TOWARD A
SCIENTIFIC PRACTICE
OF SCIENCE EDUCATION

TOWARD A
SCIENTIFIC PRACTICE
OF SCIENCE EDUCATION

Edited by
MARJORIE GARDNER
University of California, Berkeley

JAMES G. GREENO
Stanford University
Institute for Research on Learning

FREDERICK REIF
ALAN H. SCHOENFELD
ANDREA DISESSA
ELIZABETH STAGE
University of California, Berkeley

LEA

LAWRENCE ERLBAUM ASSOCIATES, PUBLISHERS
1990   Hillsdale, New Jersey   Hove and London

Lawrence Erlbaum Associates, Inc., Publishers
365 Broadway
Hillsdale, New Jersey 07642

Library of Congress Card Number 90-3867

Printed in the United States of America
10 9 8 7 6 5 4 3

Contents

Foreword

James G. Greeno
Stanford University and Institute for Research on Learning

Marjorie Gardner
Lawrence Hall of Science, University of California, Berkeley

This book reflects a vision of a field that is in the process of development. We believe that a revised and advanced field of science education can emerge from the convergence and synthesis of several current scientific and technological activities. This book includes some examples of research progress of the kind that we hope will form the integrated discipline of science education.

The papers in this volume were presented at a conference that was an effort toward this revision and advancement. At a previous meeting in 1986, members of the communities of science educators, cognitive scientists, and educational technologists met to discuss and formulate a research agenda for science education. In addition to a report of the group's conclusions (Linn, 1987), the meeting accomplished a step toward forming an inclusive community of research and development for science education.

The participants in the 1986 meeting agreed that there is an important agenda for research in science education and that the communities of science educators, science-education researchers, cognitive scientists, and technologists bring important perspectives and capabilities to that scientific activity. They did not completely agree on every point that should be on the agenda or on the relative importance of the points, but that is as it should be. The community should not try to work in a single-minded way, but rather should pursue a collection of overlapping but nonidentical goals and thereby discover which directions are most productive. The shared sense of the group, however, was that important programs of research and development are being pursued, and that some of the community's effort should be directed toward bringing these various activities into closer contact. This led to our decision, along with our colleagues, to hold a conference in 1988, at which the papers in this volume were presented. We invited individuals working on the social context of science learning, in addition to technology, cognitive science, and science education researchers.

The conference that this volume presents was, in part, a test of the hypothesis developed at the 1986 meeting, namely, that there is an important agenda for research in science education and that the various communities of researchers are engaged in work that is significant for the development of a new integrated field. We decided to test this hypothesis directly by bringing together individuals from the various communities to present their work and encourage discussion among the participants.

The first condition for developing a new intellectual field is the existence of research problems that are productive and about which the community can interact meaningfully. We believe that this condition is met, and we present this book as our evidence. These are not the only examples of work that would be synthesized in the field of science education; any meeting represents a partial sample. But the point we wish to make is that significant examples exist, and we hope that our colleagues agree that these papers definitely establish that.

Another condition for developing this field is that individuals working in its various subcommunities interact productively about each other's problems as well as their own. This is harder to demonstrate in a volume of research papers, but on the basis of our experience in the two meetings, we are optimistic about that as well. The discussions were mutually engaging and spirited, and participants' comments about the meetings were positive. Many individuals at the meetings met each other for the first time and apparently were favorably impressed. Most of the final versions of papers that you can read here differ significantly from the versions that were presented, reflecting comments and questions that were given by other participants. The shared sense of engagement, including agreements as well as significant unresolved issues, is reflected in the summary section that Linn contributed to this book. The development of a genuine scientific community is a long-term process, of course, but we see the success of these meetings as a positive sign.

Organization of the Book

The papers in this book are in four sections, reflecting four research traditions that we feel can come together in a scientific practice of science education.

First, there is a community of science-education researchers whose intellectual homes are in the study of curriculum and teaching of scientific disciplines. Discipline-based research and development was the main activity of the science-education field during the important period of curriculum reform in the 1950s and 1960s and continues to play a major role.

A second community of researchers in cognitive science studies general principles of learning, knowing, understanding, and reasoning. Cognitive science is, itself, a field in the process of development, forming as a convergence of parts of artificial intelligence, cognitive psychology, linguistics, philosophy, and other disciplines. The research in this developing field differs from earlier research, especially in psychology, in a way that is important for science education. Modern cognitive science attends to the content of information that people learn, know, understand, and reason with. Earlier research on cognition was abstract and content-free; however, in cognitive science beginning in the late 1950s, simulation models of cognitive structures and processes include hypotheses about the specific information structures that are known and understood and the specific reasoning operations that are applied to those structures.

Until about 10 years ago, the communities of discipline-based educators and cognitive scientists had very little in common. Since the late 1970s, however, there has been an increasing tendency for cognitive scientists to be concerned with problem solving, knowing, and learning in subject-matter domains, especially in mathematics and science. And simultaneously, there has been an increasing tendency for scientific discipline-based researchers to make use of theoretical and empirical methods developed in cognitive science in their research and development of instruction. Both of these trends are evident in the papers in the first two sections of this volume. Much work remains before the science of cognition and discipline-based educational research and development are well integrated, but there is a strong and growing intellectual basis for that integration, if the communities of researchers choose to develop it.

The third section of papers is concerned with the social context of learning, a topic on which a body of interdisciplinary research and development is beginning to grow. Studies of cognition in everyday settings are shedding interesting new light on the capabilities of individuals to reason successfully about quantities and causal relations in the world, and relations between this everyday reasoning and school learning are just beginning to be examined. Investigations of social organization of schools, including socially determined attitudes toward schooling and participation in group activities, benefit strongly from use of concepts and methods developed in the social sciences. We are hopeful that a convergence of methods and concepts of social science, cognitive science, and discipline-based educational study can develop productively to broaden the scientific base of science education.

The final section of this book presents discussions of educational technology in science and mathematics education. Development of advanced technology for education has had somewhat disconnected components, with some efforts related primarily to discipline-based concerns, some to cognitive studies, a few to social concerns, and several to general concerns of artificial intelligence. The development of complex technological systems can serve as a vehicle for further integration of these various intellectual strands as papers in this volume indicate.

The Idea of a Scientific Practice

The title we chose for this volume is a coined term, and it may bear a brief discussion. As we envision the developing field of science education, it would become an integrated disciplinary activity including development of resources and materials for science education as well as development of ideas about learning, knowing, and reasoning in science. The field would also be engaged in continuing evaluation, refinement, and restructuring of these resources and ideas. We believe that the model of basic research by a group of scientists, with results that inform practice by a group of educators, is misconceived. The search for knowledge and understanding and the development of educational resources must be concurrent concerns and interactive activities. The alternative vision, which we prefer, has inquiry coupled with development of resources so that development is guided by and informs the growth of scientific principles and concepts, and scientific inquiry addresses questions that are important in practice. Such a melding of inquiry and practice might well be called either a practical science or a scientific practice of science education. By either name, we hope that these papers contribute to its development; we'll hope and work for its continued progress.

ACKNOWLEDGMENT

Support of the conference on which this book was based was provided by the National Science Foundation under grant MDR-8550921, the Lawrence Hall of Science; the Graduate School of Education, University of California, Berkeley; and by the Institute for Research on Learning.

1
VIEW FROM
THE DISCIPLINES

Marjorie Gardner
Elizabeth Stage
University of California, Berkeley

Whether from the natural or from the synthetic world, science is a whole fabric, a beautifully interwoven tapestry. Humans split it into disciplines for study purposes. We compartmentalize in order to handle its many subtle complexities, yet we yearn to integrate as evidenced by so many efforts toward interdisciplinary science and mathematics.

For the opening session of the Research Conference, active researchers from each of the four traditional areas of science instruction—biology, chemistry, mathematics, and physics—were asked to summarize recent research results, current trends, and recommendations for important research projects for the future. The purpose was to set the framework for the more interdisciplinary sections to follow. James Stewart from the University of Wisconsin at Madison reports on biology; Dudley Herron from Purdue University reports on chemistry; Jack Lochhead from the University of Massachusetts at Amherst reports on mathematics; and Lillian McDermott from the University of Washington reports on physics. Their chapters and reference lists provide the reader with a useful summary, a wealth of ideas and sources.

McDermott notes that there has been more research on learning and teaching of physics than in any other science discipline. She discusses physics educational research from three perspectives, that of the cognitive psychologist, the physics instructor, and the science educator. Major attention is then given to research efforts directed toward elucidating students' understanding of physics concepts, scientific representations, and the reasoning required for the development and interpretation of both concepts and representations. Questions for future studies are identified for each of the three areas she discusses.

Herron takes the constructivist point of view as he reviews recent research in chemical education and looks to the future. Citing research done in the United

States and internationally, he critiques research efforts related to problem solving and conceptual understanding. In surveying research in these two major areas, Herron explores misconceptions, experts versus novices, and representations. The chapter concludes with a section that looks to the future by summarizing our current knowledge and identifying research that is needed.

Stewart begins by noting that the biological sciences are the most commonly taught sciences at all levels as well as the most rapidly changing due to the current biological "revolution." The first half of the chapter is concerned with the current state of biological sciences educational research; the second part deals with the future and identifies some of the important research that needs to be done. Stewart notes that much of the research to date has been of the correlation studies type as he surveys results of these studies at the elementary, secondary, and university levels. More sophisticated studies concerned with genetics and evolution are then reviewed. Studies of the uses of advanced technology including the computer are surveyed. In looking to the future, he calls for a research consortium in biological science education. The research for such a consortium might include continuation of descriptive research studies, problem-solving research, and research related to the findings of cognitive scientists.

Lochhead describes the recent, rapid, almost explosive advancements in the mathematical sciences as well as the heavy demands on mathematics education for advances in research. He identifies needed changes throughout the chapter and calls for flexibility, and the capacity to respond to rapid change. He also examines some of the predictable changes in terms of the curriculum and instructional materials, modes of instruction and student learning strategies (e.g., problem solving, metacognition). The role and use of calculators and computers are explored in terms of current research. Lochhead turns near the end of the chapter specifically to recommended areas for future research.

As the "View from the Disciplines" was unveiled, the current somewhat fragmentary nature of research became more apparent and elevated awareness of the need for longitudinal studies and team efforts. Three common threads are identifiable in the four chapters: attention to problem solving, the constructivist view of how students learn, and the role of technology in instruction. Little cross-disciplinary work is being done. Researchers identify themselves as mathematicians, chemists, physicists, biologists or geologists when doing educational research. All four authors recognize that students construct knowledge for themselves and that their knowledge of rules, formulas, and algorithms is virtually useless unless they can apply what they've learned to novel situations. In the three science papers, there's further acknowledgment of the importance of understanding the origin of student misconceptions. The need for interdisciplinary collaborative effort and/or perhaps more importantly for Research Centers where resources can be garnered for in-depth and longitudinal studies become evident.

"View from the Disciplines" serves as a backdrop for the more interdisciplinary areas of Instructional Design, Science Education in the Social Context, and the Impact of Technology.

1 A View From Physics

Lillian C. McDermott
University of Washington

INTRODUCTION

There has been more research on the learning and teaching of physics than on any other scientific discipline. Until recently, most investigations have focused on mechanics, particularly on kinematics and on the relation between force and motion.[1] The field of inquiry is now considerably broader and includes several other content areas such as heat, electricity, and optics.

Physics has been chosen as a domain for investigation by cognitive psychologists, science educators, and physicists. These groups share some of the same goals, but their primary motivation for doing research is often different. As a consequence, they often do not ask the same questions and even when they do, they may interpret the same answers in different ways. The broad range in perspective is illustrated by the diagram in Fig. 1.1 In actual practice, differences among the groups are not as sharply defined as they appear in the diagram.

The nature of a paper on the status and future of research in physics education is likely to be strongly influenced by the background and orientation of the author. The point of view taken here is that of a physics instructor whose primary motivation for research is to understand better what students find difficult about physics and to use this information to help make instruction more effective.[2] The

---

[1]For an overview of research on conceptual understanding in mechanics, see McDermott (1984).

[2]For examples of the author's approach to research in physics education, see Trowbridge & McDermott (1980, 1981); Goldberg & McDermott (1986, 1987); McDermott, Rosenquist, & van Zee (1987); Lawson & McDermott (1987). For examples of the application of this research to curriculum development, see Hewson (1985); Rosenquist & McDermott (1987).

3

PERSPECTIVES ON RESEARCH IN PHYSICS EDUCATION

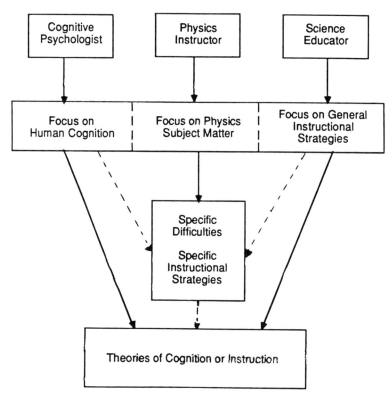

FIG. 1.1. Physics has been chosen as a domain for investigation by cognitive psychologists, science educators, and physicists.

direction and methods for research are derived from an interest in physics for its own sake and an interest in teaching that particular subject. The emphasis in the research is to identify specific difficulties and to develop instructional strategies to address these difficulties. This focus is not meant to imply a lack of interest on the part of the author in the general theoretical and instructional issues that concern the cognitive psychologist and science educator; rather, the approach reflects a pragmatic attitude toward instruction that is common among physicists who teach the subject. The empirical emphasis is also a consequence of the belief that the most effective way to improve instruction is by first concentrating on specific instances and generalizing only at later stages.

Some physicists hold a contrasting point of view.[3] As shown in the diagram in

---

[3]For a discussion by a physicist with a more theoretical perspective, see Reif (1986, 1987 a & b); Hestenes (1987).

Fig. 1.1, this perspective is closer to that of cognitive psychologists. Physics has proved to be an appealing domain for studies that focus on problem-solving. The interest that drives the research of these investigations is often less on specific subject matter and more on underlying thought processes. An important goal for cognitive psychologists is the development of theoretical models of human cognition that can be used as a basis for planning instruction.[4]

Still another approach toward research in physics education characterizes the work of science educators. The title, as used in this paper, does not refer to the science instructor who is a subject matter specialist, but is reserved for those who are directly concerned with the education of teachers or with curriculum and instruction in the schools. As indicated in Fig. 1.1, science educators are usually more broadly interested in teaching science in general than physics in particular. Although physics may provide the context, the focus for research is often on the development of instructional strategies and theories of instruction that extend beyond the teaching of physics.[5]

The particular view that is presented in this paper has evolved over several years and has been influenced by the experience of the Physics Education Group at the University of Washington. The group, which is an integral part of the Physics Department, is actively involved in teaching physics to students with a wide variety of preparation. The instructional environment provides a setting for conducting research and curriculum development from a strong disciplinary perspective. We have found it useful to organize these activities into categories that correspond to various aspects of student understanding in physics. Our investigations are directed toward elucidating the following aspects of student understanding: the concepts of physics, scientific representations (e.g., diagrams, graphs, equations), and the reasoning required for the development and interpretation of both concepts and representations. We make use of problems primarily to gain insight into conceptual and reasoning difficulties rather than to examine problem-solving capability as an end in itself. There is a major emphasis in our research on the ability of students to make connections among concepts, representations, and real world phenomena.

In this paper, the organizational structure for discussion of research will be provided by a loose classification scheme consisting of four categories: (a) concepts, (b) representations, (c) reasoning, and (d) problem solving. These are not mutually exclusive. An investigation may fit equally well into more than one category. The choice has been determined by the aspect of research that a particular study is used to illustrate. To call attention to recent work outside of mechanics, the illustrations have been drawn from other content areas whenever possible.

---

[4]For a discussion by a cognitive psychologist about implications from research for physics instruction, see Larkin (1980).

[5]For a discussion by a science educator about applications of research results to physics instruction, see Gilbert & Watts (1983); Champagne, Gunstone, & Klopfer, (1985).

CONCEPTS

The discussion in this section focuses on a line of research in which qualitative interpretation of a concept is required. The task presented to the students may involve real objects and actual events or deal with a hypothetical situation. Most investigations in which actual equipment is used involve one-on-one interviews or small group activities in which there is dialogue between the investigator and students. Sometimes a laboratory demonstration provides the basis for written questions simultaneously administered to a large group. In other investigations, the task is presented only in written form and student response is entirely in writing.

Criteria for Understanding

The determination of what constitutes adequate conceptual understanding depends on the type of study and on the point of view of the investigator. In investigations based on actual phenomena that the student observes or can easily imagine, the emphasis is on the ability of students to use a concept (or set of concepts) correctly in performing a specified task. The criteria may include some or all of the following: (a) The ability to apply the concept to the situation observed and to describe the reasoning used; (b) the ability to recognize circumstances under which the concept is or is not applicable; and (c) the ability to distinguish clearly between the concept under scrutiny and similar but different concepts that might apply to the same situation. In some investigations, the emphasis may be on student facility with different ways of representing the concept (e.g., diagrams, graphs, equations) or with the ability to make connections among these representations and the real world.

Many studies do not involve actual apparatus. Questions about a physical situation may be described on paper or on a computer screen. There may or may not be supplementary interviews. In cases in which the student responds only in writing or by typing on a keyboard, it is much more difficult and often impossible to extract the amount of conceptual detail that the interview situation allows. On the other hand, mass testing by questionnaire or computer allows the investigator to estimate the prevalence of a particular response.

Some studies place less emphasis on the ability to apply concepts than on the ability to relate a set of concepts that may be applicable under certain general conditions. The students are encouraged to think about the concepts from a theoretical perspective. For example, there have been a number of studies in which students are asked to draw "maps" showing relationships among concepts. From the ways in which students group the concepts, indicate a hierarchy, and show connections, inferences are drawn about the level of conceptual understanding. In such cases, the criterion for understanding refers to the accuracy and level of sophistication that the student demonstrates in drawing the diagram.

Misconceptions

Although the methods of research are diverse, some generalities emerge. Students have certain incorrect ideas about physics that they have not learned through formal instruction, or at least that they were not intentionally taught. Some have resulted from misinterpretation of daily experience; others are of a different origin. To the degree that these ideas are in conflict with the formal concepts of physics, the physicist considers them to be "misconceptions." The term misconceptions will be used here although it is recognized that some investigators would rather refer to alternate conceptions.

It has been shown by a number of studies that students often complete a physics course with some of the same misconceptions with which they began. Furthermore, certain errors are characteristic of student responses to certain types of questions (see footnote 1). These observations have led some investigators to hypothesize that students bring to the study of physics a strongly held system of beliefs about how the world operates (McCloskey, 1983). A contrasting point of view is that students' knowledge of the world is fragmentary and unstable, with a tendency to shift according to the context (di Sessa, 1988). There is disagreement about whether certain observed regularities in response occur because students have a mental model for cause and effect or for some other reason. For example, perhaps the similar features among answers are simply elicited by the way in which the questions are asked (Viennot, 1985a, 1985b).

Although there is a difference of opinion about whether or not students have a consistent system, there is no doubt that there are some common misconceptions that do not disappear spontaneously as the relevant material is taught. To bring about conceptual change, it is frequently necessary to make a conscious effort to help students reject certain ideas and accept others (Strike & Posner, 1982). The way such instruction is designed may be influenced by the inferences made about how students think.

Constructivist Epistemology

The results from research are consistent with the view that the mind is not a blank slate upon which an instructor may write correct statements that the student can learn passively. It is also clear that, whatever their origin, incorrect ideas that are well entrenched in the student mind may interfere with the ability to learn what is being taught. These circumstances have led to an interest in constructivist epistemology among science educators. Basic to this approach are the beliefs that (a) Each individual must actively construct his or her own concepts, and (b) that the knowledge that a person already has will determine, to a large extent, what he or she can learn. The implications for instruction that can be derived from these tenets may be used to guide the design of curriculum from precollege through undergraduate levels (Driver & Bell, 1986; Schuster, 1987).

7

Linguistic Complications

It is not only common experience with the physical world that leads students to develop ideas that contradict those of the physicist. Linguistic elements also play an important role. Often the picture conjured up in a student's mind is different from the meaning the words are intended to convey. For example, a physics student who reads a problem about a ball that is "dropped" in an ascending elevator may not realize that in this case the ball initially moves upward with respect to the ground. When words have both a technical and colloquial meaning, the concepts that are associated with them may be muddled. Terms like *force* and *energy* that are understood in an unambiguous way by physicists are often interpreted by students in a context-dependent manner (Touger, Dufresne, Gerace, & Mestre, 1987).

Quite apart from the problems caused by differences in the everyday and technical use of a word, other linguistic complications may be introduced in the course of defining scientific terms. For example, Kenealy (1987) examined how various populations interpreted the statement: "Acceleration is the time rate of change of velocity." The definition is from one of the most widely used high school physics textbooks in the United States (Williams, Trinklein, & Metcalfe, 1984, p. 48). Participants in the survey included students in eighth grade through college and high school science teachers. A significant fraction of answers identified acceleration as an amount of time required to change a velocity.

Examples of Research

Theoretical Constructions: Concept Mapping in Electricity

An example of research in which a theoretical construction by the student constitutes the primary source of data is provided by the concept-mapping studies of Moreira (1987). One study involved engineering students in an introductory physics course at a Brazilian university. The students were asked to draw maps showing relationships among the main physical concepts that they had studied in electricity. They were also asked to write key words along the lines linking the concepts to make explicit the relationship between them. Upon completion, the maps were discussed on an individual basis with the students who drew them.

The map shown in Fig. 1.2 is a copy of one drawn by a student. The student has selected electric charge as the most important concept and linked it to electric current, electric field, and electric potential. However, the field and the potential are not linked to each other. (These links and the others shown as dotted lines were added during discussion of the map.) Electric force and potential difference did not appear on the original drawing. The ensuing discussion revealed that the student made no distinction between the concepts of potential and potential difference.

8

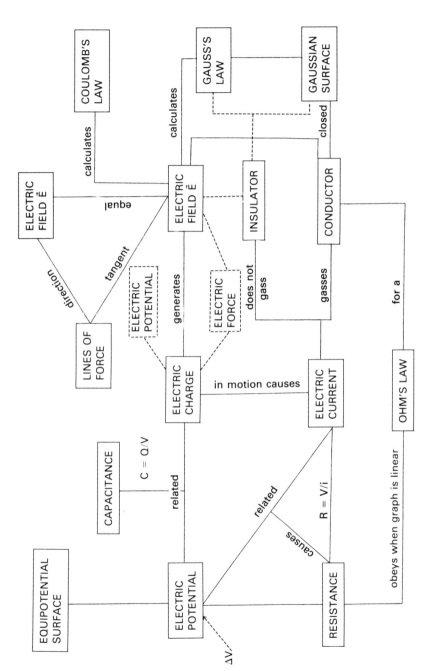

FIG. 1.2. Concept map in electricity drawn by a Brazilian engineering student taking introductory physics (Moreira, 1987).

9

Real Phenomena: Light and Image Formation in Geometrical Optics

Student observation, or visualization, of real phenomena forms the basis of much of the research on conceptual understanding in physics. To illustrate how different investigations can make a cumulative contribution to our knowledge of how students think about physical phenomena, we review briefly some of the research involving geometrical optics. Other topics (e.g., dynamics, electric circuits, or heat and temperature) could also have been used for illustration.

Children's Ideas about Light. A number of studies have identified some incorrect ideas about light that are common among children and adolescents (and sometimes among adults) who have not studied the topic formally.[6] It appears that before about the age of twelve children do not usually recognize light as an entity independent of its source or its effects. In the early teens, children begin to identify light as an entity that can travel in space and that can be obstructed and reflected. Their understanding of how light propagates is limited, however. Many believe that light travels farther from its source at night than during the day. They do not separate the idea of light from how bright it is. They also may think of light as a force acting on an object. Often seeing is considered an activity of the observer rather than the result of the reception of light by the eye.

From studies such as the foregoing, we can gain some insight about the state of knowledge with which many students begin formal study of optics. As a result of instruction in optics in high school or college, most of these naive ideas are superseded by concepts the physicist uses to explain how light is transmitted from a source to an observer and how objects can be seen. The vestiges of some of these ideas may remain, however, and may interfere with the development of a student's understanding of how an image is formed and seen.

Formal study of geometrical optics typically begins with the study of image formation by mirrors and lenses. Students learn how a lens or mirror can form an image of an object and how the location and size of the image can be predicted. They often do experiments with mirrors and lenses in the laboratory and almost always work problems involving images.

Student Understanding of Real Images. Documentation from research is beginning to bring about more awareness on the part of high school and college teachers that the ability to solve standard physics problems is no indication that a sound conceptual understanding has been achieved. Problems in geometrical optics are no exception, even though this topic is generally considered one of the

---

[6]For sources for the statements in the summary, see Piaget (1974); Tiberghien, Delacote, Ghiglione, & Matalon (1980); Guesne (1984); Stead & Osborne (1980); Watts (1985); Andersson, & Kärrqvist, (1983); Eaton, Anderson, & Smith, (1984); Feher & Rice (1985; Jung (1987).

simplest in a physics course. The following example illustrates how little we sometimes know about what students really understand if we look only at their ability to solve standard problems.

The illustration is taken from research conducted in collaboration with Fred Goldberg during the 2-year period he spent with the Physics Education Group at the University of Washington (Goldberg & McDermott, 1986, 1987). The work described is based on a task from an investigation on student understanding of the real image formed by a single mirror or lens. The students involved were volunteers from the introductory physics sequence required for majors in engineering, physics, and other physical sciences. Calculus is required for this course. Most of the data were collected from individual interviews in which students were asked a series of questions about a simple demonstration that they could observe. Each was shown the same demonstration and asked the same questions. The demonstration was a simple optical system consisting of a lens, a light bulb, and a screen, all mounted on an optical bench. A real, inverted image of the lighted filament of the bulb was visible on the screen, as can be seen in Fig. 1.3.

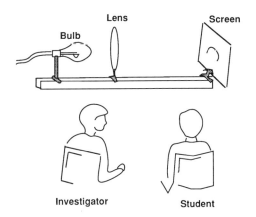

FIG. 1.3. Investigator asks student: "Suppose I were to cover the top part of the lens, leaving the bottom half uncovered. Would anything change on the screen?" The table shows the percentage of students who gave the correct answer both before and after instruction (Goldberg & McDermott, 1987).

Interview Data Summary

	Pre (N=36)	Post (N=23)
Complete image (correct)	0%	35%
Half of image	95%	55%
Other	5%	10%

Before discussing a question that caused the students difficulty, we first consider a task that they could perform. In exploratory interviews, we found that students who had completed geometrical optics could generally use the thin-lens formula to solve the following problem: Given the focal length and the object distance, predict the location, characteristics, and magnification of the image. The students could also solve the problem by drawing an appropriate ray diagram. Furthermore, they were able to check their solutions by using laboratory apparatus and could make the proper connections between the numbers from their algebraic solutions and the corresponding distances on an optical bench.

Let us now contrast what the students could do with what they could not do. During the individual demonstration interviews, the investigator asked the following question: "Suppose I were to cover the top part of the lens, leaving the bottom half uncovered, would anything change on the screen?" The results in Fig. 1.3 indicate that many students did not realize that the complete image could still be seen with only part of the lens.

In reporting the results, we refer to the students who had taken physics in high school but not yet at the university as prestudents, and those who had completed the optics portion of the university course as poststudents. None of the prestudents gave the correct response. About one third of the poststudents made a correct prediction. In spite of the fact that these students knew how to use the thin-lens formula, many did not know how to answer a basic question that they had not been asked before. By far the most common response was that only half the image would be seen if the upper half of the lens were blocked. Most students claimed that the bottom half of the image would disappear, a prediction consistent with their knowledge that the image in this situation is inverted.

It is not only the mistakes that students make that are of interest. The explanations they give in support of their answers can give us some insights into their thinking. A particularly interesting form of incorrect reasoning on the lens task is illustrated by the explanation offered by a student who drew an essentially correct ray diagram, similar to the one shown in Fig. 1.4.

The student drew two rays from the top of the object: (a) one parallel to the principal axis (ray #1), and (b) the other toward the center of the lens (ray #2). After passing through the lens, ray #1 was drawn so that it passed through the focal point and ray #2 was shown undeviated. The image was located at the point where the two rays intersected. The student described the ray-tracing procedure correctly, but then went on to say: "Now if you block off the top part of the lens, that would block off rays #1 and #2 from getting through, so the bottom of the image would be blocked. The bottom part of the object, which corresponds to the upper part of the image, would still be there."

Thus we have a situation in which a student was able to do all that is usually required on a typical examination but seemed to have totally missed a crucial concept in geometrical optics: From each point on an object, there are an infinite number of rays which, to close approximation, will converge at a single image

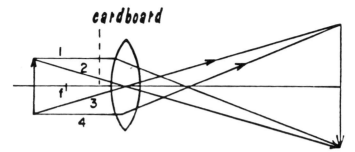

FIG. 1.4.   A student was able to draw this essentially correct ray diagram even though the reasoning that half the lens would produce half the image was incorrect (Goldberg & McDermott, 1987).

point after passing through the lens. It is unlikely that a complicated numerical problem involving several applications of the lens formula would have revealed as much about conceptual understanding as the simple qualitative question asked. It is also worth noting that the belief that the two rays used to locate the image are necessary, rather than merely sufficient, must have developed during the course of instruction. Unlike some misconceptions, this one cannot be attributed to misinterpretation of everyday experience.

The results on the lens task cannot be explained on the basis that the participants in the study were poor students. It has been our experience that students who participate in interviews generally receive a grade of A or B in physics. The less capable students seldom volunteer. Moreover, when a multiple-choice version of this question was asked on final examinations administered to more than 200 introductory physics students, only about one fourth recognized that the entire image would remain intact if half the lens were blocked.

Questions for Future Study

The example taken from geometrical optics illustrates the kind of conceptual detail that research can provide. As mentioned earlier, most of the research so far has involved concepts in mechanics. To guide the design of curriculum, we need answers to questions such as those below for *all* topics in introductory physics.

What ideas do students have before instruction that might interfere with developing a sound conceptual understanding? Which ideas can be built upon to promote learning and which need to be changed? Are linguistic elements of such critical importance that they need to singled out for special attention? What conceptual difficulties do students encounter during instruction? What strategies can help overcome these difficulties? How can students learn to distinguish related concepts? What instructional techniques can help students make connections between concepts and real world phenomena? We need to know more about how conceptual understanding can be developed and how conceptual change can be fostered.

REPRESENTATIONS

An inability to use and interpret scientific representations of various kinds (e.g., diagrams, graphs, equations) is quite common among physics students. A number of studies have explored this aspect of student knowledge in which elements other than conceptual understanding are involved.

Diagrams

Diagrams are a form of scientific representation frequently used in physics as an aid in the analysis of a physical situation or in the solution of a theoretical problem. Examples are free–body diagrams in mechanics, ray diagrams in optics, and circuit diagrams in electricity. Diagrams offer a way to organize information into an easily accessible form, to show conceptual relationships that may not be evident from a physical layout or verbal description, and to make predictions.

Ray Diagrams

The ray diagram drawn by the student for the lens task described in the previous section is essentially correct in form. The student knows the geometrical algorithm for construction but is unable to interpret the information the ray diagram contains and do the reasoning necessary to make a prediction. Had the student drawn the third ray that can be used to locate the image, he or she might have realized that at least one ray would emerge from the lens. (This particular ray is drawn from the head of the arrow through the focal point. After passing through the lens, it emerges parallel to the principal axis.) However, in that case, the lack of understanding of the ray diagram might have passed undetected. In spite of having learned the procedure for drawing a ray diagram, the student cannot extract from it the implicit information.

As might be expected, secondary school students also have difficulty with ray diagrams. In a study conducted in India, Ramadas (1982) found that very few students could draw correct ray diagrams for even simple situations. From an analysis of responses to written test questions, she found that the students were generally unable to abstract from the situation described the information needed to construct an appropriate diagram.

Circuit Diagrams

Electric circuit diagrams are another form of scientific representation that students often do not interpret properly. Difficulties occur both in drawing dia-

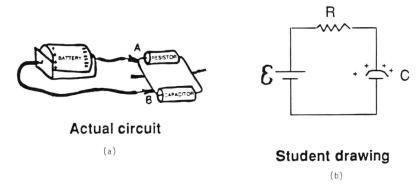

Actual circuit

(a)

Student drawing

(b)

FIG. 1.5. (a) An actual circuit shown to a student during an individual interview; (b) Circuit diagram drawn by the student, who ignores the wire *AB* that connects the resistor and capacitor across the battery (Fredette & Clement, 1981).

grams to represent real circuits and in interpreting diagrams to answer questions about hypothetical circuits.

When Fredette and Clement (1981) asked students to draw circuit diagrams of actual circuits, they found that students frequently did not represent on their diagrams wires that "shorted out" elements in the circuit. The students seemed to think that shorting wires do not merit inclusion in a circuit diagram because they "don't really do anything." An example is provided by the circuit shown in Fig. 1.5a. In the diagram in Fig. 1.5b, which was drawn by a student, the wire *AB* that connects the resistor and capacitor across the battery is ignored. The failure to represent this wire may indicate one or more related problems. The student may not recognize that virtually all of the current will be in the shorting wire and may not interpret the situation as eliminating electrically the resistor and capacitor from the circuit. The student may not understand that the purpose of a circuit diagram is to show electrical connections as clearly and explicitly as possible.

Johsua (1984) found that responses by high school and university students to questions about identical electric circuits depended upon the way in which the circuit diagrams were drawn. When asked to describe the current between points *A* and *B* in the circuit of Fig 1.6, approximately 60% of the students answered correctly if the circuit was drawn as in Fig. 1.6a. However, only 25% answered correctly if the circuit was drawn as in Fig. 1.6b. Johsua found that students tended to view the lines on circuit diagrams as a "system of pipes" through which fluid can flow. In trying to decide how the current would be distributed, the students did not analyze the diagram to determine the potential difference between points *A* and *B*. Of course, the students' difficulties were not purely representational. As in most cases, difficulty with a scientific representation cannot be viewed apart from difficulty with the concepts involved.

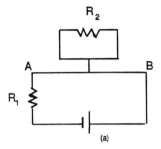 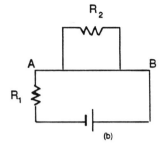

Circuit	Correct	Wrong	No answer
a	62	31	7
b	26	68	4

FIG. 1.6. When asked to describe the current in the circuit between points *A* and *B,* approximately 60% of the students answered correctly if the figure was drawn as in (a), but only 25% answered correctly if the circuit was drawn as in (b) (Johsua, 1984).

Motion Graphs

Several recent investigations on scientific representation have been devoted to motion graphs. Similar types of errors have been found among students at all levels. Common difficulties include drawing and interpreting graphs as if they were spatial pictures and trying to use the height of a graph to extract information contained in the slope.

Microcomputer-based laboratories (MBLs), which were developed at the Technical Education Research Center (TERC), allow students to watch a graph being generated as an object moves. In particular, they can see an instantaneous graph of their own motion (Thornton, 1987). In one study 52 undergraduates, who were enrolled in a physics course for students majoring in the humanities, participated in a single MBL session. These students performed significantly better than calculus-level students on examination questions requiring interpretation of motion graphs (Thornton, 1987).

Graph-as-a-picture and slope/height confusion were the most prevalent difficulties identified by Mokros and Tinker (1987) during clinical interviews with 25 seventh and eighth graders. Mokros and Tinker examined the development of graphing skills among 125 students who participated in a series of MBL lessons, in which they made real-time graphs of their own motion. A multiple-choice quiz was administered as a posttest. The students were asked to match verbal descriptions and pictures of various motions with a set of motion graphs. The increased success in choosing correct responses on the posttest compared with preinstructional performance suggests that there was an improvement in ability to distinguish the graph of a motion from its physical appearance.

When a motion was described in words, 75% of the students selected an appropriate *position* versus *time* graph. However, when a motion was both described in words and sketched in a diagram, the students were less successful in choosing between the correct *velocity* versus *time* graph and one that resembled a picture of the motion. Another recent study suggests that the simultaneous movement of the student and production of the graph may be an important factor in the gains reported for MBL instruction. Even a short delay in feedback seems to be disadvantageous (Brasell, 1987).

In another investigation on graphing, students in a calculus-level physics course at the University of Washington were given the diagram of the ball and track shown in Fig. 1.7, as well as the following description: The ball moves with steady speed along the level segment, accelerates down the incline, and then continues at a higher constant speed along the last segment (McDermott, Rosenquist & van Zee, 1987; van Zee & McDermott, 1987). The students were asked to sketch position, velocity, and acceleration versus time graphs for the motion of the ball. The only correct response from 118 students is shown in Fig. 1.7.

From the types of errors that were made, it was possible to identify some specific difficulties. All but one student neglected the fact that each segment of the motion takes place in a shorter interval of time than the preceding one. There were many other more serious errors. A relatively common one was the drawing of two or more nearly identical graphs. More frequent was the apparent attempt to emulate the appearance of the track in the shape of the graphs. For example, half of the students represented the motion along the straight inclined track by a straight line on the x versus t graph instead of a curved line. Almost as many drew parallel lines for the first and third segments of that graph, perhaps because the corresponding track segments were parallel in space.

In an extension of this study, individual interviews were conducted to identify whether there were generalizable differences in approach between students who could sketch correct graphs and those who could not. It was found that the "experts" (successful students) differed from the "novices" (unsuccessful students) in several ways. Among the more striking contrasts in procedure were the following: (a) Experts generally began by defining the axes; novices started by drawing a line; (b) experts tried to match the shape of the graph to the way the variable was changing in time; novices often tried to match the shape of the graph to the shape of the path of the motion; (c) experts used a line to represent a constant value of x, v, or a during a time interval; novices sometimes represented a constant value with a single point; and (d) experts checked for consistency in slopes and heights among graphs; novices seemed to ignore or reject such relationships.

**Let x = The position of ball rolling along a track
as shown below:**

Sketch graphs of this motion below:

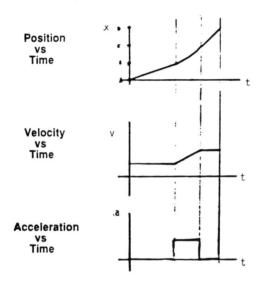

Position
vs
Time

Velocity
vs
Time

Acceleration
vs
Time

FIG. 1.7.   Only one of the 118 calculus-level physics students was able
to sketch these correct graphs for the motion of the ball on the track
(van Zee & McDermott, 1987).

The generally poor performance on this task demonstrates a widespread inability even among mathematically able students to relate graphs to actual events.
As in the case of the ray diagram and circuit diagrams previously discussed, there
is a lack of understanding of the motion graph as a way of representing and
analyzing real world phenomena.

Questions for Future Study

The examples above illustrate some of the difficulties students have with
diagrams and graphs. The problems are not solely conceptual in nature, although
lack of such understanding may play a critical role. To develop appropriate
instructional strategies, we need to identify the specific difficulties students have
with various representations. Diagrams, graphs, and equations all involve differ-

ent ways of thinking. The nature of the problems encountered is different in each case.

An important question is the role of various representations in the development of conceptual understanding. Because different representations emphasize different aspects of a concept, the more ways one can represent a concept, the deeper one's understanding is likely to be. What type of instruction can help students make connections between a concept and various representations of that concept, between one representation and another, and between various representations and the real world?

Diagrams, graphs, and equations are useful in contexts other than physics. The ability to construct and interpret these representations is a valuable skill that is worth developing in its own right. Results from research indicate that the ability to use representations does not evolve spontaneously during instruction but must be specifically cultivated. What type of instruction can promote such development? How can students learn to transfer facility with a particular form of representation from one context to another?

REASONING

Many physicists would maintain that one of the most important benefits that can be derived from the study of physics is development of scientific reasoning skills. They see problem solving as contributing to this goal. However, there is no convincing evidence that reasoning ability improves as students work standard problems in an introductory course. Arons (1976, 1982, 1983, 1984 a, b) has written extensively about the necessity of designing instruction to promote development of the capacity to reason.

Several kinds of reasoning processes needed for scientific work could be developed in introductory physics. Among them are *proportional, ratio, analogical,* and *hypothetico-deductive* (model based) reasoning. The list is far from complete and the terms lack sharpness. However, they are sufficiently descriptive to convey the nature of certain reasoning skills that many physicists consider important.

Proportional and Ratio Reasoning

Although most students who survive a physics course can reason with proportions to some extent, many do not fully understand the meaning of the number obtained by carrying out the division specified in the statement of a proportion (Arons, 1976, 1983). For example, students often do not know how to interpret the meaning of the result obtained by dividing the mass of a substance by its volume. By referring to the formula they may recognize the result as the density, but they do not identify this number as the number of units of mass for each unit of volume. In other words, they do not picture a cubic centimeter of the substance as having a mass in grams numerically equal to the density.

Students have even more difficulty in reasoning with ratios when more than a simple proportion is involved. For example, unless numbers are supplied, many students cannot tell what happens to the electrostatic force between two charges when each charge is increased by a factor of four and the distance between them is halved. They cannot conclude that the force increases by a factor of 64.

Analogical Reasoning

The ability to reason by analogy is very important in physics. Physicists regularly use analogies to analyze unfamiliar systems in terms of systems they understand. Physics instructors make frequent use of analogies in teaching new concepts. For example, angular velocity and angular momentum are often introduced as analogous to the corresponding linear quantities. Relatively little attention is devoted to these topics in an introductory course partly because of time constraints, but also because the student is expected to understand the material by analogy to the linear situation. However, instructors know from experience that student understanding of dynamics is much poorer when rotations are involved.

There has been some research on the use of analogies for teaching concepts from physics. Gentner and Gentner (1983) examined how students used "flowing water" and "teeming crowd" analogies in making predictions about the current in an electric circuit. They were interested in determining whether the analogy had only a surface effect, that is, affected only the language used in speaking about the circuit or whether the analogy generated ideas that the students used in making predictions. It was found that there was a difference in the predictions that depended on which analogy was involved. The analogies seemed to have influenced the way the students thought about the circuits.

In another study, Clement (1987) found that high school students could reason with analogies if the corresponding quantities and relationships were made explicit and a great deal of time was devoted to consideration of the analogy. The focus of the research was not on analogical reasoning but on conceptual change in mechanics. Clement explored the effectiveness of using analogies to help students overcome some common misconceptions that seem to be firmly held. One of these involves the normal force exerted by a table on a book. Many students are unwilling to accept the idea that a table, an inert object, can exert a force upward on a book. In trying to address this difficulty, Clement used an approach similar to one used by Minstrell but placed greater emphasis on reasoning by analogy (Minstrell, 1982).

To make the existence of an upward force plausible, Clement introduces an anchoring situation to which the target situation (the book on the table) can be compared. For example, he might ask the students if a hand placed under a book exerts an upward force. Students usually admit that the hand exerts a force but

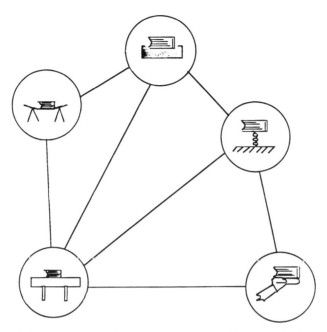

FIG. 1.8. An example of anchor, target, and bridging analogies that might be used to convince students that a table exerts an upward force on a book (Clement, 1987; Murray, Schultz, Brown, & Clement, 1987).

may not believe that an analogy between a hand and a table is valid. One is alive and one is not. To make this analogy more acceptable, Clement suggests one or more bridging analogies. A possible sequence is shown in Fig. 1.8. In this instance, a book on a coiled spring serves as an intermediate analogy. Although students usually recognize that the spring can exert an upward force when compressed from above, they often do not see the table and spring as analogous. Other bridging analogies may then be proposed, such as a book on foam rubber that sags or a book on a thin board that bends slightly. These analogies have the advantage that the deformation of the foam rubber or thin board suggests a mechanism that could account for the ability of the table to exert an upward force on the book.

When this teaching experiment was tried in several high school classes, there was a significant difference in favor of experimental over control groups in acceptance of the physicist's interpretation that the table exerts an upward force on the book. Similar results were obtained when analogies were used to help students understand frictional forces and Newton's third law. In each case, the instructors found that the students needed to participate in many discussions before they would accept the validity of the analogies that were suggested. When a belief is strongly held, it is particularly difficult to convince students that an analogy exists if acceptance of the analogy requires giving up the belief.

Clement and his associates have found that sometimes a long chain of bridging analogies is necessary to convince some students that the target and anchor systems are indeed similar with respect to the feature under consideration. To individualize instruction, the group has designed a prototypical computer program for an analogy-based tutor. The computer can generate a series of bridging analogies that can be selected to make long or short steps on the basis of student response (Murray, Schultz, Brown, & Clement, 1987).

Hypothetico-Deductive Reasoning

The construction of a scientific model involves many steps of inductive and deductive reasoning. The building of a model usually begins with an observation that may trigger an analogy. The analogy may suggest a hypothesis. The hypothesis is formulated with as few assumptions as possible. Deductions that follow are tested against other observations. The process is repeated. When necessary, new assumptions are made and new hypotheses generated. Constantly tested by observation, the model grows in complexity. It is continuously being verified and its predictive capability tested.

The process summarized above is important in physics. However, students in a traditional physics course seldom have the opportunity to become actively engaged in the type of thinking required. As a consequence, even good students who are mathematically able are often unable to reason from a scientific model and may not even understand what a scientific model is.

There are topics in introductory physics that can provide opportunities for students to gain direct experience in model-building (Arons, 1982). The study of electric circuits is one. There is some evidence that students who have developed a coherent model for an electric circuit from their own observations can remember and use this model to solve qualitative circuit problems that are difficult for students who have not had this experience (McDermott & van Zee, 1984).

Questions for Future Study

In addition to content, students taking physics are assumed to be learning the processes of science. It is not clear that in a typical introductory course that the ability to do scientific reasoning is consciously developed. To design instruction to accomplish this goal, we need to identify the specific difficulties students have with different types of reasoning.

To what extent is reasoning a critical element in conceptual understanding? Often students are expected to memorize the definition of a concept, such as velocity or acceleration, but are not expected to demonstrate that they can do the reasoning by which the concept is constructed. They may not be able to give a clear operational definition (or in Reif's terms, the procedural specification) that gives an unambiguous meaning to the concept (Heller & Reif, 1984; Reif, 1985; Trowbridge & McDermott, 1980, 1981). Does going through the step-by-step reasoning involved in the construction of a concept enhance a student's ability to apply the concept, especially in situations that have not been expressly taught?

How capable are students of using a suggested analogy if they are not specifically shown how to make the necessary correspondences (as they are by Clement). Can students be taught how to generate their own analogies for new situations? What disadvantages are there in teaching new ideas by having students reason by analogy? Can students learn how to recognize the limitations or are they likely to develop new misconceptions by making correspondences that are not valid? Would examining many instances in which a concept is applicable and helping students abstract a common feature be pedagogically wiser than suggesting analogies to them?

What is the most effective way to help students learn to use a scientific model to predict and explain simple phenomena? Is it sufficient to present a model as a set of rules that students can memorize and apply deductively, or is the ability to use a model best developed by requiring students to engage in the deductive and inductive reasoning that are part of the model-building process?

How can the study of physics contribute to the development of higher order thinking skills? What kinds of instruction can help students develop the ability to ask questions of themselves that can help them recognize what they do or do not understand? How can we elicit from students the type of qualitative reasoning that can guide them toward coherent understanding of a topic? What role does awareness of one's own thinking play in developing the conceptual understanding and scientific skills needed to do well in physics?

PROBLEM-SOLVING

The precision with which concepts are defined and the formal reasoning required to use and interpret them make physics a fertile field in which to investigate problem-solving. The primary objective in some studies in this area is to understand human thought processes. In others, the goal is to identify the knowledge and procedures needed to solve physics problems successfully. Some investigations are directed toward both of these outcomes.

Many studies on problem-solving focus on identifying differences between novices and experts. Often a major emphasis is to determine the nature of expertise and to use this knowledge to develop procedures to effect transition from the novice to the expert state. The research often has a strong theoretical element and the construction of performance models may play an important role.

Descriptive Performance Models

In some research projects, the emphasis is on describing differences in what novices and experts actually do. For example, Chi, Feltovich, and Glaser (1981) have shown that novices and experts classify physics problems into types in very different ways. Whereas experts consider general underlying principles, such as the conservation of energy, novices tend to concentrate on surface features, such as an inclined plane or a pulley.

Observation of a novice or expert in the process of solving a problem may lead to other generalizations about differences. A task analysis can be carried out that describes the procedures that were followed. Larkin (1983) found that novices typically work physics problems backwards in a linear fashion, identifying the unknown quantity and then searching for equations that contain it. Experts typically work forward, constructing a representation of the problem from general physics principles and then writing the appropriate equations.

By characterizing the differences between novice and expert behavior, it is hoped that techniques can be developed to teach novices suitable procedures that will help them make the transition from novice to expert. For example, Gerace and Mestre have developed a computer program, the Hierarchical Analysis Tool, that leads students to analyze problems qualitatively in the manner typical of experts (Dufresne, Gerace, Hardiman, & Mestre, 1987).

The computer may be used to simulate the differences between novice and expert behavior and to construct a dynamic performance model for the transition from novice to expert. For example, Larkin (1981) has designed a program (ABLE) that can "learn" from solving successively more complicated problems in mechanics and thus develop into a more expert program (MORE ABLE). A more recent program (FERMI) incorporates general problem-solving procedures that can be applied in different topics in physics, such as fluid statics or electric circuits (Larkin, Reif, Carbonell, & Gugliotta, 1988). It is anticipated that such computer models may guide development of effective, intelligent tutoring systems.

Prescriptive Performance Models

Another approach to developing a model for good problem-solving performance is theoretical and involves a hypothetical task analysis. From the determination of what is necessary in the way of tacit knowledge and implicit procedures to solve a problem successfully, Reif (1985, 1987) constructs a prescriptive model. In this case, there is no requirement that the model replicate what an expert actually does. It is recognized that an expert may use intuitive knowledge that may not be accessible to a novice. The important feature is that the procedures lead effectively to a solution. The expectation is that by learning these procedures, students will become better problem-solvers. The instructional materials that are developed on the basis of the research are evaluated in terms of the problem-solving performance of students.

From his prescriptive model, Reif has formulated guidelines for teaching scientific concepts. These include teaching an explicit procedure for specifying a concept as well as descriptive knowledge about the concept, asking the student to apply this procedure systematically in various specially selected cases, guiding the student to summarize knowledge acquired through examining these special cases, and teaching the student to detect, diagnose and correct errors.

Rule-based Problem-Solving Models

In research motivated by Siegler's rule-based studies (1976), Maloney (1985) analyzed responses on multiple-choice tests to infer the rules students used to compare the behavior of two systems. In a study that involved carts rolling on inclined planes, he identified patterns in student responses and noted that the strategies employed often seemed to depend on the order of the questions.

A different approach to problem-solving research is illustrated by the work of White and Frederiksen (1987). A goal of the research was the development of an effective method for use on a computer to teach students how to troubleshoot electric circuits. The procedures that experts appear to use to solve circuit problems were analyzed and put into the form of rules. The rules were arranged into sets (student models) that increase in size and complexity as they approach the level of the expert. Instruction on the use of this problem-solving model begins with the presentation of a simple set of rules sufficient for solving simple circuit problems. The students gradually progress to more difficult problems that require use of increasingly larger numbers of rules for solution. The rules are taught as the need for them arises. On the basis of their ability to solve more difficult problems, the students are described as moving from a novice to a more expert state.

Questions for Future Study

The ability to solve standard problems is frequently taken as a measure of student understanding in physics. It is often assumed that successful problem-solving involves all the other aspects of understanding that have been discussed.

How much does instruction in how to solve problems contribute to student understanding of concepts and representations? Does practice in problem-solving promote the development of scientific reasoning ability so that a student can reason successfully about new situations? When students follow prescribed procedures, are they thinking of the physics involved or is their attention devoted to following directions? What happens when problems are presented that cannot be solved by the patterns taught? To what extent do students transfer problem-solving techniques learned in one context to new areas and to domains outside physics?

How can the computer help improve problem-solving performance? Should it be used to calculate, to simulate, to provide drill and practice, to tutor? Is there sufficient similarity between computers and the human mind to gain useful insights for instruction from models of novice/expert behavior or from models of transition from novice to expert? How should intelligent tutoring systems be designed?

Is the fact that mathematical complexity does not have to limit the selection of problems for a physics course an advantage or a disadvantage? Removal of this constraint allows the use of real-world problems that may be quite complicated. For example, the inclusion of air resistance and other nonlinear phenomena makes possible consideration of more realistic situations in mechanics. Are students sufficiently motivated by real-world examples to warrant using them in place of problems that are conceptually simpler and more readily understood?

CONCLUSIONS

It is a consequence of the broad scope of activity in research in physics education that the brief overview presented here has omitted so much of what has been done in the last few years. Only a few examples of recent work in mechanics have been cited and even fewer illustrations taken from optics and electricity. Some topics, such as heat, have not been included. Suggestions for future study have been limited to questions for which a foundation was laid in the discussion.

The emphasis on subject matter reflects the disciplinary orientation from which the paper was written. Underlying the discussion is the belief that many of the difficulties students encounter in learning physics are a consequence of the nature of the material and must be addressed in that context. Other difficulties that may cut across subject matter boundaries are also often best treated in the same way since the ability to transfer reasoning skills from one context to another seems to develop slowly. Our knowledge about how students think is still too incomplete to provide a firm foundation for constructing useful theories of general applicability. Thus it seems prudent for the present to continue acquiring data rich in conceptual detail and to concentrate on developing instructional strategies that are demonstrably effective with specific content.

If the major goal of research is to improve instruction, then the ultimate test of its validity must involve students and teachers. We need to consider carefully what students should be expected to know and be able to do as a result of studying a particular body of material. It is important that the objectives for teaching introductory physics represent some sort of consensus among instructors. We must recognize that we cannot make realistic recommendations for improving instruction without consulting those who teach the subject at the level involved. To influence practice in the classroom, the results from research should be reported in journals that physics instructors read and in language that they can understand. The vocabulary used should be straightforward and not require familiarity with the psychological and educational literature to be comprehensible.

In working toward a scientific practice of science education, we must be sure to maintain continual blending of research with curriculum development and instruction. The three components reinforce one another and their joint presence helps insure that a project is kept relevant to the needs of students and teachers.

ACKNOWLEDGMENTS

The substantive contributions by Emily H. van Zee to the preparation of this paper are deeply appreciated as is the invaluable assistance provided by Joan Valles. Support by the National Science Foundation for the work of the Physics Education Group at the University of Washington is also gratefully acknowledged.

REFERENCES

Andersson, B., & Kärrqvist, C. (1983). How Swedish pupils, aged 12–15 years, understand light and its properties. *European Journal of Science Education, 5* (4), 387–40.

Arons, A. B. (1976). Cultivating the capacity for formal reasoning: Objectives and procedures in an introductory physical science course. *American Journal of Physics, 44* (9), 834–838.

Arons, A. B. (1982). Phenomenology and logical reasoning in introductory physics courses. *American Journal of Physics, 50* (1), 13–20.

Arons, A. B. (1983). Student patterns of thinking and reasoning. Part One. *The Physics Teacher, 21* (12), 576–581.

Arons, A. B. (1984). Student patterns of thinking and reasoning. Part Two. *The Physics Teacher, 22* (1), 21–26.

Arons, A. B. (1984). Student patterns of thinking and reasoning. Part Three. *The Physics Teacher, 22* (2), 88–93.

Brasell, H. (1987). The effect of real-time laboratory graphing on learning graphic representations of distance and velocity. *Journal of Research in Science Teaching, 24* (4), 385–395.

Champagne, A. B., Gunstone, R. F., & Klopfer, L. E. (1985). Instructional consequences of students' knowledge about physical phenomena. In L. H. T. West & A. L. Pines (Eds.), *Cognitive structure and conceptual change* (pp. 61–90). Orlando, FL: Academic Press., Inc.

Chi, M. T. H., Feltovich, P. J., & Glaser, R. (1981). Categorization and representation of physics problems by experts and novices. *Cognitive Science, 5,* 121–152.

Clement, J. (1987). Overcoming students' misconceptions in physics: The role of anchoring intuitions and analogical validity. In J. Novak (Ed.), *Proceedings of second international seminar; Misconceptions and educational strategies in science and mathematics III* (pp. 84–97). Ithaca, NY: Cornell University.

di Sessa, A. (1988). Knowledge in pieces. In G. Forman & P. Pufall (Eds.), *Constructivism in the computer age* (pp. 49–70). Hillsdale, NJ: Lawrence Erlbaum Associates.

Driver, R., & Bell, B. (1986). Students' thinking and the learning of science: a constructivist view. *School Science Review, March*, 443–456.

Dufresne, R., Gerace, W., Hardiman, P., & Mestre, J. (1987). Hierarchically structured problem solving in elementary mechanics: Guiding novices' problem analysis. In J. Novak (Ed.), *Proceedings of Second International Seminar: Misconceptions and Educational Strategies in Science and Mathematics III.* (pp. 16–130). Ithaca, NY: Cornell University.

Eaton, J., Anderson, C., & Smith, E. (1984). Students' misconceptions interfere with science learning: Case studies of fifth-grade students. *Elementary School Journal, 84* (4), 365–379.

Feher, E., & Rice, K. (1985). Development of scientific concepts through the use of interactive exhibits in a museum. *Curator, 28*, 35–46.

Fredette, N. H., & Clement, J. (1981). Student misconceptions of an electric circuit: What do they mean? *Journal of College Science Teaching, 11*, 280–285.

Gentner, D., & Gentner, D. R. (1983). Flowing waters or teeming crowds: Mental models of electricity. In D. Gentner & A. L. Stevens (Eds.), *Mental models* (pp. 99–129). Hillsdale, NJ: Lawrence Erlbaum Associates.

Gilbert, J. K., & Watts, D. M. (1983). Concepts, misconceptions and alternative conceptions: Changing perspectives in science education. *Studies in Science Education, 10*, 61–98.

Goldberg, F. M., & McDermott, L. C. (1986). Student difficulties in understanding image formation by a plane mirror. *The Physics Teacher, 24* (8), 472–480.

Goldberg, F. M., & McDermott, L. C. (1987). An investigation of student understanding of the real image formed by a converging lens or concave mirror. *American Journal of Physics, 55* (2), 108–119.

Guesne, E. (1984) Children's ideas about light. In E. J. Wenham (Ed.), *New trends in physics teaching IV* (pp. 179–192). Paris: UNESCO.

Heller, J. I., & Reif, F. (1984). Prescribing effective human problem-solving processes: Problem description in physics. *Cognition and Instruction, 1* (2), 177–216.

Hestenes, D. (1987). Toward a modeling theory of physics instruction. *American Journal of Physics, 55* (5), 440–454.

Hewson, P. W. (1985). Diagnosis and remediation of an alternative conception of velocity using a microcomputer program. *American Journal of Physics, 53* (7), 684–690.

Johsua, S. (1984). Students' interpretation of simple electrical diagrams. *European Journal of Science Education, 6* (3), 271–275.

Jung, W. (1987). Understanding student's understandings: The case of elementary optics. In J. Novak (Ed.), *Proceedings of Second International Seminar: Misconceptions and Educational Strategies in Science and Mathematics III* (pp. 268–277). Ithaca, NY: Cornell University.

Kenealy, P. (1987). A syntactic source of a common "misconception" about acceleration. In J. Novak (Ed.), *Proceedings of Second International Seminar: Misconceptions and Educational Strategies in Science and Mathematics III* (pp. 278–292). Ithaca, NY: Cornell University.

Larkin, J. H. (1980). Teaching problem solving in physics: The psychological laboratory and the practical classroom. In D. T. Tuma & F. Reif (Eds.), *Problem solving and education: Issues in teaching and research* (pp. 111–125). Hillsdale, NJ: Lawrence Erlbaum Associates.

Larkin, J. H. (1981). Cognition of learning physics. *American Journal of Physics, 49* (6), 534–541.

Larkin, J. H. (1983). The role of problem representation in physics. In D. Gentner & A. L. Stevens (Eds.), *Mental models* (pp. 75–98). Hillsdale, NJ: Lawrence Erlbaum Associates.

Larkin, J. H., Reif, F., Carbonell, J., & Gugliotta, A. (1988). FERMI: A flexible expert reasoner with multi-domain interfacing. *Cognitive Science, 12* (1), 101–138.

Lawson, R. A., & McDermott, L. C. (1987). Student understanding of the work-energy and impulse-momentum theorems. *American Journal of Physics, 55* (9), 811–817.

Maloney, D. (1985). Rule-governed approaches to physics: Conservation of mechanical energy. *Journal of Research in Science Teaching, 22* (3), 261–278.

McCloskey, M. (1983). Naive theories of motion: In D. Gentner & A. L. Stevens (Eds.), *Mental models* (pp. 299–324). Hillsdale, NJ: Lawrence Erlbaum Associates.

McDermott, L. C. (1984). Research on conceptual understanding in mechanics. *Physics Today, 37,* 24–32.

McDermott, L. C., Rosenquist, M. L., & van Zee, E. H. (1987). Student difficulties in connecting graphs and physics: Examples from kinematics. *American Journal of Physics, 55* (6), 503–513.

McDermott, L. C., & van Zee, E. H. (1984). Identifying and addressing student difficulties with electric circuits. In R. Duit, W. Jung, & C. von Rhöneck (Eds.), *Proceedings of an international workshop: Aspects of understanding electricity* (pp. 39–48). Ludwigsburg, W. Germany: Pädagogische Hochschule Ludwigsburg.

Minstrell, J. (1982). Explaining the "at rest" condition of an object. *The Physics Teacher, 20* (1), 10–14.

Mokros, J. R., & Tinker, R. F. (1987). The impact of microcomputer-based labs on children's ability to interpret graphs. *Journal of Research in Science Teaching, 24* (4), 369–383.

Moreira, M. (1987). Concept mapping as a possible strategy to detect and to deal with misconceptions in physics: In J. Novak (Ed.), *Proceedings of Second International Seminar: Misconceptions and Educational Strategies in Science and Mathematics III* (pp. 352–360). Ithaca, NY: Cornell University.

Murray, T., Schultz, K., Brown, D., & Clement, J. (1987). *Remediating physics misconceptions using an analogy-based computer tutor. The Journal of Interactive Learning Environments, 1* (1), (In press: January 1990).

Novak, J.(Ed.). (1987). *Proceedings of Second International Seminar: Misconceptions and Educational Strategies in Science and Mathematics III.* Ithaca, NY: Cornell University.

Piaget, J. (1974). *Understanding causality.* New York: Norton.

Ramadas, J. (1982). Use of ray diagrams in optics. *School Science, 10,* 1–8.

Reif, F. (1985). Acquiring an effective understanding of scientific concepts. In L. H. T. West & L. Pines (Eds.), *Cognitive structure and conceptual change* (pp. 133–151). Orlando, FL: Academic Press, Inc.

Reif, F. (1986). Scientific approaches to science education. *Physics Today, 39,* 48–54.

Reif, F. (1987a). Instructional design, cognition, and technology: Applications to the teaching of scientific concepts. *Journal of Research in Science Teaching, 24* (4), 309–324.

Reif, F. (1987b). Interpretation of scientific or mathematical concepts: Cognitive issues and instructional implications. *Cognitive Science, 11,* 395–416.

Rosenquist, M. L., & McDermott, L. C. (1987). A conceptual approach to teaching kinematics. *American Journal of Physics, 55* (5), 407–415.

Schuster, D. G. (1987). Understanding scientific derivations: A task analysis and constructivist learning strategy: In J. Novak (Ed.), *Proceedings of Second International Seminar: Misconceptions and Educational Strategies in Science and Mathematics III,* (pp. 448–456). Ithaca, NY: Cornell University.

Siegler, R. S. (1987). Three aspects of cognitive development. *Cognitive Psychology, 8,* 481–522.

Stead, B. F., & Osborne, R. J. (1980). Exploring science students' concepts of light. *Australian Science Teaching Journal, 26* (3), 84–90.

Strike, K. A., & Posner, G. J. (1982). Conceptual change and science teaching. *European Journal of Science Education, 4* (3), 231–240.

Thornton, R. K. (1987). Access to college science: Microcomputer-based laboratories for the naive science learner. *Collegiate Microcomputer, V* (1), 100–106,

Thornton, R. K. (1987). Tools for scientific thinking—microcomputer-based laboratories for physics teaching. *Physics Education, 22* (4), 230–238.

Tiberghien, A. (1983). Revue critique sur les recherches visant à élucider le sens de la notion de lumière pour les élèves de 10 à 16 ans. English translation in book: "Critical review of research aimed at elucidating the sense that notions of light have for students aged 10 to 16 years." *Research on Physics Education: Proceedings of the First International Workshop* (pp. 125–134). La Londe les Maures, France.

Tiberghien, A., Delacote, G., Ghiglione, R., & Matalon, B. (1980). Conceptions de la lumière chez l'enfant de 10–12 ans. English translation in book: "Conceptions of light in children aged 10–12." *Revue Française de Pedagogie, 50,* 24–41.

Touger, J. S., Dufresne, R., Gerace, W., & Mestre, J. (1987). Hierarchical organization of knowledge and coherent explanation in the domain of elementary mechanics. In J. Novak (Ed.), *Proceedings of Second International Seminar: Misconceptions and Educational Strategies in Science and Mathematics III.* (pp. 517–530). Ithaca, NY: Cornell University.

Trowbridge, D. E., & McDermott, L. C. (1980). Investigation of student understanding of the concept of velocity in one dimension. *American Journal of Physics, 48* (12), 1020–1028.

Trowbridge, D. E., & McDermott, L. C. (1981). Investigation of student understanding of the concept of acceleration in one dimension. *American Journal of Physics, 49* (3), 242–253.

van Zee, E. H., & McDermott, L. C. (1987). Investigation of student difficulties with graphical representations in physics. In J. Novak (Ed.) *Proceedings of Second International Seminar: Misconceptions and Educational Strategies in Science and Mathematics III.* (pp. 531–539). Ithaca, NY: Cornell University.

Viennot, L. (1985a). Analyzing students' reasoning in science: A pragmatic view of theoretical problems. *European Journal of Science Education, 7* (2), 151–162.

Viennot, L. (1985b). Analyzing students' reasoning: Tendencies in interpretation. *American Journal of Physics, 53* (5), 432–436.

Watts, D. M. (1985). Student conceptions of light: A case study. *Physics Education, 20,* 183–187.

White, B., & Frederiksen, J. (1987). *Causal model progressions as a foundation for intelligent learning environments.* (Report No. 6686). Cambridge, MA: BBN Laboratories Inc.

Williams, J., Trinklein, F., & Metcalfe, H. C. (1984). *Modern physics.* New York: Holt, Rhinehart and Winston.

2 Research in Chemical Education: Results and Directions

J. Dudley Herron
Purdue University

INTRODUCTION: THE CONSTRUCTIVIST PERSPECTIVE

This review is a reflection of my thinking as much as it is a reflection of existing research. My beliefs about what is important in chemical education, what I know about teaching and learning, and the idiosyncratic thought patterns that I use to make sense out of what I read and observe have inevitably influenced what I reviewed and how I interpreted it. This is self-evident and has undoubtedly been understood for centuries, but there is a new awareness of this constructive nature of learning that has shaped recent research in science education. Whether one traces this view of learning to Piaget as Bodner (1986) does in his discussion of constructivism, views it in terms of a new alternative conceptions-research paradigm as do Gilbert and Swift (1985), or in terms of artificial intelligence and cognitive science in general, as does Resnick (1987), it is a view that is shaping current research in education, and chemical education is no exception. If we can learn how, it will shape the teaching of chemistry as well.

RESEARCH ON PROBLEM SOLVING

Importance of Problem Solving

As Resnick (1987) points out in the recent monograph, *Education and Learning to Think,* the so called "higher order" thinking skills associated with problem solving have always been an important goal of education, but they have not always been a goal for all students. Education has been an elitist tradition.

Chemical education, and science education in general, has been a part of this elitist tradition. Whether acknowledged or not, we have operated on the assumption that some people are born bright and some are born stupid; only the bright ones should succeed in chemistry. Our beliefs about intelligence and the nature of our discipline have shaped chemical education and limited it. So have the beliefs of students.

Student Beliefs

Carter (1987) analyzed the beliefs of undergraduate chemistry students and the influence of belief on problem solving. Carter interviewed nine students enrolled in the first semester of a year-long general chemistry course for science and engineering majors. Students were presented with a variety of traditional and nontraditional problems in chemistry and nonchemistry contexts in a series of interviews throughout the semester.

Carter described the beliefs held by four students, Lorna, Chadd, Maureen, and Rob, to illustrate the range of beliefs observed in the study. Lorna and Maureen see chemistry as abstract and alien. Their job is to absorb and reproduce knowledge presented by the teacher, an authority from another world. Problems are tasks that require calculations and an answer, but not a question, and the only purpose of solving problems is to get an answer that "they" want. The way to do problems is to reproduce algorithms and recognize problem types; there is no role for creativity in chemistry. The way to succeed is to work the same problems over and over until they are memorized.

Rob and Chadd hold quite different beliefs. To them, chemistry is a creative way of understanding concepts and problems. They see themselves as the source of knowledge and they see their role as putting concepts together and applying them to solve problems. The teacher is there to motivate, answer questions, and explain when necessary. Problems are tasks where one must think creatively and synthesize ideas; problems are not algorithmic. The goal of problem solving is to understand ideas and apply them to new contexts, and the way to get good at it is to work problems, think about the concepts involved, and relate ideas to previous knowledge.

> It would be comforting to believe that students like Lorna and Maureen are just "dumb", or "lazy", and that their difficulties in learning chemistry are not related to their formal instruction in chemistry. Lorna and Maureen are neither dull nor lazy, however. They [are] bright, hardworking students trying as best they [can] to make sense out of chemistry as taught in their high school and college courses. They put in many hours on what they believed was productive study. They tried to get good grades—which they believed measured understanding and success. The idea that chemistry may be understood in ways other than their instrumental methods would have astonished Lorna and Maureen. (Carter, 1987, p. 315)

Students like Lorna and Maureen are estranged from chemistry as a discipline. Chemistry is "out there" and the people who do it are some undefined "they" who want students to perform in mysterious ways that don't make much sense. The student's job is to figure out what they want done, to recall how they say a problem should be solved, and to apply their procedure to generate an answer they expect. There is little consideration of whether the answer to a problem or the problem itself makes sense. It isn't supposed to.

This impression that chemistry is largely arbitrary and meaningless is not confined to students in college courses. It seems to be shared by the 13–16 year-olds in Australia who were asked by Ellerton and Ellerton (1987) to make up questions that would be difficult to answer and to indicate how they would respond if the question were on a test. Most of the questions emphasized memory, facts, and formal treatment of symbols, presumably what the students thought chemistry is about. But the most interesting result was the students' indication of how they would respond to the questions they had written. They frequently said that they wouldn't or couldn't respond, that they would skip the question, or would simply panic.

Effect of Problem Complexity

Student beliefs are only one of many factors that influence student performance in problem solving. The complexity of the problem certainly plays a role. When operating in a relatively familiar domain such as stoichiometric calculations, students are able to solve one-step problems such as finding the mole mass of a compound or calculating the number of moles corresponding to a given mass of the compound, but they are quite unsuccessful when a problem requires the stringing together of such steps to solve a more complex task. Frazer and Sleet (1984) gave tests consisting of three complex problems along with separate items over the subproblems involved to 76 sixth-form students (age 17–18) in four English schools. For the first problem 77% were successful on all of the subproblems, but only 37% succeeded on the main problem; for problem two the respective percentages were 83% and 52%; for problem three they were 73% and 57%. Lazonby, Morris, and Waddington (1985) report similar findings.

There are several possible explanations for this kind of difficulty. One might argue that in one-step problems students aren't solving problems at all but are simply applying well practiced algorithms. When the more complex problem is presented, deeper understanding of the relationships among concepts and skills is required to string steps together to produce a sensible result. Alternatively, failure on the more complex tasks may be due to poorly developed metacognitive skills that govern the organization of work, sequencing of tasks, and checking of results. We are not certain what is going on. It is certain that the number of elements to be considered plays a role in success.

M-demand and Working Memory

It is generally agreed that solving problems or performing any other intellectual task involves thinking in what we call working memory and that the capacity of working memory is limited. Pascual-Leone describes that capacity as *M-space* and has proposed a neo-Piagetian model of intellectual development that adds the idea of M-space to Piaget's well-known theory of intellectual development (Pascual-Leone & Smith, 1969). According to the model, intellectual performance is influenced by both the intellectual level of the task and the M–demand of the task.

Niaz and Lawson (1985) examined the influence of intellectual development, M-space, and field dependence-field independence on success at balancing chemical equations by trial and error. They concluded that balancing equations by trial and error requires formal thought and that success on more difficult equations (those that require several steps) depends on M-space. (Field dependence had little effect.) In a subsequent study Niaz (1987) analyzed chemistry test items for their *M-demand* (i.e., the number of "bits" of information that must be held in memory in order to solve the problem) and measured the M-space of students. Niaz observed that the greater the M-demand of the items, the lower the success rate, and the greater the M-space of the students, the higher the success rate. In an earlier study Johnstone and El-Banna (1986) obtained similar results; they found a sharp drop in the proportion of students getting test items correct when the M-demand exceeds five or six.

So what shall we make of this research which shows that problems with a high M-demand are not solved by our students? Such problems remain, and experts in chemistry have somehow learned to solve them. Have the experts enlarged their M-space or have they managed in other ways to deal with complex tasks?

Although work by Pascual-Leone (1970), Case (1972), and Scardamalia (1977) indicates that M-space increases with age, it seems to level off in the midteens. There is little evidence that the M-space of experts is appreciably greater than that of high school or college chemistry students. However, experts organize information in chunks that can be manipulated with less demand on M-space than would be the case for the novice.

Conceptual Understanding

Experts and novices differ in other important ways. Proportional reasoning, hypothetico-deductive reasoning, and the ability to think in terms of atoms and molecules—entities that must be imagined rather than observed—play an important role in chemistry. Such reasoning is characteristic of formal operational thought—thought patterns that are poorly developed in a substantial fraction of our students (Chiappetta, 1982; Herron, 1975).

It seems reasonable that *formal operational* students would be successful at solving complex chemistry problems whereas *concrete operational* students would not. Sue Nurrenbern and I designed a study of chemistry problem solving at the high school level to test the idea (Nurrenbern, 1980). The results were disappointing. Essentially none of the students in that study were successful, and individual interviews of students suggested that the fundamental problem was that students didn't understand the concepts required to solve the problems.

Clearly, how well concepts are understood is an important factor in problem-solving success and it seems reasonable that how concepts are organized—what concepts are connected—will also be important. Kempa and Nicholls (1983) studied the level of concept attainment for acid, alkali, element, ion, and salt among students ages 14–15 taking GCE *O*-level chemistry in England and the effect of concept attainment on solving examination problems related to the concepts. They concluded that: (a) The cognitive structures of good problem solvers are more complex and contain more associations than those of poor problem solvers, (b) the links between concepts do not appear to be different for good and poor problem solvers, but that the strength of those links may differ, and (c) that the deficiencies in cognitive structures of poor problem solvers are predominantly for concepts of higher levels of abstraction.

Other Difficulties

Although conceptual understanding is essential for successful problem solving, there are other issues to consider. Tom Greenbowe (1984) interviewed 30 chemistry students and 1 chemistry professor as they solved a series of stoichiometry problems. The students ranged from new college freshmen to graduate students, and they differed considerably in their ability to solve problems.

Several interesting things came out of the study. First, all of the subjects, including the professor, had difficulty with the more complex problems even though all were taken from introductory texts. Second, there was a substantial difference between the way "experts" in the study solved problems and the solutions outlined in texts. Solutions provided in texts represent straightforward, efficient algorithms that cut cleanly and move swiftly to solution. They may represent how experts approach the task after it is clearly understood and well-practiced, but they do not represent the way experts approach the task when it is an unfamiliar problem.

This difference between the problem-solving performance of experts and text-book solutions is significant because the examples must convey to students an unrealistic idea about how problems are actually attacked. The examples provide no indication of the false starts, dead ends, and illogical attempts that characterize problem solving in its early stages, nor do they reveal the substantial time and effort expended to construct a useful representation of a problem before the systematic solution shown in examples is possible.

Representation

Clear and careful representation of the problem seemed to be the key to success in Greenbowe's study, and successful problem solvers represented problems in a variety of ways. They thought about the chemistry, and they made connections between symbolic representations such as formulas and equations, the microscopic entities symbolized, and macroscopic events. The less successful students looked for cues in the problem statement and recalled algorithms that seemed to fit. They often focused on an inappropriate balanced equation and applied the factor-label algorithm to arrive at an answer that made no sense. Consider Sue, one of the students in Greenbowe's study (Herron & Greenbowe, 1986).

The following problem (Mahan, 1975) was difficult for most of the students in the study (only 32% solved it), and Sue's errors are typical.

A 1.00-g mixture of cuprous oxide, Cu_2O, and cupric oxide, CuO, was quantitatively reduced to 0.839 g of metallic copper by passing hydrogen gas over the hot mixture. What was the mass of CuO in the original mixture? (p. 29)

Sue wrote the following equations and then applied the factor-label algorithm to find the mass of CuO in the original mixture.

$$CuO + Cu_2O \rightarrow Cu_3O_2$$
$$Cu_3O_2 + 2 H_2 \rightarrow 3 Cu + 2 H_2O$$

Not only did Sue assume a 1:1 molar ratio of the two oxides in writing the equations, but she reported that there was 1.05 g of CuO in the 1.00-g mixture. She was oblivious to both errors and seemed confident in her answer.

Behavior such as Sue's is observed among high school students as well. Anamuah-Mensah (1986) studied the strategies used by chemistry students in Canada to solve titration problems. Students were taken into the laboratory where they titrated HCl against standardized NaOH and calculated the molarity of the HCl. When they finished, they were asked to repeat the calculations: (a) assuming a 2:1 molar ratio of acid to base, (b) assuming that the acid was H_2SO_4, rather than HCl, and (c) assuming that Na_2CO_3 was the base rather than NaOH. Although the majority of the high- and average-ability students solved the original problem, students at all ability levels seemed to respond instrumentally rather than relationally. Performance on the prediction problems seems to support this inference; only 13 of the 49 students in the study got all three of them correct. Studies by Gabel and Sherwood (1984) and Gabel, Sherwood and Enochs (1984) also suggest that the majority of high school students solve chemistry problems using only algorithmic methods and do not understand the chemical concepts on which the problems are based.

Problem Solving Instruction

Approaches used to teach problem solving vary from highly structured approaches, where students are given an explicit set of steps to follow, to approaches that emphasize general strategies that may help in solving some problems while affording little help in solving others.

One of the best examples of the structured approach is that employed by Mettes and his associates in The Netherlands (Kramers–Pals, Lambrechts, & Wolff, 1983; Mettes, Pilot, & Roossink, 1981; Mettes, Pilot, Roossink, & Kramers–Pals, 1980, 1981). In their approach, students are given a general Program of Actions and Methods (PAM), which is divided into four phases: (a) read and analyze the problem, (b) transform the problem into a standard problem, (c) carry out the routine operations of the standard problem, and (d) review the result. Although these phases are similar to recommendations made in less structured approaches, here the steps are explicit and sequential.

The PAM has been translated into a Systematic Approach to Solving Problems Chart (SAP), a flowchart which spells out in detail the steps students must take in solving a thermodynamics problem. Students solve problems individually or in small groups using the SAP worksheets and are expected to design their own Key Relations (KR) charts listing important laws, definitions, equations, and the like. Students taught to use the SAP charts in a university thermodynamics course outperformed other students on course examinations.

Bunce and Heikkinen (1986) were less successful with their structured approach to problem-solving instruction. Bunce used a worksheet when solving mathematical chemistry problems in college lectures. The worksheet contained: (a) a sketch of the situation described in the problem, (b) all information (rules, equations, definitions, etc.) needed to solve the problem, (c) labels for information given or requested, (d) a qualitative statement regarding what the problem asked, (e) a breakdown of the problem into subproblems, (f) a list of steps needed to solve the problem, and (g) the solution. This technique had little effect, partly because students didn't use it. According to students' reports, it was too time consuming.

A less structured approach to problem solving instruction is illustrated by Frank's (1986) study in an introductory chemistry course for science and engineering majors at Purdue. Each of the two graduate students taught two recitation sections. Problem solving was stressed in one section, but not in the other. The instruction was loosely based on Polya's problem-solving model: (a) understanding the problem, (b) designing the plan, (c) carrying out the plan, and (d) looking back (Polya, 1957). However, the first two stages were combined into one called "planning the solution," partly because Yackel's (1984) work had shown that these stages are not done separately or sequentially.

Students were encouraged to ask questions such as, "What is the unknown? What are the conditions? What do these substances look like? How do the atoms and molecules involved here interact? How could you symbolize what you see?" Instructors asked similar questions when they modeled problem solutions, and they pointed out when a student had neglected one of the stages such as quitting a problem without judging whether the answer makes sense.

Much of the class time was spent with students working on novel problems in small groups. The instructors monitored the progress of the small groups and gave help only when it was needed. The class period concluded with a discussion of strategies used and what was learned about problem solving. Students in the treatment group outperformed students in a control group on the four major exams in the course, and think–aloud interviews conducted at the end of the semester showed that the experimental students made more generalizable representations of problems, had fewer uncorrected math errors, were more persistent, and evaluated their work more frequently than did students from the control group.

The improvements in performance observed in Frank's study were modest. We believe that greater improvement would require intervention in all parts of the course. That is what Fasching and Erickson (1985) did in their study. One half of each lecture period was spent in unstructured, small-group problem-solving activity. In addition, a research project was required, and students engaged in open–ended laboratory activities.

Fasching and Erickson apparently did not conduct class discussions of general problem-solving strategies, an important component of Frank's intervention, but otherwise their procedures appear to be similar. Unfortunately, Fasching and Erickson were unable to conduct a controlled evaluation of their treatment, but their qualitative analysis of student performance suggests noticeable improvement in problem solving by about 40% of their students.

If we believe that knowledge is constructed and that we begin that construction by analyzing events that can be experienced directly, it may be useful to introduce problems in a familiar context before moving to more abstract, formal contexts. Some support for this notion is found in Gabel and Samuel's (1986) study of high school students' ability to solve molarity problems. Tests were constructed to include common molarity problems and analogous problems involving making lemonade. The results suggest that learning analog problems might help students understand molarity. It was also found that problems involving fractions were more difficult than those involving whole numbers and that problems involving dilution or evaporation of solvent were more difficult than those involving dissolving solids. These differences in difficulty may be due to greater difficulty in constructing meaningful representations for less familiar situations or to differences in M-demand.

Visualization

The ability to visualize the situation described in a problem probably aids problem solving by providing a representation that guides solution. In Greenbowe's study more successful problem solvers frequently referred to the macroscopic conditions implied by a problem statement or described what atoms and molecules must be doing as they solved problems; less successful students did not. These visualizations seem to provide useful information about what is sensible to do.

Kleinman, Griffin, and Kerner (1987) investigated the kind of images called forth by key words in chemistry (bond, energy, equilibrium, functional group, mole, etc.) by undergraduates, graduate students, and faculty. Images were classified as *associative* if the response was a word association such as K_{sp} (solubility product constant) in response to solubility. Images were *real* if they were concrete or real-world such as "sun" for energy. *Model* images were abstractions such as "electron density" for orbital.

The total number of images recalled and the level of abstraction of the images increased with the experience of the subjects (faculty had better recall than graduate students and graduate students had better recall than undergraduates). Almost 50% of the images reported by undergraduates (average of 9.1 out of 19.8) were associative whereas only 10% (average of 3 out of 31) of the images reported by faculty were associative. Twenty four out of 31 images reported by faculty were model images. This difference in the kind of images evoked undoubtedly accounts for many differences in the way novices and experts represent problems.

If images are important, perhaps we should try to encourage students to form them. McIntosh (1986) investigated this proposition in 9th-grade physical science classes, which studied gases for three days. One group of students was asked to create an image of a typical gas as it responded to pressure, temperature, or volume and to draw that image in their notes. Another group was asked to write a rule and say it aloud. Following the three days of instruction, tests were given—a 6-item recall test and a 6-item essay test calling for application of the rules to solve novel problems. In addition, students were asked to rate, on a 5-point scale, the degree to which they used imagery when answering each question. On the recall questions, those who wrote the rules and said them aloud scored higher than those who recalled an image and drew a picture, but there was no main effect on the transfer test. There was, however, an interesting interaction. Those who wrote the rules and claimed that they did not use images to answer the questions scored considerably lower than the other students. The highest scoring students were those who wrote the rules and also claimed that they used images when answering the test questions. In view of the research reported in the section on conceptual understanding, it seems likely that the 3-day treatment requiring students to draw pictures of gases was insufficient to produce meaningful images of gas behavior, but that students who had constructed and spontaneously used images (regardless of treatment) were aided by those images when solving problems.

For many problems in organic chemistry the ability to visualize molecular structures is clearly important. These structures are normally represented by 2-dimensional line drawings, and many students have difficulty identifying identical structures represented after rotation. Seddon and Tariz (1982) found that between 55% and 78% of English students ages 16–17 and 36% to 72% of Portuguese students ages 18–21 were able to identify representations of molecular structures after various spatial changes took place. In a subsequent study, Seddon and Moore (1986) reported an unexpected result in a study aimed at improving students' ability to recognize molecular structures after rotation. Students were given models that they compared with profiles of line drawings of the same structure presented on a television screen. No short-term learning occurred. However, when the models were eliminated and students simply focused on the line drawings as they were transformed, the instruction was effective.

In another study of spatial visualization, Bodner and McMillen (1986) tested the spatial ability of college chemistry students and calculated correlations between scores on the spatial test and achievement on stoichiometry ($R = .29$), crystal structure measured by multiple-choice test items ($r = .32$) and free response items ($r = .35$); and a comprehensive final exam ($r = .30$). The authors suggest that the correlations are probably due to cognitive restructuring required for success on both the spatial test items and the achievement test items.

RESEARCH ON CONCEPTUAL UNDERSTANDING

Understanding Chemical Symbols

There is a substantial body of research documenting misconceptions about fundamental chemistry concepts. For example, balancing chemical equations has been a standard part of the chemistry curriculum for years, but there is evidence that students don't understand what equations should be telling them. Yarroch (1985) gave 14 students in the top third of two classes four simple equations to balance. As expected, all of the students were successful at balancing the equations, but only five could draw diagrams that accurately depicted the molecules and groups of molecules represented. Similarly, in a series of studies, Nurrenbern and Pickering (1987) found that students could solve routine problems concerning gas laws and stoichiometry reasonably well, but when asked to identify diagrams that would represent the distribution of gas molecules in a container or draw diagrams to represent the changes in molecules taking place in the stoichiometry problems, student performance dropped drastically. Many others have documented common difficulties that students have "thinking atoms and molecules" (Ben-Zvi, Eylon, & Silberstein, 1982, 1986; Bleichroth, 1965; Cros, & Maurin, 1986; Glassman, 1967; Mitchell & Kellington, 1982; Novick & Nussbaum, 1978, 1981; Pfundt, 1981, 1982).

Gorodetsky and Gussarsky (1986, 1987), Wheeler and Kass (1978) and others have documented common misconceptions concerning equilibrium; still others have studied common misconceptions about mole, volume, heat, temperature, energy, and a multitude of other concepts commonly taught in chemistry (Albert, 1978; Andersson, 1980; Barke, 1982; Ericson, 1979, 1980; Johnstone, Mac-Donald & Webb, 1977; Novick & Mannis, 1976; Osborne & Cosgrove, 1983; Shayer & Wylam, 1981).

Misconceptions and Alternative Explanations

If all we were learning from the research on misconceptions was that students have them, it probably wouldn't be worth mentioning. Fortunately, there is more. The better work in this area aims at understanding how student conceptions evolve, how their "naive" conceptions interfere with learning more powerful constructs, and how we might build on students' existing concepts to teach those more powerful concepts.

It is useful to divide consideration of misconceptions into two categories. One category of misconceptions deals with what actually happens in the physical world. Students believe that heavy objects always sink in water, that the bubbles seen in boiling water are hydrogen or oxygen, that rapidly boiling water is at a higher temperature than gently boiling water, and that mass changes when matter melts or boils. These ideas are simply contrary to empirical facts.

Other misconceptions deal with students' explanations of what happens in the natural world. For example, several studies have been done concerning students' conceptions of heat (Albert, 1978; Erickson, 1979, 1980; Hewson, 1984; Shayer & Wylam, 1981). In general, students understand heat in terms of the caloric theory; heat is a substance that flows between bodies. This is not the scientist's view and, as a consequence, it is labeled as a misconception. However, students can explain what happens in the natural world in terms of a caloric theory of heat and, as Fuchs (1987) points out, it is possible to develop an internally consistent theory of thermodynamics around the caloric idea. The student's conception "works;" it explains what happens in the natural world, but the explanation differs from the one accepted in science. The point is that many ideas labeled as misconceptions are not misconceptions about what happens but alternative explanations of what happens. In most cases those explanations are quite logical from the student's point of view, are consistent with their understanding of the world, and are resistant to change.

Factors that Shape Conceptual Understanding

Exactly how students arrive at naive conceptions is not known, but culture, everyday experience, and language all seem to play important roles. Hewson (1984, 1986) examined the role of language and culture on student understanding of heat and of floating and sinking. Common metaphors in Sesotho, the native language of the South African students involved in the study, appeared to affect the students' conceptualization of heat. Sesotho does not contain a word for volume as understood in science and words for mass/weight are rather imprecise. Furthermore, precise measurements of volume are seldom made in everyday experience. These factors seemed to influence the students' attempts to explain why some objects float and others sink, and Hewson points out that the formal instruction students received on mass, volume, and density generally did not have a significant impact on their explanations.

In another study of the role of language and culture in student conceptions, Lynch, Chipman, and Pachaury (1985) investigated understanding of several words related to matter (length, mass, area, volume, element, compound, mixture, atom, electron, etc.) in India and Australia. Although total test scores were comparable in the two cultures, some words were much better understood in one culture than the other. Gardner (1972) in Australia and Cassells and Johnstone in Scotland (Cassells, 1976, 1980; Cassells & Johnstone, 1980) have looked at students' understanding of ordinary English words frequently used in science. The essential outcome of these studies is that a substantial number of students do not understand many common words used to teach and test students in science, and this lack of understanding contributes to misconceptions as well as interferes with efforts to test students' understanding of science concepts.

Even scientists do not agree on the precise meaning of words. In an informal survey conducted several years ago, college chemistry students, graduate students, and professors were shown the following series of equations and asked which processes were properly described as ionization (Driscoll, 1978; Herron, 1977, 1978).

$$Na(g) \rightarrow Na^+(g) + 1 e^-$$

$$Cl(g) + 1 e^- \rightarrow Cl^-(g)$$

$$Fe^{2+}(g) \rightarrow Fe^{3+}(g) + 1 e^-$$

$$F^-(g) \rightarrow F(g) + 1 e^-$$

$$HCl(g) \rightarrow H^+(aq) + Cl^-(aq)$$

$$NaCl(s) \;\; \rightarrow \;\; Na^+(aq) + Cl^-(aq)$$

$$HAc(aq) \;\; \rightarrow \;\; H^+(aq) + Ac^-(aq)$$

The results indicated considerable variation in respondents' conceptions of ionization.

In a more recent article, Weninger (1982) calls attention to inconsistencies in the way various symbols such as "=" and "→" are used in chemistry classes and makes an appeal for more precise use of symbolic language. Laudable though this appeal may be, it seems unlikely that we will achieve consistent, precise use of language within the chemistry classroom or that inconsistency and imprecision will cease to be a source of confusion for students.

Changing Conceptual Understanding

According to the conceptual change model outlined by Strike and Posner (1982), the first condition for conceptual change is personal dissatisfaction with existing conceptions. To produce that dissatisfaction, educators suggest that we confront students with some discrepant event—some observation that cannot be explained in terms of the students' current understanding. However, a study by Rowell and Dawson (1983) casts doubt on the utility of this strategy in precipitating conceptual change. They sum up their research on the use of laboratory counterexamples by saying:

> the results . . . together with those of Karmiloff-Smith and Inhelder (1974) detail the problem that empirical counterexamples may not provide the information needed by many students to correct their misconceptions, certainly not in the short term, and possibly not at all. (p. 212)

These results should not be taken to mean that counterexamples are not important, only that they are not sufficient. In the study by Karmiloff-Smith and Inhelder cited above, the authors discuss several factors that influence conceptual change. In the study, children tried to solve the task of balancing blocks on a rod. When the children simply tried to get the blocks to balance without considering why, they would persist for a long time if they were successful, but they did not learn much. It was only when they failed repeatedly and were forced to explore the reasons for failure by adopting some "theory-in-action" such as the block balances at the geometric center that they learned. Although negative responses were a necessary condition for progress, they were not sufficient. Furthermore, as a theory-in-action became consolidated—that is, it seemed to be working consistently—information that tended to contradict the theory (counterexamples) was not easily assimilated.

The difficulty of assimilating information that contradicts a consolidated theory is seen in a case study of "Sharon," reported by Scott (1987). Sharon, a 14-year–old of average intelligence, was studying solids, liquids, and gases in Scott's class. Sharon started with a continuous model of matter but adopted a particle model as a result of instruction. However, it took several weeks for Sharon to accept that particles could be far apart with nothing between. A story which Scott told about a goldfish that died in distilled water because there was no air seemed to prove to Sharon that there couldn't be air between particles of water, as she had supposed, but even after accepting the proof, she didn't give up the idea until much later. Scott uses the incident to call into question the utility of critical experiments in casting doubt on a student's current understanding.

Conceptions of Matter

It is clearly not enough to confront students with information that conflicts with their theories-in-action. We need to know more about their conceptions, how they develop, and how they change over time. Studies are beginning to provide some insight here.

Andersson (1986) has provided a thoughtful synthesis of several studies that examined students' understanding of chemical reactions. He places student explanations in five categories:

1. *It is just like that*. This, of course, is no explanation; the individual just accepts that it happens.

2. *Displacement*. Displacement explanations suggest that something in one place has simply moved to another. For example, in explaining how dark coating appears on copper pipes, students suggested that "hot water makes steam which forms a coating on the pipe."

3. *Modification*. Modification explanations suggest that what appear to be different substances are actually the same substance in a different form. For example, in describing burning alcohol, "When you burn alcohol, there's alcohol vapour; that's what you'd expect, its normal."

4. *Transmutation*. Here the student suggests that a chemical change has produced totally different materials. "The steel wool that has burnt has turned into carbon. Carbon weighs more."

5. *Chemical interaction*. This category contains those explanations that would be acceptable to a chemist; for example, "Copper and oxygen have reacted." as an explanation for the dark coating that forms on a copper pipe:

It is important to note that these explanations are all rational and are correct for many events. Even scientists have no better explanation for opposite charges attracting than "it is just like that," dark coatings can appear on copper pipes because substances such as paint are "displaced," rock candy and powdered sugar look very different but are viewed as a simple "modification" without chemical change, nuclear reactions are commonly described as "transmutations," and or course many changes are attributed to "chemical interaction." Rather than criticizing students for giving such "naive" explanations, we need to understand how they derive them so we can suggest more powerful explanations in an intelligible manner.

Stavy (1987) investigated the development of conservation of matter in children in first to ninth grades (ages 6–15) in Israel. At about the second grade children conserve matter when confronted with the standard Piagetian task where clay is deformed, but Stavy's work shows that they do not conserve when confronted with other situations.

Several tasks were used in the study. In one, identical candles were shown and one was then melted. Two ice cubes were shown and then one was melted. In other tasks, drops of acetone were placed in two closed tubes and one was evaporated; similarly, crystals of iodine were placed in tubes and one was sublimed. In another task two cups of water and two spoons filled with sugar were observed and in one of the cups, the spoonful of sugar was dissolved. Finally, two water thermometers were prepared and one was subsequently heated, causing the water to rise in the narrow tube. In each task students were questioned to determine whether they thought the amount of matter present was still the same after the transformations. The proportion of students at each age level who were successful on the tasks varied considerably from task to task. Of particular interest was the difference in performance on the two evaporation tasks. Up to Grade 7, students were considerably more successful on the iodine task, where the gas was visible, than on the acetone task. The author suggests that in teaching science concepts, careful selection of examples that provide perceptible clues concerning what is taking place is likely to help.

Reustrom (1987) interviewed 20 students in Grades 7 through 9 to determine how they conceptualize matter and how those conceptions develop over time. Based on extensive interviews, the author suggests that matter is first seen as a continuum and then as a continuum with atoms within the continuum. (Note that at this stage matter is still continuous and that the atoms are simply objects within it rather than the basic building blocks of the matter.) Next, matter is seen as made up of discontinuous units with a "shell" or skin that holds some kind of nucleus such as "taste," "odor," or atoms. Matter may then be seen as consisting of particles, but there may be some other substance between the particles. Finally, the atomic view of the chemist is adopted. The students' understanding of the atoms themselves emerges through stages as well. First, atoms are simply words with no apparent meaning at all, then names for very small things. Next, atoms are seen as being made up of other atoms, and ultimately the chemists' model is accepted. However, the author was never certain whether students accepted this final view with real conviction or simply responded with the right words.

The understanding of matter expressed by students in Reustrom's study was certainly influenced by their study in science, and it is unlikely that the particular developmental sequence described would be the same for all students in all classes. What is interesting about the study is the gradual change in the students' ideas from a conception that seems reasonable in the face of everyday experience to the conception accepted by chemists. This gradual evolution of the atomic model seems to support the constructivist view that new ideas are built from existing ideas rather than accepted as they are.

WHAT WE SEEM TO KNOW; WHAT WE NEED TO LEARN

At a time when the ability to address novel tasks using chemical and other science knowledge appears to be increasingly important for economic and social survival, we are becoming increasingly aware that a large percentage of our students are not acquiring the understanding needed to do so. The simple fact is that for many—perhaps most—of our students, chemistry just doesn't make sense. Even those who earn good grades often perform dismally when asked to apply what they have learned to solve novel problems.

Through research such as that reviewed here, we are beginning to understand many of the reasons for such poor performance: Students have learned to apply algorithms to solve routine problems, but they have not learned general strategies that they can use to solve novel tasks. Many believe that they are incapable of understanding the ideas presented in chemistry, they see chemistry as something that "they" do "out there" to produce information that must be received without question. Students manipulate symbols according to memorized rules without connecting the symbols with the macroscopic events and the microscopic models that the symbols represent. They are often overwhelmed by complex problems because the M-demand of the task exceeds the capacity of their working memory, and they have not developed organizational strategies that enable them to handle the complexities. Many have not developed the logical operations used in science to rationalize and explain experience, and the arguments based on formal operations reasoning do not make sense. When working problems, they either do not see the need to verify their logic and results, have not developed effective verification strategies, or have such a weak knowledge base that common verification strategies can't work. The ideas that they have developed to explain experience often conflict with and interfere with the ideas we ask them to accept.

Through the research on conceptual understanding we are beginning to learn how students view the world and how those views develop over time, and we are beginning to test some promising models for conceptual change. Still, much research is needed before we will be able to make the changes in chemical education that we need. Few teachers at either the pre-college or college level are aware of "children's science" and how it differs from "school science." Furthermore, most teachers operate on the assumption that knowledge is transmitted, intact, from teacher to learner and little consideration is given to how students construct knowledge that is sensible enough to use. For those teachers who are aware of students' conceptions of science and believe that knowledge is constructed rather than received, there are few specific guidelines to suggest how they should proceed.

Rowell and Dawson (1983) suggest that we should first find out how students think about a situation. We need research to show efficient ways that teachers can do this. Will simple questioning in the context of class discussion suffice, or are other strategies required? What questions should teachers ask? How should teachers interpret student responses?

We are told to ask students to hold their ideas and tell them that we are going to teach another possibility to be evaluated later, but the research suggests that powerfully-held theories-in-action make it difficult for students to attend to ideas that conflict with their theories. Can students actually "hear" the new idea? We need research to show that this suggestion is viable.

We are told to teach the new idea, linking it, where possible, to ideas already held. How is this done? Are some kinds of linkages more effective than others?

We are told that after the new idea has been taught, we should recall the old explanations for comparison with the new idea and for comparison with reality. But the chemists's reality is not easily presented for comparison. Atoms and molecules cannot be seen, and inferences from what can be seen are not straight—forward. How can we show the reality that acetone is the invisible gas produced when it is evaporated, but not the invisible gas produced when it burns? We need to learn how we can present the reality of the chemist in a convincing manner.

In addition to research that is needed to show how we can implement the conceptual change models that have been proposed, we have other problems to address. There are powerful traditions to contend with, texts to be covered, and statewide competency exams to be passed. All of this in the face of students who are apathetic and distracted by extra-curricular activities and jobs. How do we solve those problems?

Constructivists see learners as "mentally active agents struggling to make sense of their world" (Pines & West, 1986, p. 584). Perhaps if we restructure the curriculum around problems to be solved rather than content to be covered, the apathy that plagues the classroom will disappear, but that is yet to be demonstrated, and the idea has been around at least since Dewey.

Clearly, much remains to be done.

REFERENCES

Abraham, M., & Renner, J. (1985). The sequence of learning cycle activities in high school chemistry. *Journal of Research in Science Teaching, 22,* 121–143.

Albert, E. (1978). Development of the concept of heat in children. *Science Education, 62,* 389–399.

Alcorn, F. L. (1985). The relationships between problem solving style as measured by the Myers Briggs type indicator and achievement in college chemistry at the community college (Doctoral dissertation, Virginia Polytechnic Institute and State University, 1984). *Dissertation Abstracts International, 46,* 939A.

Allen, J., Barker, L., & Ransden, J. (1986). Guided inquiry laboratory. *Journal of Chemical Education, 63,* 533–534.

Anamuah-Mensah, J. (1986). Cognitive strategies used by chemistry students to solve volumetric analysis problems. *Journal of Research in Science Teaching, 23,* 759–769.

Anamuah-Mensah, J., Erickson, G., & Gaskell, J. (1987). Development and validation of a path analytic model of students' performance in chemistry. *Journal of Research in Science Teaching, 24,* 723–738.

Andersson, B. (1980). Some aspects of children's understanding of boiling point. In U. A. Archenhold (Ed.), *Cognitive development research in science and mathematics* (pp. 252–259). Proceedings of an International Seminar. The University of Leeds, Leeds, England.

Andersson, B. (1986). Pupils' explanations of some aspects of chemical reactions. *Science Education, 70,* 549–563.

Arzi, H., Ben-Zvi, R., & Gamiel, U. (1986). Forgetting versus savings: The many facets of long-term retention. *Science Education, 70,* 171–188.

Barke, H. (1982). Problems bei der verwendung von symbolen im chemieunterricht. Eine empirische untersuchung an schulern der sekundarstufe 1 [Problems associated with the application of symbols in chemistry teaching: An empirical study of secondary school children]. *Naturwissenschaften im Unterricht Ph/Ch, 30,* 131–133.

Beasly, W. (1985). Improving student laboratory performance: How much practice makes perfect. *Science Education, 69,* 567–576.

Ben-Zvi, R., Eylon, B., & Silberstein, J. (1982). Students vs. chemistry: A study of student conceptions of structure and process (Unpublished technical report), Rehovot, Israel: Weizmann Institute, Dept. of Science Teaching.

Ben-Zvi, R., Eylon, B., & Silberstein, J. (1986). Is an atom of copper malleable? *Journal of Chemical Education, 63,* 64–66.

Bleichroth, W. (1965). Was wissen unsere volksschulkinder vom atom? [What do our elementary school children know about the atom?] *Zeitschrift fur Naturlehre unter Naturkunde, 4,* 89–94.

Bodner, G. (1986). Constructivism: A theory of knowledge. *Journal of Chemical Education, 63,* 873–878.

Bodner, G. (1987). The role of algorithms in teaching problem solving. *Journal of Chemical Education, 64,* 513–514.

Bodner, G., & McMillen, T. (1986). Cognitive restructuring as an early stage in problem solving. *Journal of Research in Science Teaching, 23,* 727–737.

Bunce, D. M. (1985). The effects of teaching a complete and explicit problem solving approach on mathematical chemistry achievement of college students (Doctoral dissertation, University of Maryland, 1984). *Dissertation Abstracts International, 46(3),* 665A.

Bunce, D. M., & Heikkinen, H. (1986). The effects of an explicit problem solving approach on mathematical chemistry achievement. *Journal of Research in Science Teaching, 23,* 11–20.

Cachapuz, A., & Martins, I. (1987). High school students ideas about energy of chemical reactions. *Proceedings of the Second International Seminar on Misconceptions and Educational Strategies in Science and Mathematics. Vol. III* (pp. 60–68). Ithaca, NY: Cornell University.

Carter, C. (1987). *The role of beliefs in general chemistry problem solving.* Unpublished doctoral dissertation, Purdue University.

Case, R. (1972). Validation of a neo-Piagetian capacity construct. *Journal of Experimental Child Psychology, 14,* 287–302.

Cassels, J. R. T. (1976). *Language in chemistry: The effect of some aspects of language on 101 grade chemistry candidates.* Unpublished Master's thesis, University of Glasgow, Glasgow, Scotland.

Cassels, J. R. T. (1980). *Language and thinking in science: Some investigations with multiple choice questions.* Unpublished doctoral dissertation, University of Glasgow, Glasgow, Scotland.

Cassels, J. R. T., & Johnstone, A. H. (1980). *Understanding of non-technical words in science.* London: Royal Society of Chemistry.

Chadran, S., Treagust, D., & Tobin, K. (1987). The role of cognitive factors in chemistry achievement. *Journal of Research in Science Teaching, 24,* 145–160.

Chiappetta, E. (1982). A review of Piagetian studies relevant to science instruction at the secondary and college level. *Science Education, 66,* 85–93.

Choi, B., & Gennaro, E. (1987). The effectiveness of using computer simulated experiments on junior high school students' understanding of the volume displacement concept. *Journal of Research in Science Teaching, 24,* 539–552.

Cobb, P. J., & Steffe, L. P. (1983). The constructivist researcher as teacher and model builder. *Journal of Research in Mathematics Education, 14,* 83–94.

Craney, C. L., & Armstrong, R. W. (1985). Predictors of grades in general chemistry for allied health students. *Journal of Chemical Education, 62,* 137–129.

Cros, D., & Maurin, M. (1986). Conceptions of first year university students of the constituents of matter and the notion of acids and bases. *European Journal of Science Education, 8,* 305–313.

De Boer, G. (1987). Predicting continued participation in college chemistry for men and women. *Journal of Research in Science Teaching, 24,* 527–538.

Dombrowski, J., & Hagelberg, R. (1985). The effects of a safety unit on student safety knowledge and behavior. *Science Education, 69,* 527–533.

Drake, R. F. (1985). Working backwards is a forward step in the solution of problems by dimensional analysis. *Journal of Chemical Education, 62,* 414.

Driscoll, D. (1978). Comments on ionization. *Journal of Chemical Education, 55,* 465.

Ehindero, O. (1985). Differential cognitive responses to adjunct study questions in the learning of chemistry textual materials. *European Journal of Science Education, 1,* 423–429.

Ellerton, N., & Ellerton, H. (1987). Mathematics and chemistry problems created by students. *Proceedings of the Second International Seminar on Misconceptions and Educational Strategies in Science and Mathematics. Vol. III* (pp. 131–136). Ithaca, NY: Cornell University.

Ericson, G. (1979). Children's conceptions of heat and temperature. *Science Education, 63,* 221–230.

Ericson, G. (1980). Children's viewpoints of heat: A second look. *Science Education, 64,* 323–336.

Fasching, J., & Erickson, B. (1985). Group discussions in the chemistry classroom and the problem-solving skills of students. *Journal of Chemical Education, 62,* 842–846.

Frank, D. (1986). Implementing instruction to improve the problem-solving abilities of general chemistry students (Doctoral dissertation, Purdue University, 1985). *Dissertation Abstracts International, 47,* 141A.

Frazer, M., & Sleet, R. (1984). A study of students' attempts to solve chemical problems. *European Journal of Science Education, 6,* 141–152.

Freeman, W. A. (1984). Relative long-term benefits of a PSI and a traditional-style remedial chemistry course. *Journal of Chemical Education, 61,* 617–619.

Fuchs, H. (1987). Thermodynamics: A "misconceived" theory. *Proceedings of the Second International Seminar on Misconceptions and Educational Strategies in Science and Mathematics, Vol. III* (pp. 160–167). Ithaca, NY: Cornell University.

Gabel, D. L. (1981). *Facilitating problem solving in high school chemistry* (NSF Technical Report No. RISE-SED-79-20744). Bloomington: Indiana University, School of Education. (ERIC Document Reproduction Service No. ED 210 192)

Gabel, D. L., & Enochs, L. (1987). Different approaches for teaching volume and students' visualization ability. *Science Education, 71,* 591–597.

Gabel, D. L., & Samuel, K. (1986). High school students' ability to solve molarity problems and their analogue counterparts. *Journal of Research in Science Teaching, 23,* 165–176.

Gabel, D. L., & Sherwood, R. D. (1984). Analyzing difficulties with mole-concept tasks by using familiar analogue tasks. *Journal of Research in Science Teaching, 21,* 843–851.

Gabel, D. L., Sherwood, R. D., & Enochs, L. (1984). Problem-solving skills of high school chemistry students. *Journal of Research in Science Teaching, 21,* 221–233.

Gardner, P. L. (1972). *Words in science.* Melbourne: Australian Science Education Project.

Garrett, J. (1983). Teaching factor-label method without sleight of hand. *Journal of Chemical Education, 60,* 962.

Genyea, J. (1983). Improving students' problem-solving skills: A methodical approach for a preparatory chemistry course. *Journal of Chemical Education, 60,* 478–482.

Gilbert, G. L. (1980). How do I get the answer? Problem solving in chemistry. *Journal of Chemical Education, 57,* 79–81.

Gilbert, J., & Swift, D. (1985). Towards a Lakatosian analysis of the Piagetian and alternative conceptions research programs. *Science Education, 69,* 681–696.

Glassman, S. (1967). High school students' ideas with respect to certain concepts related to chemical formulas and equations. *Science Education, 51,* 84–103.

Gorodetsky, M., & Gussarsky, E. (1986). Misconceptualization of the chemical equilibrium concept as revealed by different evaluation methods. *European Journal of Science Education, 8,* 427–441.

Gorodetsky, M., & Gussarsky, E. (1987). The roles of students and teachers in misconceptualization of aspects in "chemical equilibrium." *Proceedings of the Second International Seminar on Misconceptions and Educational Strategies in Science and Mathematics. Vol. III* (pp. 187–193). Ithaca, NY: Cornell University.

Gorodetsky, M., & Hoz, R. (1985). Changes in the group cognitive structure of some chemical equilibrium concepts following a university course in general chemistry. *Science Education, 69,* 185–199.

Greenbowe, T. (1984). An investigation of variables involved in chemistry problem solving (Doctoral dissertation, Purdue University, 1983). *Dissertation Abstracts International, 44,* 3651A.

Greeno, J. G. (1980). Trends in the theory of knowledge for problem solving. In D. T. Tuma & F. Reif (Eds.), *Problem solving and education: Issues in teaching and research* (pp. 9–23). Hillsdale, NJ: Lawrence Erlbaum Associates.

Hackling, M., & Garnett, P. (1985). Misconceptions of chemical equilibrium. *European Journal of Science Education, 7,* 205–214.

Hall, J. (1973). Conservation concepts in elementary chemistry. *Journal of Research in Science Teaching, 10,* 143–146.

Herron, J. D. (1975). Piaget for chemists. *Journal of Chemical Education, 52,* 146–150.

Herron, J. D. (1977). Are chemical terms well defined? *Journal of Chemical Education, 54,* 758.

Herron, J. D. (1978). Response to 'Are chemical terms well defined?' *Journal of Chemical Education, 55,* 393–394.

Herron, J. D., & Greenbowe, T. (1986). What can we do about Sue: A case study of competence. *Journal of Chemical Education, 63,* 528–531.

Hewson, M. (1984). The influence of intellectual environment on conceptions of heat. *European Journal of Science Education, 6,* 245–262.

Hewson, M. (1986). The acquisition of scientific knowledge: Analysis and representation of students' conceptions concerning density. *Science Education, 70,* 159–170.

Hodson, D. (1984). Some effects of changes in question structure and sequence on performance in a multiple choice chemistry test. *Research in Science and Technological Education, 2,* 177–185.

Hodson, D. (1984). The effect of changes in item sequence on student performance in a multiple-choice chemistry test. *Journal of Research in Science Teaching, 21,* 489–495.

Horton, P. B., Fronk, R. H., & Walton, R. W. (1985). The effect of writing assignments on achievement in college general chemistry. *Journal of Research in Science Teaching, 22,* 535–541.

Isom, F., & Rowsey, R. (1986). The effect of a new pre-laboratory procedure on students' achievement in chemistry. *Journal of Research in Science Teaching, 23,* 231–235.

Johnstone, A., & El-Banna, H. (1986). Capacities, demands and processes: A predictive model for science education. *Education in Chemistry, 23,* 80–84.

Johnstone, A., MacDonald, J., & Webb, G. (1977). Misconceptions in school thermodynamics. *Physics Education, 12,* 248–251.

Karmiloff-Smith, A., & Inhelder, B. (1974). If you want to get ahead, get a theory. In P. N. Johnson-Laird & P. C. Wason (Eds.), *Thinking: Readings in cognitive science* (pp. 293–306). Cambridge: Cambridge University Press.

Kempa, R., & Nicholls, C. (1983). Problem-solving ability and cognitive structure: An exploratory investigation. *European Journal of Science Education, 5,* 171–184.

Klainin, S. (1987). Learning achievement in upper secondary school chemistry in Thailand: Some remarkable sex reversals. *International Journal of Science Education, 9,* 217–227.

Kleinman, R., Griffin, H., & Kerner, N. K. (1987). Images in chemistry. *Journal of Chemical Education, 64,* 766–770.

Kramers-Pals, H., Lambrecht, J., & Wolff, P. (1983). The transformation of quantitative problems to standard problems in general chemistry. *European Journal of Science Education, 5,* 275–287.

Lawson, A. (1985). A review of research on formal reasoning and science teaching. *Journal of Resarch in Science Teaching, 22,* 569–617.

Lazonby, J., Morris, J., & Waddington, D. (1985). The mole: Questioning format can make a difference. *Journal of Chemical Education, 62,* 60–61.

Lehman, J. R., Koran, J. J., & Koran, M. L. (1984). Interaction of learner characteristics with learning from three models of the Periodic Table. *Journal of Research in Science Teaching, 21,* 885–893.

Lynch, P., Chipman, H., & Pachaury, A. (1985). The language of science and the high school student: The recognition of concept definitions. A comparison of Hindi speaking students in India and English speaking students in Australia. *Journal of Research in Science Teaching, 22,* 675–686.

Mahan, B. (1975). *University chemistry* (3rd Ed.). Reading, MA: Addison-Wesley.

Mathews, G., Brook, V., & Kran-Gandapus, T. (1984). Cognitive structure determination as a tool in science teaching. Part I: A new method of creating concept maps. *European Journal of Science Educaion, 6,* 169–177.

Mathews, G., Brook, V., & Kran-Gandapus, T. (1984). Cognitive structure determination as a tool in science teaching. Part II: The measurement of Piaget-specific levels. *European Journal of Science Education, 6,* 289–297.

Mathews, G., Brook, V., & Kran-Gandapus, T. (1984). Cognitive structure determination as a tool in science teaching. Part III: Results. *European Journal of Science Education, 7,* 263–279.

McIntosh, W. (1986). The effect of imagery generation on science rule learning. *Journal of Research in Science Teaching, 23,* 1–9.

Mettes, C. T. C. W., Pilot, A., & Roossink, H. J. (1981). Linking factual and procedural knowledge in solving science problems: A case study in a thermodynamics course. *Instructional Science, 10,* 303–316.

Mettes, C. T. C. W., Pilot, A., Roossink, H. J., & Kramers-Pals, H. (1980). Teaching and learning problem solving in science. Part I: A general strategy. *Journal of Chemical Education, 57,* 882–884.

Mettes, C. T. C. W., Pilot, A., Roossink, H. J., & Kramers-Pals, H. (1981). Teaching and learning problem solving in science. Part II: Learning problem solving in a thermodynamics course. *Journal of Chemical Education, 58,* 51–55.

Mikkelson, A. (1985). Computer assisted instruction in remedial teaching in first year chemistry. *Education in Chemistry, 22,* 117–118.

Mitchell, H., & Kellington, S. (1982). Learning difficulties associated with the particulate theory of matter in the Scottish Integrated Science Course. *European Journal of Science Education, 4,* 429–440.

Mulder, T., & Verdonk, A. H. (1984). A behavioral analysis of the laboratory learning process: Redesigning a teaching unit on recrystallization. *Journal of Chemical Education, 61,* 451–453.

Niaz, M. (1985). Evaluation of formal operational reasoning by Venezuelan freshmen students. *Research in Science and Technological Education, 3,* 43–50.

Niaz, M. (1987). Relation between M-space of students and M-demand of different items of general chemistry and its interpretation based upon neo-Piagetian Theory of Pascual-Leone. *Journal of Chemical Education, 64,* 502–505.

Niaz, M., & Lawson, A. (1985). Balancing chemical equations: The role of developmental level and mental capacity. *Journal of Research in Science Teaching, 22,* 41–51.

Novick, S., & Mannis, J. (1976). A study of student perception of the mole concept. *Journal of Chemical Education, 53,* 720–722.

Novick, S., & Nussbaum, J. (1978). Junior high school pupils' understanding of the particulate nature of matter: An interview study. *Science Education, 62,* 273–281.

Novick, S., & Nussbaum, J. (1981). Pupils' understanding of the particulate nature of matter: A cross-age study. *Science Education, 65,* 187–196.

Nurrenbern, S. C. (1980). Problem solving behaviors of concrete and formal operational high school chemistry students when solving chemistry problems requiring Piagetian formal reasoning skills (Doctoral dissertation, Purdue University, 1979) *Dissertation Abstracts International, 40,* 4986A.

Nurrenbern, S., & Pickering, M. (1987). Concept learning versus problem solving: Is there a difference? *Journal of Chemical Education, 64,* 508–510.

Okebukola, P. (1987). Students' performance in practical chemistry: A study of some related factors. *Journal of Research in Science Teaching, 24,* 119–126.

Osborne, R., & Bell, B. (1983). Science teaching and children's views of the world. *European Journal of Science Education, 5,* 1–14.

Osborne, R., & Cosgrove, M. (1983). Children's conceptions of the changes of state of water, *Journal of Research in Science Teaching, 20,* 825–838.

Pankiewicz, P. R. (1985). The effects of a self-designed introductory junior high school organic chemistry module on selected student charateristics (Doctoral dissertation, Clark University, 1984). *Dissertation Abstracts International, 45,* 2057A.

Pascual-Leone, J. (1970). A mathematical model for the transition rule in Piaget's developmental stages. *Acta Psychologica, 63,* 301–345.

Pascual-Leone, J., & Smith, J. (1969). The encoding and decoding of symbols by children: New experimental paradigm and a neo-Piagetian model. *Journal of Experimental Child Psychology, 8,* 328–355.

Perez, D., & Torregrosa, J. (1983). A model of problem-solving in accordance with scientific methodology. *European Journal of Science Education, 5,* 447–455.

Pfundt, H. (1981). The atom: The final link in the division process or the first building block? Pre-instructional conceptions about the structure of substances. *Chemica Didactica, 7,* 75–94.

Pfundt, H. (1982). Pre-instructional conceptions about transformation of substances. *Chemica Didactica, 8,*

Pickering, M. (1986). Laboratory education as a problem in organization, *Journal of College Science Teaching, 16,* 187–189.

Pines, A. L., & West, L. (1986). Conceptual understanding and science learning: An interpretation of research within a sources-of-knowledge framework. *Science Education, 70,* 583–604.

Polya, G. (1957). *How to solve it: A new aspect of mathematical method* (3rd ed.). Princeton, NJ: Princeton University Press.

Rafel, J., & Mans, C. (1987). Alternative framework about the learning of changes of state of aggregation of matter: Sorting of answers into models. *Proceedings of the Second International Seminar on Misconceptions and Educational Strategies in Science and Mathematics. Vol. III* (pp. 392–399). Ithaca, NY: Cornell University,

Raines, S. J. (1984). Problem-similarity recognition and problem-solving success in high school chemistry. *Dissertation Abstracts International, 45,* 800A.

Reif, F. (1981). Teaching problem solving: A scientific approach. *The Physics Teacher, 19,* 310–316.

Reustrom, L. (1987). Pupils' conceptions of matter: A phenomenographic approach. *Proceedings of the Second International Seminar on Misconceptions and Educational Strategies in Science and Mathematics, Vol. III* (pp. 400–419). Ithaca, NY: Cornell University,

Resnick, L. (1987). *Education and learning to think.* Washington, D. C.: National Academy Press.

Roberts, W., & Sutton, C. (1984). Adults' recollections from school chemistry—facts, principles & learning. *Education in Chemistry, 21,* 82–85.

Rowell, J., & Dawson, C. (1983). Laboratory counterexamples and the growth of understanding in science. *European Journal of Science Education, 5,* 203–215.

Ryan, M., Robin, D., & Carmichael, J. (1980). A Piagetian-based general chemistry laboratory program for science majors. *Journal of Chemical Education, 57,* 642–645.

Scardamalia, M. (1977). Information processing capacities and the problem of horizontal decalage: A demonstration using combinatorial reasoning tasks. *Child Development, 48,* 28–37.

Schmidt, H. (1987). Secondary school students' learning difficulties in stoichiometry. *Proceedings of the Second International Seminar on Misconceptions and Educational Strategies in Science and Mathematics, Vol. I* (pp. 396–404). Ithaca, NY: Cornell University.

Scott, P. (1987). The process of conceptual change in science: A case study of the development of a secondary pupil's ideas relating to matter. *Proceedings of the Second International Seminar on Misconceptions and Educational Strategies in Science and Mathematics. Vol. II* (pp. 404–418). Ithaca, NY: Cornell University,

Seddon, G., & Moore, R. (1986). An unexpected effect in the use of models for teaching the visualization of rotation in molecular structure. *European Journal of Science Education, 8,* 79–86.

Seddon, G. M., & Shubber, K. E. (1984). The effects of presentation mode and colour in teaching the visualization of rotation in diagrams of molecular structures. *Research in Science and Technological Education, 2,* 167–176.

Seddon, G., & Tariz, R. (1982). The visualization of spatial transformation in diagrams of molecular structures. *European Journal of Science Education, 4,* 409–420.

Shayer, M., & Wylam, H. (1981). The development of the concept of heat and temperature in 10–13 year olds. *Journal of Research in Science Teaching, 18,* 419–434.

Solomon, J. (1982). How children learn about energy or Does the first law come first? *The School Science Review, 63,* 415–422.

Solomon, J. (1983). Learning about energy: How pupils think in two domains, *European Journal of Science Education, 5,* 49–59.

Staver, J. R., & Halsted, D. A. (1985). The effects of reasoning, use of models, sex type, and their interactions on post-test achievement in chemical bonding after constant instruction. *Journal of Research in Science Teaching, 22,* 437–447.

Stavy, R. (1987). Acquisition of conservation of matter. *Proceedings of the Second International Seminar on Misconceptions and Educational Strategies in Science and Mathematics, Vol. I* (pp. 456–465). Ithaca, NY: Cornell University.

Streitberger, H. E. (1985). College chemistry students' recommendations to high school students. *Journal of Chemical Education, 62,* 700–701.

Strike, K., & Posner, G. (1982). Conceptual change and science teaching. *European Journal of Science Education, 4,* 231–240.

Vermont, D. F. (1985). Comparative effectiveness of instructional strategies on developing the chemical mole concept. *Dissertation Abstracts International, 45,* 2473A.

Voelker, A. (1975). Elementary school children's attainment of the concepts of physical and chemical change: A replication. *Journal of Research in Science Teaching, 12,* 5–14.

Waddling, R. E. L. (1983). Titration calculations—a problem-solving approach. *Journal of Chemical Education, 60,* 230–232.

Wainwright, C. L. (1985). The effectiveness of a computer-assisted instruction package in supplementing teaching of selected concepts in high school chemistry: Writing formulas and balancing chemical equations. *Dissertation Abstracts International, 45,* 2473A.

Weninger, J. (1982). Fundamental remarks on so-called reaction equations. *European Journal of Science Education, 4,* 391–396.

Wheeler, A., & Kass, H. (1978). Student misconceptions in chemical equilibrium. *Science Education, 62,* 223–232.

Widing, R. W., Jr. (1985). Systematic Analysis of chemistry topics. *Dissertation Abstracts International, 45,* 2057A.

Wienekaup, H., Jansen, W., Fickenfrerichs, H., & Peper, R. (1987). Does unconscious behavior of teachers cause chemistry lessons to be unpopular with girls. *International Journal of Science Education, 1,* 281–286.

Williams, H., Turner, C., Debreuil, L., Fast, J., & Berestiansky, J. (1979). Formal operational reasoning by chemistry students. *Journal of Chemical Education, 56,* 599.

Wilson, A. (1987). Teaching use of formal thought for improved chemistry achievement. *International Journal of Science Education, 9,* 197–202.

Yackel, E. (1984). Characteristics of problem representation indicative of understanding in mathematics problem solving (Doctoral dissertation, Purdue University, 1984). *Dissertation Abstracts International, 45,* 2021A.

Yarroch, W. (1985). Student understanding of chemical equation balancing. *Journal of Research in Science Teaching, 22,* 449–459.

Zitzewitz, B., & Berger, C. (1985). Application of mathematical learning models. *Journal of Research in Science Teaching, 22,* 775–777.

3 Biology Education Research: A View from the Field

Jim Stewart
University of Wisconsin, Madison

INTRODUCTION

During the last several years there have been many calls for reforming science education in the United States (for example, the National Academy of Science & National Academy of Engineering, 1982). If these are to be answered, attention needs to be directed toward the biological sciences, including biology education research. The basis for this claim is found in elementary, middle, and high school classrooms throughout the country. Biology and life science content are, and likely will remain, the most commonly taught at all levels. One reason is that elementary teachers tend to be most comfortable teaching life science topics. In addition, for students at both the elementary and secondary levels, learning about living things, including themselves, has an attraction that is difficult for other sciences to match. Thus research that sheds light on teaching and learning in biology has the potential to improve the science education that students receive. This may be particularly important given that we are in the midst of a biological "revolution"—one that will continue to present significant political, economic, ethical, and educational issues for our society to grapple with.

The first section of this chapter will be an overview of current biology education research that I think has the most promise for improving teaching and learning. The second section of this chapter contains my impressions of what most needs to be done to insure that biology education research continues to contribute to the improvement of science education in the United States.

THE CURRENT STATE OF AFFAIRS

The biology education research and instructional development projects described in this section represent the emergence of science education research from an era of unproductive correlation studies into a more productive one, if the measure of productivity is significant insights into the four commonplaces of education (the student, the teacher, the content, and the milieu) and their intersections (Schwab, 1964). Studies will be examined in the areas of elementary school life science, high school and university biology, and technology and biology education research.

Elementary School Life Science

There is a rich literature of descriptive research on elementary school students' conceptions of living and nonliving; plant and animal, including differences between the two; classification schemes for members of particular animal groups such as vertebrates; food; plants as producers; and the human body, particularly the structure and function of various organs and organ systems. Researchers have reported that students have knowledge about the content, that this knowledge is often different from that which teachers expect them to learn, and that it is often difficult to change. These descriptive studies could lead to a better education in the biological sciences for prospective elementary school teachers and their students. Unfortunately, this research has had little impact in either of these two areas, possibly because the individual studies have not been part of larger, multifaceted research programs. An important exception, which could be a model for educational research in the biological sciences, is the program of Smith and Anderson at Michigan State University. Within the context of learning as "conceptual change," they have focused on "four aspects of the teaching-learning situation: the curriculum materials (especially the text and teacher's guide),the teacher's planning, actual classroom interaction, and student conceptions of the topics covered." (1984, p. 686)

In a case study of elementary science teaching (fifth-grade students studying the Science Curriculum Improvement Study "Plants as Producers" unit), Smith and Anderson (1984) identified three significant factors that influence the interactions between teacher and students, and the results of those interactions on learning. The factors that they identified were: (a) the teacher's conceptions of the actual content of instruction, which differed from those that the students held; (b) the beliefs that the teacher held about how students learn, which was a view of learning by accretion rather than as a process of conceptual change; and (c) the teacher's epistemological beliefs about the relationship between evidence and scientific theories. The teacher held a positivistic view that theories are generated by following an inductive logic, rather than being inventions used to guide investigation. The significance of this research, in addition to its contribution to

56

the understanding of student learning about plant growth and development, is that Smith and Anderson looked at many complex interactions that take place in classrooms. In addition, their research has led to the development of revised teachers' guides to help teachers recognize and address the alternate conceptions of their students.

High School and University Biology

Genetics, particularly transmission genetics, has been a primary focus for research at the high school and university levels. This research has provided important insights into students' problem-solving performance as well as into their conceptual knowledge. These studies have included both empirical and theoretical research (task analyses and analyses of the learning outcomes that may result from problem-solving instruction). In addition to the research on genetics, there has been a limited amount of research on students' understanding of evolution.

Transmission Genetics

Much of the research on transmission genetics has been done at the University of Wisconsin at Madison (Albright, 1987; Collins, 1989; Slack & Stewart, 1988; Stewart, 1983, 1988; Stewart & Dale, 1981, 1988; Thomson & Stewart, 1985). Related studies have been conducted at other universities (Peard, 1983; Smith, 1986; Smith & Good, 1984; Tolman, 1982; Walker, Mertens, & Hendrix, 1980). This research has been most influenced by the alternate conceptions research in science education and problem-solving research, including general studies of problem solving (Newell & Simon, 1972) and the studies of expert and novice performance in physics (Chi, Feltovich, & Glaser, 1981; Reif, 1983). For example, at Wisconsin, we have been developing models of desired performance, by studying the problem-solving performance of geneticists and by analyzing the structure of transmission genetics (Collins, 1989; Collins & Stewart, 1988). We have also studied college students (Albright, 1987) and high school students (Slack & Stewart, 1988; Stewart, 1983; Stewart & Dale, 1988) to understand novice performance. The results of this research are being used to develop a model of instruction for transmission genetics. An example of this approach can be found in the *MENDEL* tutoring system (Streibel, Stewart, Koedinger, Collins, & Jungck, 1987) to be defined later in this chapter.

We began by studying the procedural and conceptual knowledge of genetics and meiosis that high school students used to solve typical textbook problems. These problems require students to reason from causes (known information on inheritance patterns, such as which variation of a trait is dominant) to effects (offspring genotype or phenotype data). Because solutions can be obtained by using algorithms, correct answers do not necessarily measure students' understanding of genetics (Stewart, 1983). For example, we found that students have multiple mental models of meiosis, including erroneous ones, that they use to account for their problem solutions (Stewart & Dale, 1988). Based on these results, and on analyses of the procedural and conceptual knowledge that underlie "meaningful" problem solving, we designed instruction that did help students connect the two (Thomson & Stewart, 1985).

While the results of this research on students' conceptual knowledge were important, they were less valuable in terms of insights into procedural knowledge, since the procedures most used by students were genetics-specific algorithms (Stewart, 1983). However, genetics problem-solving research has been done using realistic problems that provide more meaningful insights into the procedural knowledge of problem solvers. These problems require that at some point in a solution, solvers reason from effects (phenotype data from crosses) to causes (an explanation for the phenotype data in terms of underlying genetics phenomena). Therefore, they are more likely than typical textbook problems to promote the acquisition of important learning outcomes, including genetics-independent and genetics-specific problem-solving heuristics; conceptual knowledge of genetics; and a better understanding of science as a human problem-solving activity.

Smith and Good (1983, 1984) studied problem solving on textbook effect-to-cause problems that require the analysis of a limited set of phenotypic data in order to infer inheritance patterns. They identified 32 general tendencies that only successful problem solvers used. They also reported that the strategies that best characterized the performance of successful solvers included seeking a solution rather than an answer, checking for consistent logic, approaching a problem working forward, checking for one variable at a time, and looking for evidence that would invalidate previous assumptions. In related research, Smith (1986) observed that unsuccessful solvers tended to model an instructor' solution without understanding basic concepts, were dependent on expected ratios, failed to consider alternative hypotheses, were unable to assign genotypes to individuals according to their hypotheses, and changed hypotheses without reason.

We have studied the solving of effect-to-cause problems by using problems produced by the microcomputer program, *GENETICS CONSTRUCTION KIT,* (GCK) (Jungck & Calley, 1985). These problems differ from Smith's in that solvers need to generate and interpret data. Collins (1987) studied the performance of experts solving GCK problems, and reported that experts' knowledge was organized by inheritance pattern schemata that allowed them to generate hypotheses about the genetics operating in the problem even when they had very little data. They then used the selected hypothesis as a basis for generating and evaluating additional data. Collins' description of the problem-solving performance of genetics experts is similar to descriptions of expert performance in physics. Genetics experts' knowledge was organized hierarchically around schemata that include basic concepts and principles of genetics. These schemata allowed experts to quickly recognize patterns in data and to construct theoretical representations of problems in terms of genotype-to-phenotype mappings. In addition, these schemata directed their crossing (data generation) strategy of working back and forth between data and hypotheses.

Albright (1987) and Slack and Stewart (1988) have studied the performance of novices on GCK problems and found, as might be expected, that they are less efficient than Collins' experts. The experts had well-developed schemata that they used to generate and run mental models of problem solutions. Novices lacked such schema and did not perceive genetics problem solving to be a process of generating and testing hypotheses. Some of the major characteristics of novices were as follows:

1. They did not think of genetics as involving patterns across generations.
2. They viewed each cross as a distinct problem with no necessary connection to past or future crosses.
3. They worked at the level of phenotype rather than genotype and made little connection between the two levels.
4. They tended to work backwards in the sense that they tried to explain crosses rather than working forward by making predictions about upcoming crosses.

Other Genetics

In addition to problem solving in transmission genetics, there have been other studies involving genetics content. In one of these, Fisher (1985) reported on a persistent problem that university biology students had understanding the process of translation in the synthesis of proteins. She provided possible explanations as to why students persisted in believing, in spite of excellent instruction, that amino acids are products of the process of translation: (a) there is a strong word association between the two terms because of the way the content is taught; (b) students are drawn to the answer that requires the fewest steps to verify; (c) students do not have a mental model of the process of protein synthesis that they can use to determine a correct answer to a question; and (d) students experience a conflict because activating enzymes (the correct answer to the question) are products of the process of translation and, at the same time, play a role in catalyzing the translation reaction.

Hildebrand (1986) studied cognitive representations of the basic biological entities: (a) DNA, (b) gene, and (c) chromosome. She provided detailed descriptions of experts' and novices' mental models of these basic biological entities which underlie meaningful understanding and problem solving in genetics. This study represents one of the first times in the biological sciences that the metaphor of "running a mental model" was employed to characterize what individuals, particularly experts, do when they think about the content of, or solve problems in, their domains. For Hildebrand, studying cognitive representations is important because

> mental model reasoning is a privileged level of reasoning about biological processes because it concerns the fundamental, but "invisible" entities and processes that give rise to higher level (observable) characteristics among living organisms (Hildebrand, 1986, p. 5).

Evolution

Even though there have been several studies on students' conceptions of various aspects of evolution, only one was done in this country. In that study Bishop and Anderson (1986a) described student conceptions of natural selection and its roles in evolution, and consequently developed a module for teaching the content to high school or introductory college students (Bishop & Anderson, 1986b). By using a combination of multiple–choice and essay questions along with interviews of college biology students, they described student conceptions of evolution by natural selection both before and after instruction. They reported the following results:

1. Although students had taken an average of 1.9 years of biology courses, performance on the pretest was low,
2. There was no relationship between the amount of previous biology exposure and pretest or posttest performance,
3. A belief in the truthfulness of evolutionary theory was also unrelated to pretest or posttest performance,
4. Instruction using specially designed materials was moderately successful in improving students' understanding of evolution.

The most significant features of this research are the rich descriptions of student conceptions and the instruction that was developed which took into account specific student misconceptions. These misconceptions can be summarized as follows. Students:

1. fail to consider that selection has two components, a source of population variation and a process of selection that acts on that variation,
2. have a *Lamarckian*-like view of evolution (that acquired characteristics are inherited) that involves need, use, and disuse,
3. do not view evolutionary change as a phenomenon that happens to populations, and
4. misunderstand the concepts of adapt/adaption and fitness.

Technology and Biology Education Research

Research and development efforts in biology education often employ advanced technologies. Three such efforts will be described, each of which includes research as well as software development.

BioQUEST

BioQUEST (Peterson & Jungck, 1988) is a consortium of university biology faculty from six colleges and universities who are developing a one-year introductory biology course centered around 12 *strategic simulation* programs. The programs are termed strategic simulations because they motivate students to develop problem-solving strategies and because they are simulations of actual experimental behavior of natural biological systems. Each of the strategic simulations, including ones in transmission genetics, microbial genetics, molecular genetics, physiology, and evolution, take full advantage of the computer in terms of: (a) novelty of problems each time a program is run, (b) realistic outcomes for each experiment performed, and (c) infinite opportunities to perform experiments.

Underlying the efforts of BioQUEST is a view of biology education that focuses on the nature of the interactions among teachers and students. Although BioQUEST has as its primary purpose the improvement of introductory university biology courses through the use of computer technologies, there is considerable transfer to high school biology and to life science instruction at any grade level, even if computer technologies are not involved. The transfer occurs because a goal of BioQUEST is to help students develop a new set of expectations about learning science by addressing several problems in biology teaching, including:

1. a lack of integrative thinking and problem-solving skills among students,
2. the perpetuation of misconceptions about learning science,
3. fragmented and poorly understood (by teachers) teaching materials and methods.

The view of education that BioQUEST is developing is called post-Socratic (Jungck & Calley, 1985), or design learning (Peterson, Jungck, Sharpe, & Finzer, 1987) and stresses the open-ended problem-posing and problem-solving aspects of science. This approach is far removed from the pre-posed problems with finite solution spaces presented in textbooks. A significant feature of design learning is that it " . . . demands active attention to both the discovery method and the discovery itself." (Peterson et al., 1987, p.113) By offering students environments that require strategy development, the members of the BioQUEST consortium intend that students will be continuously challenged to evaluate three structures of their knowledge: (a) methods (strategies), (b) justification (how you know that you know), and (c) the components that are under investigation (what is the real problem?). By doing this the BioQUEST consortium will help students focus on the three p's of science education: (a) problem posing, (b) problem solving and (c) persuading peers about the correctness and utility of a solution.

MENDEL

The MENDEL project at the University of Wisconsin (Streibel et al., 1987) has used research findings on genetics learning and problem solving to develop an intelligent tutoring system for transmission genetics. The primary objective of the MENDEL project is to create a computer environment for transmission genetics in which students can develop model-based problem-solving strategies by conducting realistic genetics experiments and receiving appropriate problem-solving advice. The MENDEL software includes: (a) a problem "GENER-ATOR"; (b) a "TUTOR" that includes a model-based tutoring strategies; (c) a student "MODELER" that includes a model of student performance; (d) an "ADVISOR" that questions students and provides them with tutorial advice; and (e) a hypothesis "CHECKER" that evaluates students' hypotheses. The research that has led to the development of this software includes studies of expert and novice problem-solving performance, tutoring strategies of genetics teachers, student use of MENDEL's notational tools, and effective advising strategies to help novices understand model-based problem solving.

SemNet

The third project, SemNet (Fisher, et al., 1987) builds on research on human memory and computer science, particularly on how knowledge is represented, and on the use of knowledge representation techniques such as concept maps and semantic networks to enhance student learning. Software has been developed that permits users to create and to easily modify semantic networks of their knowledge. Fisher et al. (1987) list the following as strengths of SemNet:

1. Instructors can create multidimensional externalizations of their knowledge. When analyzing a topic for a lecture, an instructor can use SemNet to manipulate ideas without worrying about linear order of expression.
2. Instructors can convey their knowledge to students in a nonlinear, highly interconnected semantic network.
3. Students can browse through multidimensional representations in any order that their individual learning needs require, thus assimilating information in a highly interconnected form.
4. Students can create semantic networks of topics they are studying, re-

organize or elaborate upon networks provided by an instructor, and use masking features of networks for reviewing material.

The authors have just begun to gather data on the effect of SemNet on student learning, and have developed a set of research questions that have the potential to improve biology instruction. In addition, the SemNet software is not limited to the disciplines researched by Fisher et al. (1987). Thus it could be incorporated into other projects such as BioQUEST or MENDEL.

THE FUTURE: A RESEARCH CONSORTIUM

Given the current valuable educational research in the biological sciences, there is reason to believe that over the next few years practical benefits may occur. But there are steps that must be taken in order for this potential to be realized.

A parallel situation exists in the biological sciences, where researchers are planning one of the most ambitious projects ever undertaken: mapping or even sequencing the human genome. Although a key element for the success of this project is the commitment of federal money, success will also depend on careful planning and coordinating of efforts to insure that the appropriate research questions are asked. For such planning and coordinating to occur, political, institutional, and research group differences need to be acknowledged and resolved.

Biology education needs a similarly ambitious and well-articulated program of research. A national consortium, providing direction for biology education research would be one way to meet this need. For such a consortium to be successful, the input of many individuals and organizations with different research interests would be required. The initial leadership for such a consortium could come from institutions that have active programs in biology education research. In addition, the consortium would need to establish ties with other organizations that are committed to improving biology education, such as the National Association of Biology Teachers and the Biological Sciences Study Commission. These organizations would complement a research consortium because of their interest in issues of biology curriculum and instruction and because of their close work with teachers.

A significant outcome of such a collaboration would be an emphasis on evaluating the purpose and content of biology education. This interest in curriculum development in the biological sciences is essential to the success of a consortium for biology education research because research should reflect the most significant learning outcomes that we have for students. It also should have a direct effect on teacher planning, curriculum development, instructional materials development, teacher education, student evaluation, and student learning and problem solving. For those things to occur, federal funding agencies, partic-

ularly the National Science Foundation and the Department of Education, will
need to be involved. Funds from these agencies will be essential to provide:

1. Support that will lead to the development of the consortium;
2. Opportunities for the individuals within the consortium to continue to
   meet once the agenda for the consortium has been set;
3. Scholarships and fellowships that would allow universities to attract the
   brightest most able graduate students and to attract outstanding teachers as
   fellows (The existence of a research program and the availability of financial support should be incentives, also indicating the legitimacy and importance of the undertaking.);
4. Support for the research that will be done by the consortium;
5. Incentives to insure that the research the consortium does benefits from the
   involvement of biologists, precollege teachers and administrators, and
   psychologists;
6. Support to insure that the results of the consortium's research efforts will
   reach the classroom.

Once a research consortium is established, it would have the responsibility for
developing a research agenda. It is my feeling that the following areas are ones
that should be given serious consideration.

Continuation of Descriptive Research

Some of the most useful educational research in the biological sciences—descriptive reports of students' knowledge of particular content—needs to be continued. Although changes in instruction can be based on past descriptive work,
there will continue to be a need for rich descriptions of students' knowledge, both
before and after instruction. At the high school level, research could be done in
areas of social significance related to students' understandings of the biological
basis of race and gender. For example, to examine the extent to which students
use biological determinist views to interpret the world could have important
consequences for instruction.

Another area in which additional descriptive research is needed is on students'
understanding of the interconnections among closely related content areas. At the
high school level, for example, even though we have a reasonable understanding
of students' knowledge of basic genetics phenomena, we do not understand how
students relate transmission genetics to the genetics of gene expression. Nor do
we know how they understand events at the level of DNA to be related to events
at the level of chromosomes, the organism, or the population. Teachers expect
that students will leave a biology course with an integrated understanding, yet we
don't know whether or not this happens. We do know that students lack connections within certain content areas, such as transmission genetics, so there is

reason to expect that their understandings across content areas are even less well-developed.

Continuation of descriptive research is also important at the elementary school level. One thing that needs to be done is to extend that research to new content areas such as interactions between plants and animals and diversity and continuity within species and individuals.

Problem-Solving Research

Continuation of research on problem solving in biology is also needed in order to understand how students solve problems that require more than the application of algorithms. For example, by using programs like GENETICS CONSTRUCTION KIT (Jungck & Calley, 1985), it is possible to investigate: (a) the problem-solving strategies of students; (b) the general and genetics-specific heuristics that they use; (c) how their problem-solving performance is related to their conceptual knowledge; (d) how differences in conceptual knowledge may be correlated with differences in performance; and (e) specific differences between less successful and more successful solvers. Research on problem solving needs to be extended to other areas of biology, such as evolutionary biology, ecology, and physiology.

The typical question asked in problem-solving research takes for granted that the ability to solve problems is the outcome of instruction in which students have learned the appropriate conceptual and procedural knowledge. This approach to problem-solving research has been useful, but more needs to be done. For example, the question, "What are the gains in conceptual and procedural knowledge that emerge from solving problems?" should be addressed. The reason that we solve problems in our own areas of specialty is not to exercise our problem-solving ability or to show off our conceptual understanding, but rather to develop new conceptual and/or procedural understandings. Sometimes this new knowledge is simply the reorganization of existing knowledge, although it may also be the addition of knowledge in the form of new propositions, new concepts, or new problem-solving procedures. Presumably, if these are the reasons that we as researchers solve problems, then it would also be educationally important to understand what students learn from solving problems.

Transfer of problem-solving procedures to content areas other than those in which they were taught also should be given attention. For example, do problem-solving strategies taught and learned in transmission genetics carry over to the solving of population genetics or microbial genetics problems?

Develop New Lines of Research

A research consortium would be an ideal vehicle through which to promote lines of research that would complement those described above. For example, research in biology education has not taken full advantage of constructs developed in cognitive psychology. There is a great deal known about the structure and function of human memory that has not made its way into the biology education research. This includes information on the limits of short-term memory; processes by which information is stored in and retrieved from long-term memory; and the importance of knowledge structure on intellectual performance.

Another area that holds promise for biology education researchers is similar to Resnick's (1987) *cognitive apprenticeships*. Her view emphasizes that teaching and learning cannot be fully represented by a one-way flow of information from teacher to learner. Although I think this is an area that has the potential to influence teaching and learning in biology, I am not sure that "student as apprentice" is the best metaphor to guide biology education. In apprentice relationships, the apprentice learns procedures and skills that the master has already acquired, by solving problems identical to those that the apprentice will face. Biology, and science in general, is much more an open-ended problem-solving activity (Laudan, 1977). A more useful view is one of teacher and student joined in a cognitive collaboration. In this view, classrooms would be transformed into places where the language of instruction is biology, and teachers and students work together to solve realistic problems. The teacher is the group leader, someone that the junior collaborators will learn from (because of the problem-solving heuristics that they have developed rather than by virtue of having previously worked out solutions to the problems); however, just as in a research lab, the senior researcher/teacher will also learn, through interactions with the junior reseachers/students. Research needs to be done that both examines the interactions between students and teachers and among students, and that assesses the learning and problem-solving outcomes that occur in environments that promote cognitive collaborations.

For this to happen, researchers, teachers, and agencies that support educational research must work together to provide opportunities for teachers and students to be able to work and learn in radically different ways. For example, we need to institute and evaluate the student-as-collaborator idea in classrooms with 9 to 10 students, rather than in ones with 24 to 30. There is more to be gained by looking at ideal arrangements than at typical classroom arrangements. We cannot continue to think of educational research as having to take the structure of schools as a given. All parties involved in research, or who may benefit from it, need to be willing to experiment on changing the structure of the learning environment.

CONCLUSIONS

In this chapter I have given an overview of selected research results in biological education that I feel provide a solid basis for continued research that have the potential to positively influence teaching and learning. In addition, I have outlined some ideas that I feel must be implemented if the past research is to be meaningful. The most important of these is the establishment of a consortium for biology education research.

ACKNOWLEDGMENTS

I wish to thank Judith Van Kirk for comments on various drafts of this chapter and Robert Hafner for his help in reviewing parts of the biology education literature. I also wish to acknowledge the support of the National Science Foundation Grant # MDR 8470277 which was used, in part, to support the development of this chapter.

REFERENCES

Albright, W. C. (1987). *A description of the performance of university students solving realistic genetics problems.* Unpublished master's thesis, University of Wisconsin-Madison.

Bishop, B., & Anderson, C. (1986a). *Student conceptions of natural selection and its role in evolution* (Research Series No. 165). East Lansing: Michigan State University, The Institute for Research on Teaching.

Bishop, B., & Anderson, C. (1986b). *Evolution by natural selection: A teaching module* (Occasional Paper No. 91). East Lansing: Michigan State University, The Institute for Research on Teaching.

Chi, M. T. H., Feltovich, P. J., & Glaser, R. (1981). Categorization and representation of physics problems by experts and novices. *Cognitive Science, 5,* 121–152.

Collins, A. (1989). *A description of the strategic knowledge of experts solving transmissions genetics problems.* Manuscript submitted for publication.

Collins, A., & Stewart, J. (1988). *A description of the strategic knowledge of experts solving realistic genetics problems* (MENDEL Research Report No. 1). Madison: University of Wisconsin-Madison.

Collins, A., Stewart, J., & Slack, S. (1988). *A comparison of the problem solving performance of experts and novices on realistic genetics problems* (MENDEL Research Report No. 3). Madison: University of Wisconsin-Madison.

Fisher, K. M. (1985). A misconception in biology: Amino acids and translation. *Journal of Research in Science Teaching, 22,* 53–62.

Fisher, K. M., Faletti, J., Thornton, R., Patterson, H., Lipson, J., & Spring, C. (1987). *Computer-based knowledge representation as a tool for students and teachers.* Unpublished manuscript.

Hildebrand, A. C. (March, 1986). *Cognitive representations of basic biological entities.* Paper presented at the annual meeting of the American Education Research Association, San Francisco, CA.

Jungck, J. R., & Calley, J. N. (1985). Strategic simulations and post-Socratic pedagogy: Constructing software to develop long-term inference through experimental inquiry. *American Biology Teacher, 47,* 11–15.

Laudan, L. (1977). *Progress and its problems: Towards a theory of scientific development.* Los Angeles: University of California Press.

National Academy of Sciences & National Academy of Engineering. (1982). *Science and mathematics in the schools: A report of a convocation,* Washington, DC: National Academy Press.

Newell, A., & Simon, H. (1972). *Human problem solving.* Englewood Cliffs, NJ: Prentice-Hall, Inc.

Peard, T. L. (1983). The microcomputer in cognitive development research (or putting the byte on misconceptions). In H. Helm & J. D. Novak (Eds.), *Proceedings of the International Seminar on Misconceptions in Science and Mathematics* (pp. 112–126). Ithaca, NY.

Peterson, N., Jungck, J., Sharpe, D., & Finzer, W. (1987). A design approach to learning science, simulated laboratories: Learning via the construction of meaning. *Machine-Mediated Learning, 2,* 111–127.

Peterson, N., Jungck, J. (1988). Problem posing, problem solving and persuasion in biology education. *Academic Computing.* March/April, 14–17 and 44–50.

Resnick, L. B. (1987). Learning in school and out. *Educational Researcher, 16,* 13–20.

Reif, F. (1983). *Understanding and teaching problem solving in physics.* In *Research on physics education: Proceedings of the first international workshop,* editions du Centre National de la Researche Scientificue 15, quai Anatole-france, 75700 Paris pp. 15–53.

Schwab, J. (1964). Problems, topics, and issues. In S. Elam (Ed.), *Education and the structure of knowledge* (pp. 4–43). Fifth annual Phi Delta Kappa Symposium in Education Research. Chicago: Rand McNally,

Slack, S., & Stewart, J. (1988). High school students problem-solving performance on realistic genetics problems. (MENDEL Research Report No. 2). Madison: University of Wisconsin-Madison.

Smith, E., & Anderson, C. (1984). Plants as producers: A case study of elementary science teaching. *Journal of Research in Science Teaching, 21,* 685–698.

Smith, M. (April, 1986). Problem solving in classical genetics: Successful and unsuccessful pedigree analysis. Paper presented at the annual meeting of the American Educational Research Association, San Francisco, CA.

Smith, M., & Good, R. (1984). Problem solving and classical genetics, successful versus unsuccessful performance. *Journal of Research in Science Teaching, 21,* 895–912.

Stewart, J. (1983). Student problem solving in high school genetics. *Science Education, 67,* 523–540.

Stewart, J. (1988). Potential learning outcomes from solving genetics problems: A topology of problems. *Science Education, 72, 2 ,*237–254.

Stewart, J., & Dale, M. (1981). Solutions to genetics problems: Are they the same as correct answers? *Australian Science Teacher, 27,* 59–64.

Stewart, J., & Dale, M. (1988). High school students understanding and chromosome/gene behavior during meiosis. (MENDEL Research Report No. 4). Madison: University of Wisconsin-Madison.

Streibel, M., Stewart, J., Koedinger, K., Collins, A., & Jungck, J. (1987). MENDEL: An intelligent computer tutoring system for genetics problem-solving, conjecturing, and understanding. *Machine-Mediated Learning, 2,* 129–159.

Thomson, N., & Stewart, J. (1985). Secondary school genetics: Making problem solving instruction explicit and meaningful. *Journal of Biological Education, 19,* 53–62.

Tolman, R. (1982). Difficulties in genetics problem solving. *The American Biology Teacher, 44,* 525–527.

Walker, R. A., Mertens, T. R., & Hendrix, J. R. (1980). Sequenced instruction in genetics and Piagetian cognitive development. *The American Biology Teacher, 42 (2),*104–108.

4 Toward a Scientific Practice of Mathematics Education

Jack Lochhead
University of Massachusetts

The Need for Change

The nature of mathematics is changing rapidly—both the techniques of investigation and the areas of research interest. In addition, mathematics is becoming essential to many more disciplines. The explosive growth of new technologies has increased the number and variety of useful applications. Calculators and computers are increasing the need for mathematical knowledge by making previously qualitative disciplines (from literature to political science) more quantitative. Calculators are decreasing the need for computation and placing greater demands on analytical and thinking skills.

We have every reason to expect that the hectic pace of change in mathematics will continue through the foreseeable future. Thus, mathematics educators must learn to adjust to a situation in which each new development will be superseded before it is widely implemented. Educational research will be part of a process of continuous development and renewal. Determining appropriate goals in this uncertain environment is a daunting task. Any specific skill we choose to teach, whether computational or high-level-cognitive, could become obsolete before the learner ever uses it. It is a situation in which curriculum design deserves very careful thought and presents a challenge to even the most experienced teacher. Groups of students who in the past needed very little mathematics now must be taught to perform applications they will need in their future work and that use techniques that have not yet been invented.

In this chapter I attempt to describe the sorts of curriculum changes that many of those in mathematics education (though by no means all) are now recommending. There is wide recognition that more research needs to be done and that each innovation should be carefully monitored before it is widely implemented. After describing the general direction of change in mathematics education, I will consider relevant research and, finally, indicate some of the questions that ought to guide future investigations.

Changes That Are Needed

What are the key modifications needed in the K-12 mathematics curriculum? Some entail increased emphasis on traditional objectives such as the development of "number sense" and "symbol sense." Other changes involve the introduction of material rarely found in the current curriculum such as data analysis, graph theory, and probability. Still others involve de-emphasis of topics that aim to develop certain manipulative skills that are no longer very useful, such as long division and factoring trinomials.

But far more important than changing the content of the curriculum, is the need to encourage students to develop a spirit of inquiry, an intellectual curiosity, and a sense of mathematical power.

Mathematical Power

Education in all subjects requires a balance between developing skills and knowledge and the ability to deploy that knowledge. But computers are changing the equation, and some skills are no longer prerequisites to further study. Once it was essential to stress penmanship before there could be a focus on what had been written; word processors have shifted the balance back to the content of the writing.

Calculators mean that we can shift the emphasis in K-12 mathematics away from skill development and toward mathematical power. This means the development of the abilities to:

- understand mathematical concepts and methods,
- discern mathematical relations,
- reason logically, and
- apply mathematical concepts, methods, and relations to solve a variety of nonroutine problems.

Students who achieve a considerable degree of mathematical power during their K-12 years will be able to use mathematics in their everyday lives or in a profession or vocation and they will be able to pursue further study of mathematics or other subjects that require mathematics. But computational power itself is not enough. Students must also learn to read and understand mathematical texts and to communicate to others, orally and in writing, the results of their own mathematical investigations and problem solving. The mathematics curriculum should provide support for the teaching of reading, writing, and oral communication.

Calculator and computer technology should be used throughout the K-12 mathematics curriculum and new curriculum materials should be designed with the expectation of continuous change resulting from further scientific and technological developments.

Modern relevant applications should be a fundamental part of the curriculum to a much greater extent than at present.

"Applications" need not be constrained to "real world" problems. The significant criterion for the suitability of an application is whether it has the potential to engage the students; often this can be done with questions of purely mathematical interest such as, "What is the largest prime number?"

Mathematics Instruction

Mathematical teaching must adapt to new realities. It will no longer be appropriate for most mathematics instruction to be in the traditional mode of the teacher presenting material to a class. Thinking mathematically is an *active* conception which requires more than listening. No single teaching method or any single kind of learning experience can develop the varied kinds of mathematical abilities needed for mathematical power. The Cockcroft Report (Cockcroft, 1982) indicates some of the range of activities needed:

- *Exposition* by the teacher;
- *Discussion* between the teacher and the pupils;
- *Discussion* among pupils;
- *Practical* work;
- *Consolidation* and *Practice* of fundamental skills and routines;
- *Problem solving,* including the application of mathematics to everyday situations; and
- *Investigational* work.

The standard mode of presentation in which a teacher lectures to students may still be appropriate for the delivery of straightforward information, but more imaginative settings are needed for the development of problem solving and reasoning skills, particularly in the context of using calculators and computer technology. For example, two formats that should be used often are:

1. small group work where the class is divided into teams of, say, three to five students who work collaboratively on assigned problems (which might take anywhere from five minutes to two weeks to solve), and

2. true class discussion in which the teacher plays the role of moderator rather than leader.

In both of these formats, the teacher can be a catalyst who helps students learn to think for themselves rather than the teacher who acts as a trainer and shows the "right way" to do something. Both formats also allow the teacher to use technology interactively with students.

A useful metaphor is that of the teacher as a sort of intellectual coach. At various times, this will require the teacher to be:

1. a role model who demonstrates not just the right way, but also the false starts and the higher-order thinking skills that lead to the resolution of problems;
2. a consultant helping individuals, small groups, or the whole class to decide if their work is keeping to the subject and making reasonable progress;
3. a moderator who poses questions for the class (or individuals or groups) to consider, but leaves most of the decision-making to the class; or
4. an interlocutor coaching the students during class presentations, encouraging them to reflect on their activities and to explore mathematics on their own, challenging them to make sure that what they are doing is reasonable and purposeful, and ensuring that students can defend their conclusions.

Research Directions

Research on teaching for higher-order thinking, (Peterson, in press), lends support to the notion that instruction needs to change from the traditional teacher-presenting-material-to-the-class mode to a less structured, indirect style of teaching. Because the development of higher-level thinking in mathematics has been shown to depend on autonomous, independent learning behavior, teachers should encourage self-reliance. One type of indirect instruction that has often proved to be effective is small group cooperative learning (Lochhead, 1985; Peterson, in press; Shavelson, in press). Noddings (in press) pointed out that among the several benefits of small group learning is that small groups allow consultation, a heuristic that we all use when we encounter difficulties. Cognitive research in other content areas (Brown & Campione, in press)—using a reciprocal teaching model that includes children taking turns playing teacher and posing questions, summarizing, clarifying, and predicting—has been effective in producing self-monitoring. Reciprocal teaching is based on the premise that the opportunity to communally construct meanings produces an internalization of the process of meaning construction (Resnick, in press).

Mathematical Thinking and Learning: Findings and Implications

We have called for new modes of teaching which stress the active role students must play in the construction of their own concepts. There is now wide agreement among researchers (Linn, 1986; Resnick, 1983) for the need to pay careful attention to student-constructed knowledge (Piaget, 1954). For example, Resnick (1976), Carpenter, Moser, and Romberg (1982), and Steffe, von Glaserfeld, Richards, and Cobb (1983) have shown that students invent "counting on from larger" for themselves (when adding two numbers, say $3 + 6$, the answer is found by counting 7,8,9). It is now clear that children come to school with a rich body of knowledge about the world around them, including well developed informal systems of mathematics (Ginsburg, 1977). Education fails when children are treated as "blank slates" or "empty jugs", ignoring the fact that they have a great deal of mathematical knowledge—some of which surpasses, and some of which may contradict, what they are being taught in school (Clement, 1977; Erlwanger, 1974; Gelman & Gallistal, 1978; Ginsburg, 1977)—that can be exploited.

Rote Learning

Probably the most controversial recommendations concern reducing the emphasis on rote-learned procedures and on algorithms used for extensive paper and pencil calculations. There is extensive evidence that algorithms alone do not aid conceptual understanding. The literature on arithmetic "bugs" (see Brown & Burton, 1978; Maurer, 1987) documents this point. Research reveals that in some 40% of the mistakes students make in subtraction, the flawed procedure that produces the student's answers can be identified, and this predicts incorrect answers students will produce on similar problems. Such consistent but mistaken procedures have a natural origin. Most of the student bugs—so named because, like bugs in computer programs, they produce consistent incorrect answers—can be explained as intelligent attempts to "patch" rote–learned algorithms that are poorly understood.

Even when correctly learned, purely procedural knowledge—the ability to implement mathematical algorithms without the underlying conceptual structures—can be extremely fragile. Clement, Lochhead, and Monk (1979) have shown that even a solid procedural knowledge of algebra, such as that held by university level engineering students, does not, in most cases (over 80%), imply an ability to interpret the meaning of algebraic symbols. One can minimize the fragility of knowledge structures by teaching mathematical concepts in a fashion that stresses the underlying conceptual models (Carpenter, et al., 1982; Davis, 1984; Hiebert, 1986; Romberg & Carpenter, 1986). In sum, we now know children are active interpreters of the world around them, including the mathematical aspects of that world. In Piaget's (1973) words, "To understand is to invent". This suggests that topics in school should be arranged to exploit intuitions and informal numerical notions students bring with them to school. Second, it indicates that predominant teaching methods must be revised to adapt to the notion of child as interpreter (and sometimes constructor of incorrect theories) as opposed to child as absorber.

Development of Subject Matter

Researchers have only just begun to construct a detailed map of the phases children can go through as they gradually build up their understanding of numbers and arithmetic (Steffe et al., 1983). Even at the early ages, the picture is quite complex. As students develop, it is most effective to engage them in meaningful, complex activities focusing on conceptual issues, rather than firmly establishing all the building blocks before going on to the next level. (Collins, Brown, & Newman, in press; Hatano, 1982; Romberg & Carpenter, 1986). In certain cases, the order of presentation is critical; early introduction of some topics can be very damaging. Wearne and Hiebert (1987) have shown that if calculational algorithms are memorized before the underlying structure is understood, "it may be difficult for semantic information to penetrate routinized rules" (Wearne & Hiebert, 1987, p. 26). In short, students who learn to calculate too early may find it more difficult to reach an understanding of the material than students who have had no such experience. But this is not always the case. In fact, Steffe et al. (1983) showed that in some cases, memorized routinized rules must precede understanding. Children must be able to recite the number words in order (one, two, three . . .) before they can develop a concept of counting or number. In contrast, there is some evidence[1] to suggest that calculation algorithms involving fractions, decimal long division, and possibly multiplication are introduced far too soon in the current curriculum. The challenge for curriculum development (and research) is to determine when routinized rules should come first, and when they should not. This is an area where more research needs to be done.

Problem Solving Strategies

There is an extensive body of literature (Charles & Silver, in press; Kulik, 1980; Mason, Burton, & Stacey, 1982; Schoenfeld, 1985; Silver, 1985) indicating that

---

[1]Evidence supporting the delayed introduction of fraction and decimal calculational algorithms comes mainly in the form of the large number of students who never learn these topics. The National Assessments indicate that a very high percentage of high school students worldwide never master these topics. This is what one would expect in a case where routinized skills are blocking semantic learning. The experience of Benazet (1935), who delayed such instruction until after 6th grade, indicates that such a postponement can be very helpful.

problem-solving strategies can be taught, and suggesting various ways to do so. The main warning from the research literature is that one should be careful not to trivialize problem-solving strategies by teaching a collection of isolated tricks (e.g., "of" means multiply or cross-cancelling factors). Problem-solving strategies, in the spirit of Polya (1945), are subtle and complex. Important strategies (such as looking for a pattern by plugging in values for $n = 1,2,3,4. \ldots \ldots$) cannot be effectively taught apart from situational cues which indicate when it is appropriate to apply them.

Metacognition and "Executive Control"

An important aspect of problem-solving competence is metacognition—the ability to know when and why to use a procedure. There is ample evidence (Brown, & Campione, in press; Collins, Brown, & Newman, in press; Schoenfeld, 1985; Silver, 1985) that students who "know" more than enough domain-specific subject matter fail to solve problems because they do not use their knowledge wisely. They may: (a) jump into problems, doggedly pursuing a particular ill-chosen approach to the exclusion of anything else; (b) raise profitable alternatives, but fail to pursue them; or (c) get sidetracked into focusing on trivia while ignoring the "big picture."

Research indicates that such executive skills can be learned, resulting in significant improvements in problem-solving performance. Effects can be obtained with interventions as simple as holding class discussions that focus on executive behaviors, and by explicitly and frequently posing questions such as:

What are you doing?
Why are you doing it?
How will it help you?

(Schoenfeld, 1985; Collins, Brown, & Newman, in press.)

Beliefs: Getting a Sense of What Mathematics is About

On the Third National Assessment of Educational Progress (Carpenter et al., 1983), a stratified nationwide sample of 45,000 students worked the following problem:

An army bus holds 36 soldiers. If 1,128 soldiers are being bussed to their training site, how many buses are needed?

Roughly 29% of the students who worked the problem wrote that the number of buses needed is "31 remainder 12," but only 23% gave the correct answer to the problem. Approximately 70% of the students who took the examination performed the right operation (1,128 divided by 36 yields "31 remainder 12"). Then, however, fewer than ⅓ of those students wrote 32 buses. How can students say that the number of buses includes a remainder?

For most students, the "school mathematics mode" includes a habit of problem solving *without* sense-making: One learns to read the problem, extract the relevant numbers and the operation to be used, perform the operation, and write down the result (Lave, in press; Reusser, in press; Schoenfeld, in press-b). Consider, for example, the following nonsense problems:

There are 26 sheep and 10 goats on a ship. How old is the captain?

There are 125 sheep and 5 dogs in a flock. How old is the shepherd? (Reusser, in press)

Reusser (in press) reports that when asked to "solve" such problems, three school children in four will produce a numerical answer. There are similar data for French school children, and the NAEP data speak for themselves. In a discussion of these and similar problems, Reusser suggests that the students work the problems compliantly and without asking that they make sense, because the students have already "learned" that school math problems do not necessarily make sense. In the context of the mathematics classroom, the expectation is that problems have an answer (Why else would the teacher pose them?) and that some reasonable combination of the numbers in the problem (usually using the most recent mathematical procedure studied) will yield it. Students learn to act in the way described in the previous paragraph.

Students constantly strive to make sense of the rules that govern the world around them, including the world of their mathematics classrooms. If the classroom patterns are perceived to be arbitrary and the mathematical operations meaningless—no matter how well "mastered" as procedures—students will emerge from the classroom with a sense of mathematics being useless. Consider the example (Reusser, in press) of a student working the shepherd problem. This solution, produced by a student solving the problem out loud, is all too typical:

$125 + 5 = 130$ . . . this is too big, and $125 - 5 = 120$ is still too big . . . while $125/5 = 25$. That works. I think the shepherd is 25 years old. (Reusser, in press)

In short, the "classroom culture" in which students learn mathematics shapes their developing understanding of the nature of mathematics which, in turn, shapes how sensibly students will use the mathematics they have learned. Research indicates (Fawcett, 1938; Lave, in press; Mason, Burton, & Stacey, 1982; Schoenfeld, in press-b) that it is indeed possible to create classroom environments that are, in essence, cultures of "sense-making" and from which students emerge with an understanding of mathematics as a discipline which helps to make sense of things. The goal of teaching sense-making via mathematics should be a central concern of our curricular efforts.

Use of Calculator and Computer Technology

At the turn of the century a social planner might have expressed concern over the consequences of widespread access to the automobile. But in the last analysis, research evidence for or against the automobile would have been irrelevant and ignored. The type of study that might have made sense would have assumed the car and considered how best to use it. We believe the same is true of computers and calculators. Whether or not we like it, they are a fact of life.

Computers have already changed the face of mathematics. New fields of inquiry, such as fractal geometry, depend in large part on the computer for their very existence. Much of modern mathematics is inaccessible and inexplicable without access to computers.

But it is still reasonable to ask whether computers and calculators in the curriculum may not pose some serious dangers. In particular, should they be introduced early, before students have mastered the basics?

Effects of Calculators

The effects of calculators in school mathematics have been studied in over 100 formal investigations during the past 15 years. Those studies have tested the impact of various kinds of calculator use—from limited access in carefully selected situations to access for all aspects of mathematics instruction and testing. There have been two major summaries of reported research on calculator usage (Hembree & Dessart, 1986; Sudyam, 1982). In almost every reported study, the performance of groups using calculators equaled or exceeded that of control groups denied calculator use.

The recent Hembree and Dessart meta-analysis of 79 calculator studies sorted out the effects of calculator use on six dimensions of attitude toward mathematics as well as on acquisition, retention, and transfer of computational skills, conceptual understanding, and problem-solving ability. The analysis led them to the following conclusions:

1. Students who use calculators in concert with traditional instruction maintain their paper-and-pencil skills without apparent harm. Indeed, a use of calculators can improve the average student's basic skills with paper and pencil, both in basic operations and in problem solving.

2. Sustained calculator use by average students in Grade 4 may be counterproductive with regard to basic skills.

3. The use of calculators in testing produces much higher achievement scores than paper-and-pencil efforts, both in basic operations and in problem solving. This statement applies across all grades and ability levels. In particular, it applies for low- and high-ability students in problem solving. The overall better performance in problem solving appears to be a result of improved computation and process selection.

4. Students using calculators possess a better attitude toward mathematics and an especially better self-concept in mathematics than noncalculator students. This statement applies across all grades and ability levels.

5. Studies with special curricula indicate that materials and methods can be developed for enhancing student achievement through instruction oriented toward the calculator. However, such special instruction has been relatively unexamined by research. (Hembree & Dessart, 1986, pp. 96–97)

These findings speak directly to a number of common concerns about the potential impact of widespread calculator use in school mathematics. For those who believe that it will continue to be important for most students to acquire some measure of skill in traditional arithmetic algorithms, the research suggests that access to calculators in a well-planned program of instruction is not likely to obstruct achievement of those skills. In fact, it might well enhance acquisition of traditional skills. More optimistically, it appears that when students have access to calculators for learning and achievement testing, they perform at significantly higher levels on both computation and problem-solving measures. In particular, students using calculators seem better able to focus on correct process analysis of problem situations.

Effects of Computers

The earliest educational use of computers was primarily to deliver computer-assisted instruction (CAI), often in a programmed learning style of instruction and, most frequently, for drill or rote skills. Several reviews of research on effectiveness of CAI (Kulik, Bangert, & Williams, 1983) have concluded that it is generally very effective, offering better results in shorter time than traditional instruction.

Lately, principles of artificial intelligence have been applied to the design of sophisticated tutors for algebra, geometry, and calculus. The designers suggest that the use of such tutors can yield dramatic increases in student achievement. However, no data is available about the use of such tutors in realistic classroom settings.

There are several kinds of computer-based strategies for giving students impressive new learning tools and exploratory environments (Hansen, 1984; Pea, 1987b; Schoenfeld, in pressa). Best known is LOGO and its turtle graphics feature to teach students concepts of geometry, algebra, and general higher-order thinking skills (Papert, 1980). Research findings have failed to confirm the strongest claims that LOGO develops high level general reasoning abilities. But a variety of studies have found positive effects on more specific instructional goals (Campbell, 1987), and thousands of classroom teachers have been convinced by firsthand experience that LOGO is a powerful instructional tool.

A somewhat different sort of computer-based exploratory tool has been provided by THE GEOMETRIC SUPPOSER (Schwartz & Yerushalmy, 1987), and by the Geodraw software developed at Wicat (Bell, 1987). In each, the idea is to give students open but guided environments for exploring the results of geometric constructions. A comparable setting for algebraic exploration is provided by the Green Globs software of Dugdale (1982). There is little formal research describing the effects of these learning and teaching tools. Yerushalmy, Chazan, and Gordon (1987) provide evidence suggesting that students using the SUPPOSER may perform as well or better than control students on traditional geometry criteria, while at the same time learning other objectives as well.

There have been some specific research studies investigating the effects of computer graphics on student understanding of mathematical concepts like *function* (Rhoads, 1986; Schoenfeld, in pressa). The curriculum development work of Demana, Leitzel, Osborne, and Waits at Ohio State University has taken particular advantage of spreadsheet-like software to develop students' numerical intuition about variables and equations in prealgebra and precalculus mathematics. In each case, the computer clearly enhances student interest and understanding of important ideas.

Most studies have focused, in one way or another, on finding better ways to reach traditional goals. There have been some noteworthy exceptions to that pattern—studies that explore daring departures from conventional curriculum priorities. Lesh (1986) and Heid and Kunkle (1988) tested the effects of algebra instruction in which students used symbol-manipulation software to perform routine tasks like solving equations. Each found that students who were freed from the traditional symbolic procedural aspects of problem solving became much more adept at the important problem formulation and interpretation phases. In two similar studies of computer-aided calculus, Heid (1988) and Palmiter (1986) found that students who learned the subject with aid of computer-tool software developed much deeper understanding of fundamental concepts than did students in traditional skill-oriented courses. Heid also found that her students picked up needed procedural knowledge in a short time period following the careful conceptual background, and Palmiter found that her students acquired their understanding much more quickly than students in conventional courses.

Each of these studies addresses the fundamental questions of technology applied to mathematics curriculum: (a) What are the essential interactions of conceptual and procedural knowledge and problem solving ability? and (b) if we diminish attention to the traditional procedural skill agendas in various branches of mathematics, will something essential to problem solving or conceptual learning be inadvertently lost? A fair test of this question involves extensive and radical curriculum development and field testing with attendant risks for students who study the new curricula. Not surprisingly, the only work of this type has been in limited numbers of classes and situations. At the University of Maryland, a computer-intensive elementary algebra program has been developed to explore the feasibility of teaching fundamental concepts and problem-solving abilities while using technology to perform nearly all of the traditional symbol-manipulation skills. Preliminary evidence indicates that students can approach algebra as the study of functions and their application as models of quantitative interactions; that they can become flexible and effective algebraic problem solvers; and that the rich conceptual background of understanding about variables, functions, equations, and inequalities they acquire provides a strong foundation for the more abstract task of learning appropriate symbol-manipulation skills later (Heid, 1988).

The research cited here indicates that access to computers and calculators need not hinder attainment of traditional curricular objectives, and that it may substantially advance it. Unfortunately, there is currently no consensus on how to investigate possible dramatically new effects such as the improvement of higher-order thinking skills. A series of articles in *Educational Researcher* (Becker, 1987; Papert, 1987; Pea, 1987; Walker, 1987) illustrates the wide diversity of opinion on this topic. A key concern is the extent to which the development of powerful reasoning can be inferred from written test performance or within the limited time spans of most research studies.

From the few attempts that have been made to measure massive changes in reasoning power, it is possible to conclude that such advancements cannot come from trivial technological fixes. It has often been proposed that the availability of computers would produce significant improvements in mathematical thinking. Repeated attempts to document such change have yet to reveal a lasting effect (e.g., Pea's study of the effect of LOGO on planning, Soloway's investigation of the impact of PASCAL on understanding of algebraic syntax and Perkin's research concerning the cognitive impact of learning programming meta-principles in BASIC). Although these results do not imply that computers cannot be used to improve mathematical thinking, they suggest that simplistic approaches are not likely to work. Thus teachers and curriculum designers probably will have a significant role to play in education far into the computer age. The computer will remain a tool for teachers and students to use, thus integration of technology into "Classroom Ecology" must remain a high priority research item.

FUTURE DIRECTIONS FOR RESEARCH

Order of Learning

Many topics can be taught earlier than they have been, others ought to be taught later. Because it is impossible to forecast the full implications of such changes, they should first be implemented in a research environment. Such studies must go beyond evaluating student progress in terms of superficial measures, they must examine deep conceptual understanding as well as the long-term effects of gaining (or not gaining) such understandings at a certain stage.

Levels of Learning

It seems clear that the new representations and computational aids made possible through computer technology now allow us to teach some concepts much earlier. Research needs to be done on the degree to which students really understand advanced concepts when they are introduced early. The goal of such research should be to find a suitable sequencing and pace for the introduction of new topics.

Modes of Learning

We have discussed the growing evidence supporting the need for new modes of instruction. The theoretical basis for many of the proposed new techniques is recognition of the active role students must take in constructing their own knowledge. Studies are now needed that can determine the long-term effects of early exposure to various modes of mathematical instruction on mathematical competence and on facility in learning new mathematical concepts. In evaluating instructional techniques it is important to avoid criteria which value maximizing current performance without regard to the extended impact.

The Roles of Arithmetic and Algebraic Manipulative Skills

Within the next decade hand-held calculators capable of performing symbolic manipulations (e.g., solving linear and quadratic equations in one variable, pairs of linear equations in two variables, etc.) will become widely available. Just as the current generation of calculators allows for the rearrangement of arithmetic topics (postponing some paper-and-pencil rote skills until appropriate), the next generation of calculators will allow much more flexibility in the order of advanced topics—for example, allowing students to do fewer algebraic symbol manipulations. We must, therefore, formulate a new set of fundamental manipulative skills and determine when each of these skills should be introduced.

The Relation Between Drill-and-Practice and Understanding Mathematics

The question of how much paper-and-pencil arithmetic to teach elementary school children and how much symbol-manipulation proficiency is desirable for secondary school students depends on the correlation of such skills with achieving an understanding of the underlying mathematics. At one time, the development of such skills could be justified on social and economic grounds. But the rapid advance of calculator and computer technology has undermined such a justification. The question of what, if any, level of manipulative skills is necessary in order to be able to understand and use mathematics in a problem-solving context, is a very difficult one on which research is badly needed.

Evaluations of the Effects of Entire Curricula: The "Transfer Problem"

We need research which can help us learn more about the ways in which mathematical experiences shape people's understanding of mathematics, understandings which often mitigate against the use of mathematics in real-world situations. We need to study ways of developing curricula that help to solve the Transfer Problem—that is to provide students with the sort of background that will encourage, rather than discourage, their ability to apply what they have learned in out-of-school contexts.

Instructional Uses of Technology

The "information explosion" has just begun to result in tools that can have significant impact on the instructional process. What kinds of mathematical comprehension can these new tools foster? Are there negative side effects? How do they fit within the context of schooling? Critical here is the question of access. If home computers become a significant part of the instructional process, what happens to those who do not have such tools?

Summary

The need for a structured scientific approach to curriculum and instructional development in mathematics has never been greater, and it is increasing at a very rapid rate. Sooner or later we may be forced to take it seriously.

ACKNOWLEDGMENTS

This paper is based directly on the work of hundreds of researchers and educators, but most particularly on a draft written jointly by Anthony Ralston, Alan Schoenfeld, and Jim Fey. I am also highly indebted to the other members of the M.S.E.B. Task Force on Curriculum Frameworks. Several organizations have created an environment to make this work possible: (a) the Mathematical Science Education Board of the National Research Council, (b) the National Science Foundation, and (c) the Exxon Education Foundation. However, the ideas generated by this paper are not necessarily endorsed by any of these organizations or individuals.

REFERENCES

Becker, H. J. (1987). The importance of a methodology that maximizes falsifiability: its applicability to research about LOGO. *Educational Researcher, 16*(5), 11–16.

Bell (1987). Microcomputer-Based Course for School Geometry. In Wirszup, I. & Streit, R. (Eds.), *Developments in school mathematics education around the world* pp. 604–622. Reston, VA: National Council of Teachers of Mathematics.

Benazet, L. P. (1935–1936). The story of an experiment. *The Teaching of Arithmetic, 24* #8 p. 241–244, *24* #9 p. 301–303, *25* #1 p. 7–8.

Brown, A. L., & Campione, J. C. (in press). On the importance of knowing what you are doing: Metacognition and mathematics. In E. A. Silver & R. I. Charles (Eds.), *The teaching and evaluation of mathematics problem solving.* Reston, VA: National Council of Teachers of Mathematics; Hillsdale, NJ: Lawrence Erlbaum Associates.

Brown, J. S., & Burton, R. R. (1978). Diagnostic models for procedural bugs in basic mathematical skills. *Cognitive Science, 2,* 155–192.

Campbell, P. F. (1987). Measuring Distance: Children's Use of Number and Unit. Final report submitted to NIMH.

Carpenter, T. P., Moser, J., & Romberg, T. A. (Eds.). (1982). *Learning to add and subtract: A cognitive perspective.* Hillsdale, NJ: Lawrence Erlbaum Associates.

Carpenter, T. P., Lindquist, M. M., Matthews, W., & Silver, E. A. (1983). Results of the third NAEP mathematics assessment: Secondary school. *Mathematics Teacher, 76*(9), 652–659.

Charles, R., & Silver, E. (Eds.). (in press). *Teaching and evaluating mathematical problem solving.* Reston, VA: National Council of Teachers of Mathematics.

Clement, J. (1977). *Quantitative problem solving processes in children.* Unpublished doctoral dissertation, University of Massachusetts at Amherst.

Clement, J., Lochhead, J., & Monk, G. (1979, April). Translation difficulties in learning mathematics. *American Mathematical Monthly, 88*(4) 287–290.

Collins, A., Brown, J. S., & Newman, S. (in press). The new apprenticeship: Teaching students the craft of reading, writing, and mathematics. In L. B. Resnick (Ed.), *Cognition and instruction: Issues and agendas.* Hillsdale, NJ: Lawrence Erlbaum Associates.

Crockcroft, W. H. (1982). *Mathematics counts* (Report of the Committee of Inquiry into the Teaching of Mathematics in Schools). London: Her Majesty's Stationery Office.

Davis, R. B. (1984). *Learning mathematics: The cognitive science approach to mathematics education.* Norwood, NJ: Ablex.

Dugdale, S. (1982). Green Globs: A Microcomputer application for graphing equations. *The Mathematics Teacher, 75,* 208–214.

Erlwanger, S. H. (1974). *Case studies of children's conceptions of mathematics.* Unpublished doctoral dissertation, University of Illinois, Champaign–Urbana.

Fawcett, H. P. (1938). *The nature of proof (1938 Yearbook of the National Council of Teachers of Mathematics).* New York: Columbia University Teachers College Bureau of Publications.

Gelman, R., & Gallistal, C. (1978). *The child's understanding of number.* Cambridge, MA: Harvard University Press.

Ginsburg, H. (1977). *Children's arithmetic: The learning process.* New York: Van Nostrand.

Hansen, V. P. (Ed.). (1984). *Computers in mathematics education.* 1984 NCTM Yearbook. Reston, VA: National Council of Teachers of Mathematics.

Hatano, G. (1982). Learning to add and subtract: A Japanese perspective. In T. Carpenter, J. Moser, & T. Romberg (Eds.), *Learning to add and subtract: A cognitive perspective.* Hillsdale, NJ: Lawrence Erlbaum Associates.

Heid, M. K. (1988). Resequencing skills and concepts in applied calculus using the computer as a tool. *Journal for Research in Mathematics Education, 19,* 3–25.

Heid, M. K., & Kunkle, D. (1988). Computer generated: Tools for concept development in elementary algebra. In A. F. Coxford & A. P. Shulte (Eds.), *1988 Yearbook of the NCTM* (pp. 170–177). Reston, VA: National Council of Teachers of Mathematics.

Hembree, R., & Dessart, D. (1986). Effects of hand-held calculators in precollege mathematics education: A meta-analysis. *Journal for Research in Mathematics Education, 17*(2), 83–89.

Hiebert, J. (Ed.). (1986). *Conceptual and procedural knowledge: The case of mathematics.* Hillsdale, NJ: Lawrence Erlbaum Associates.

Kulik, S. (Ed.). (1980). *Problem solving in school mathematics.* 1980 NCTM Yearbook. Reston, VA: National Council of Teachers of Mathematics.

Kulik, J. A., Bangert, R. L., & Williams, G. W. (1983). The effects of computer-based teaching on secondary school students. *Journal of Educational Psychology, 75,* 19–26.

Lave, J. (in press). The practice of mathematics. In R. Charles & E. Silver (Eds.), *Teaching and evaluating mathematical problem solving.* Reston, VA: National Council of Teachers of Mathematics.

Lesh, R. (1986). The evolution of problem representations in the presence of powerful conceptual amplifiers. In C. Janvier (Ed.), *Problems of representation in the teaching and learning of mathematics* (pp. 107–206). Hillsdale, NJ: Lawrence Erlbaum Associates.

Linn, M. C. (Ed.). (1986). *Establishing a research base for science education: Challenges, trends and recommendations.* (Report). Berkeley: University of California–Berkeley.

Lochhead, J. (1985). Teaching analytical reasoning skills through pair-problem solving. In Segal, J., & Chipman, S. (Eds.), *Thinking and learning skills: relating instruction to basic research,* 109–131. Hillsdale, NJ: Lawrence Erlbaum Associates.

Maurer, S. (1987). New knowledge about errors and new knowledge about learners: What they mean to educators and more educators would like to know. In A. Schoenfeld (Ed.), *Cognitive science and mathematics education* (pp. 165–187). Hillsdale, NJ: Lawrence Erlbaum Associates.

Noddings, N. (in press). Preparing teachers to teach mathematical problem solving. In E. A. Silver & R. I. Charles (Eds.), *The teaching and evaluation of mathematics problem solving.* Reston, VA: National Council of Teachers of Mathematics; Hillsdale, NJ: Lawrence Erlbaum Associates.

Palmiter, J. R. (1986). The impact of a computer algebra system on college calculus. (Doctoral dissertation, Ohio State University 1986), *Dissertation Abstracts International, 47,* 1640A.

Papert, S. (1980). *Mindstorms.* New York: Basic Books.

Papert, S. (1987). Information technology and education: computer criticism vs. technocentric thinking. *Educational Researcher, 16*(1).

Pea, R. D. (1987a). The aims of software criticism: Reply to Professor Papert. *Educational Researcher, 16*(5), 4–8.

Pea, R. (1987b). Cognitive technologies for mathematics education. In A. Schoenfeld (Ed.), *Cognitive science and mathematics education* (pp. 89–122). Hillsdale, NJ: Lawrence Erlbaum Associates.

Peterson, P. L., & Carpenter, T. L. (Eds.). (in press). Learning through instruction: The study of students' thinking during instruction in mathematics [Special issue]. *Educational Psychologist.*

Peterson, P. L. (in press). Teaching for higher-order thinking in mathematics: The challenge for the next decade. In D. A. Grouws & T. J. Cooney (Eds.), *Perspectives on research on effective mathematics teaching.* Reston, VA: National Council of Teachers of Mathematics; Hillsdale, NJ: Lawrence Erlbaum Associates.

Piaget, J. (1954). *The construction of reality in the child.* New York: Basic Books.

Piaget, J. (1973). *To understand is to invent: The future of Education* (G. Robberts, Trans.). New York: Grossman. (Original work published 1948)

Polya, G. (1945). *How to solve it.* Princeton, NJ: Princeton University Press.

Resnick, L. (1976). Task analysis in instructional design: Some cases from mathematics. In D. Klahr, (Ed.), *Cognition and instruction* (pp. 51–80). Hillsdale, NJ: Lawrence Erlbaum Associates.

Resnick, L. (1983, April). Mathematics and science learning: A new compilation. *Science,* pp. 477–478.

Resnick, L. B. (in press). Treating mathematics as an ill-mannered discipline. In E. A. Silver & R. I. Charles (Eds.), *The teaching and evaluation of mathematics problem solving.* Reston, VA: National Council of Teachers of Mathematics; Hillsdale, NJ: Lawrence Erlbaum Associates.

Reusser, K. (in Press). Problem solving beyond the logic of things. Manuscript submitted for publication.

Rhoads, C. (1986). Organization of microcomputer instruction in secondary mathematics education (Doctoral dissertation, University of Maryland, 1985). *Dissertation Abstracts International 46,* 3641A.

Romberg, T. A., & Carpenter, T. P. (1986). Research on teaching and learning mathematics: Two disciplines of scientific inquiry. In M. C. Wittrock (Ed.), *Handbook of research on teaching* (3rd ed.), (pp. 850–873). New York: MacMillan.

Schoenfeld, A. H. (1985). *Mathematical problem solving.* New York: Academic Press.

Schoenfeld, A. H. (in press-a). Mathematics, technology, and higher-order thinking. In R. Nickerson (Ed.), *Technology in education in 2020: Thinking about the future.* Hillsdale, NJ: Lawrence Erlbaum Associates.

Schoenfeld, A. H. (in press-b). Problem solving in context(s). In R. Charles & E. Silver (Eds.), *Teaching and evaluating mathematical problem solving.* Reston, VA: National Council of Teachers of Mathematics.

Schwartz, J. L., & Yerushalmy, M. (1987). THE GEOMETRIC SUPPOSER: An Intellectual Prosthesis for Making Conjectures. College Mathematics Journal, Vol. 18, 58–65.

Shavelson, R. J. (in press). Teaching mathematical problem solving: Insights from teachers and tutors. In E. A. Silver & R. I. Charles (Eds.), *The teaching and evaluation of mathematical problem solving.* Reston, VA: National Council of Teachers of Mathematics; & Hillsdale, NJ: Lawrence Erlbaum Associates.

Shumway, R. (Ed.). (1980). *Research in mathematics education.* Reston, VA: National Council of Teachers of Mathematics.

Silver, E. (Ed.). (1985). *Teaching and learning mathematical problem solving: Multiple research perspectives.* Hillsdale, NJ: Lawrence Erlbaum Associates.

Steffe, L. P., von Glasersfeld, E., Richards, J., & Cobb, P. (1983). *Children's counting types: Philosophy, theory, and application.* New York: Praeger.

Sudyam, M. N. (1982). *The use of calculators in pre-college education: Fifth annual state-of-the-art review.* Columbus, OH: Calculator Information Center. (ERIC Document Reproduction Service No. ED 220 273).

Walker, Decker, F. (1987). LOGO needs Research: A Response to Papert's Paper. *Educational Researcher, 16*(5), 9–11.

Wearne, D., & Hiebert, J. (1987, May). *Constructing and using meanings for mathematical symbols: The case of the decimal fractions.* Paper presented at the Conference on Middle School Number Concepts, Northern Illinois University.

Yerushalmy, M., Chazan, D., & Gordon, M. (1987). *Guided inquiry and technology: A year-long study of children and teachers using the geometric supposer* (ETC Final Report). Harvard University.

II INSTRUCTIONAL DESIGN

Frederick Reif
University of California, Berkeley

Good design, aiming to apply available knowledge to achieve desired goals effectively and efficiently, is the central concern of any applied or engineering field. Instructional design—to achieve desired student learning effectively and efficiently—is an equally important concern in education, but the principles of good design are far less well developed.

The four chapters in this section of the book all deal with some aspects of instructional design. They make no pretense to cover many of the complex and multifarious issues involved in such design, nor would such wide-ranging discussion be possible within the limited scope of this section. Instead, the authors of these papers focus on a few major issues which appear, at the present time, to be important for achieving significant advances in the effectiveness of science education.

The following are some of the major themes discussed by these authors from somewhat different perspectives.

Potentialities of more Scientific and Technological Approaches to Education

Recent advances in the cognitive sciences (e.g., information-processing psychology, artificial intelligence, linguistics, etc.) and information technologies (e.g., computers and video technologies) offer promising prospects of approaching education in more systematic and principled ways. These prospects are discussed in my chapter and more fully in that by Larkin, Scheftic, and Chabay.

From the vantage point of more principled and analytic approaches to education, many prevailing teaching practices seem simplistic and ignore some crucially important issues. I discuss several such issues, and point out that they are both intellectually interesting and practically important, that neglecting them can make instruction ineffective or even deleterious, and that such issues need to be addressed in any serious attempts to improve education.

Brown and Campione give concrete examples of how cognitive analyses, combined with well-designed innovative teaching methods, can lead to highly successful instruction. In particular, they discuss their method of *reciprocal teaching* which has been remarkably effective in teaching students to read with good comprehension, in spite of initially severe reading difficulties. This method, which began as a small research project in cognitive science, has now been systematically expanded to the point where it is being used successfully throughout the school system of Springfield, Illinois. Brown and Campione also discuss current attempts to extend the method to the teaching of algebra.

Educational Applications of Computers

As computers have become increasingly powerful, smaller, cheaper, and more readily available, their educational applications are increasingly apparent and amenable to realistic exploitation. Indeed, computers are providing highly useful tools for educational research and also potentially powerful means of educational distribution. However, the appealing potentialities are not easily realized in large-scale practice. The chapter by Clancey and Joerger and that by Larkin et al. both discuss current attempts to further practical realizability. Both also illustrate the fruitful interplay between pure research interests and concerns with practical implementation.

Clancey's main interest has been in the educational applications of artificial intelligence (AI) and in related deep questions about the kinds of knowledge structures underlying complex human performance. But, motivated by practical considerations, he describes in his chapter a recent attempt to produce an authoring environment which would allow teachers to produce useful computer-implemented instruction. He stresses a method that would benefit from AI without getting involved in its full complexities.

Larkin et al. describe recent efforts at Carnegie-Mellon University to develop a generally useful authoring language ("cT") which makes it easy for teachers to design and produce computer-implemented instruction even if they do not possess much programming experience. Instructional programs thus produced can also easily be transported and disseminated since they can run equally well on different types of microcomputers and workstations.

As Larkin et al. point out, the effective educational exploitation of computers could presently benefit considerably by building on two lines of work which have, until now, remained fairly disjointed—namely, the older "computer-assisted instruction" and the more recent "intelligent tutoring systems" that use artificial-intelligence techniques.

Institutional Mechanisms for Realizing Present Educational Opportunities

Attempts to realize the potentialities of promising scientific and technological educational approaches are hampered by some severe difficulties. For example, there is a shortage of good and properly trained talent, particularly with the diverse kinds of expertise needed in science, cognitive science, computation, and education; existing talent groups are mostly fragmented and below critical size; and the physical and human resources needed for significant educational projects are often lacking. Larkin et al. and I point out how such problems might be overcome by establishing and judiciously structuring some educational science and technology centers (similar to some of the science or engineering centers recently proposed by the National Science Foundation). Although they have not been thoroughly discussed, suggestions to establish such centers have repeatedly been made in the recent past. However, these suggestions have not yet led to any attempts at implementation.

5 Transcending Prevailing Approaches to Science Education

Frederick Reif
University of California, Berkeley

INTRODUCTION

The design of effective instruction is centrally important for education and can benefit from the diverse expertise of all the contributors to this book (i.e., subject-matter experts, teachers and curriculum developers, educational researchers, and cognitive scientists). Because of a somewhat unusual career, I have had a "foot in all these camps"—as a research physicist, physics professor, textbook author, and researcher in education and cognitive processes. I may, therefore, be in a good position to provide a broadly-based perspective about some salient problems of instructional design.

My focus can be summarized simply and briefly:

1. Prevailing approaches to teaching and curriculum development commonly neglect some fundamental issues. As a result, customary instruction in schools and colleges frequently fails to teach important knowledge and intellectual skills, is often ineffective or inefficient, and may sometimes even be deleterious.
2. Some such fundamental issues can be readily identified, are scientifically interesting, could be successfully addressed, and would yield significant educational benefits.
3. The time is ripe for addressing such issues systematically, instead of merely perpetuating existing educational practices.

This chapter identifies several such issues, illustrates them with some examples, and points out their implications for the improvement of education.

ADDRESSING NEGLECTED ISSUES

Prevailing Educational Approaches

Traditional approaches to science teaching and curriculum development (including the innovative science-curriculum projects supported by the National Science Foundation in the 1950s and 1960s) focus on the scientific subject matter. They try to teach the important scientific facts, principles, and methods of the discipline; they aim to present the relevant topics logically; and they attempt to reflect recent scientific advances and points of view.

However, in all these efforts the student is viewed predominantly as a "black box", i.e., as a system whose inner workings are not subjected to closer scrutiny: The input of teaching and curriculum is somehow presumed to produce desirable learning outcomes in the student, but the underlying mechanisms whereby this is supposed to happen remain largely unexamined. Paradoxically, conventional science education is thus rather unscientific; it does not reflect the usual scientific curiosity of elucidating the underlying mechanisms responsible for the functioning of various systems. For example, physicists have spent enormous efforts trying to understand the underlying mechanisms of electric conduction in metals or semiconductors, but they exhibit little interest in understanding the mental functioning of their students.

Accordingly, the instruction prevalent in most schools and colleges is based on various intuitive notions and rules of thumb, rather than upon an understanding of students' thinking and learning processes. Thus the prevailing ways of designing instruction contrast greatly with the design methods common in science or engineering (e.g., with those used to design semiconductor devices or high-energy accelerators). Indeed, in science or engineering the design process is largely principled and systematic, based on careful analysis and a good understanding of underlying mechanisms.

The results of prevailing approaches to instruction are not reassuring. Many students find science difficult to learn. Quite a few get discouraged and drop out; others are frightened away altogether. Furthermore, teaching effectiveness is dubious even for those students who apparently succeed in their science courses and get good grades in them. Indeed, recent investigations reveal that many students emerge from such courses with nominal scientific knowledge which they cannot interpret properly or use flexibly; they often exhibit gross misconceptions or revert to prescientific notions; and they frequently lack even rudimentary problem-solving skills. (Clement, 1982; Halloun & Hestenes, 1985a, 1985b; McCloskey, Caramazza & Green, 1980; McDermott, 1984).

The preceding remarks suggest that prevailing approaches to instruction are excessively simplistic in their primary emphasis on scientific subject matter. In particular, they neglect the central role of human beings in doing and learning science. After all, sciences are created by humans who make observations, invent concepts, discover principles, formulate theories, make inferences, and solve problems. To ensure that students learn the thought processes needed for such scientific activities, effective instruction needs to transcend a mere preoccupation with scientific subject matter.

A More Sophisticated Approach

The preceding remarks suggest that effective instructional design should focus central attention on the student interacting with the subject matter. Consequently, it then becomes necessary to understand adequately the student, the subject matter, and the interaction between them.

From this point of view, instruction can be regarded as a transformation process of the form

$$S_i \rightarrow S_f$$

whereby the student S is transformed from an initial state S_i to a final state S_f where the student can do things that he or she could not do initially. If the aim of instructional design is to achieve this transformation process, then the following four basic problems must be analyzed:

1. The student's behavior and capabilities in the initial state S_i, and the student's underlying knowledge and thought processes in this initial state.
2. The student's desired behavior and capabilities in the final state S_f after instruction, and the underlying knowledge and thought processes needed for such behavior and capabilities.
3. The instructional transformation process $S_i \rightarrow S_f$ whereby the student's underlying knowledge and thought processes can be changed appropriately.
4. The implementation methods whereby this instructional process can be achieved in actual practice.

These problems are analogous to those addressed in medicine, where the main problem is the transformation process $S_i \rightarrow S_f$ that can take a person from an initial state of illness to a final state of good health. Correspondingly, the four subsidiary problems requiring investigation are (1) the underlying mechanisms responsible for illness, (2) the underlying mechanisms responsible for desirable physiological functioning, (3) the therapy needed to transform a person from illness to health, and (4) the medical practices and institutions needed for implementing such therapy.

The analogous four problems in instructional design are no less complex and challenging than those in scientific medicine. Their examination reveals important issues that are commonly neglected in prevailing science education, and which would lead to substantial educational improvements if they were seriously addressed. The following sections examine briefly these four instructional problems so as to identify and illustrate several such educational issues.

INITIAL STATE OF THE STUDENT

Any systematic instructional effort must start with adequate descriptive information about the student coming to instruction, and with some understanding of the underlying knowledge and thought processes of such a student.

Students' Preexisting Knowledge

Educational approaches prevalent in most science courses and curricular materials assume implicitly that students come to instruction with blank minds ready to be filled with new knowledge provided by teachers and textbooks.

This perspective is excessively simplistic: Students do not come to a learning situation with empty minds, but with considerable preexisting knowledge previously acquired by independent discovery, by informal cultural transmission in daily life, and by formal instruction in schools. In particular, this knowledge includes various naive theoretical models about the world. Such models are often primitive and poorly articulated compared to scientific theories, but are nevertheless useful in coping with everyday living.

The existence of previous knowledge implies that instruction cannot merely focus attention on the new knowledge to be acquired by a student. Instead, it needs to *restructure* the student's preexisting knowledge and such restructuring may often be more difficult than the acquisition of new knowledge by a virgin mind. Indeed, it is often remarkably hard to change students' familiar prior ways of thinking, and many confusions can result from inadequate discriminations between new scientific concepts and preexisting everyday notions.

Hence it is important to understand adequately various aspects of a student's preexisting knowledge and to take them properly into account in the design of instruction.

Aspects of Preexisting Knowledge

Conceptions about the Physical World

Students' preexisting notions about the physical world are obviously important and have been extensively investigated in recent years (Caramazza, McCloskey, & Green, 1981; Clement, 1982; Halloun & Hestenes, 1985a, 1985b; McCloskey, Caramazza, & Green, 1980; Cohen, Eylon, & Ganiel, 1983; Driver,

Guesne, & Tiberghien, 1985; Helm & Novak, 1983; Kahnemann & Tversky, 1982; McDermott, 1984; Trowbridge & McDermott, 1980, 1981; Viennot, 1979).

Such investigations show that students have fairly definite theoretical conceptions about the physical world (e.g., the motion of objects). Such conceptions are usually primitive, involve only minimal abstractions from observable phenomena, are sometimes reminiscent of notions historically prevalent in pre-Newtonian times, reflect occasionally anthropomorphic or animistic ideas, and may appear to be misconceptions from a modern scientific perspective. Despite their limited predictive and explanatory power, such naive notions are useful in daily life and are difficult to change. Indeed, students often revert to their naive preexisting notions even after they have successfully completed science courses in which they were presumed to have learned more sophisticated scientific conceptions.

The importance of students' preexisting knowledge about the physical world has been increasingly recognized by educational researchers. However, this awareness has not yet had appreciable effect on actual teaching practices. Furthermore, as pointed out in the next paragraphs, other aspects of students' preexisting knowledge are perhaps even more important and have received scarcely any attention.

Form of Students' Knowledge

The form of students' knowledge is at least as important as its content. For example, students' knowledge, whether acquired in daily life or through prior instruction, often consists of loosely interrelated knowledge fragments (diSessa, 1983, 1987) and is thus rather incoherent. (For instance, many students know the formulas for the areas of rectangles, triangles, and circles, but they have no general conception of area from which they could derive such formulas or find the areas of other surfaces.) Such incoherent knowledge is difficult to remember, difficult to regenerate if partially forgotten, and prone to inconsistencies. It also provides a poor basis for making inferences, for problem solving, or for further learning.

Furthermore, students' knowledge is often nominal rather than functional (i.e., it may be recalled and verbalized, but cannot be flexibly applied in various specific cases).

Needless to say, fragmentation and vagueness are often characteristic of everyday knowledge. However, science deliberately strives to refine knowledge so as to attain the clarity, precision, consistency, and generality needed to achieve optimal predictive and explanatory power. Accordingly, science education needs to change not merely the content, but also the form of students' knowledge. This is a difficult task that is rarely addressed explicitly in most ordinary instruction.

Notions about Science and Cognition

Students come to instruction not only with naive models about the physical world, but also with naive notions about the nature of science and about human thought processes. Such naive notions concern issues more abstract than directly observable physical phenomena, and are thus particularly difficult to change. Yet, such notions about science and thinking can have all-pervasive effects since they crucially affect how students direct their attention and what they learn (Schoenfeld, 1983).

For example, students commonly view science as useful knowledge of facts and formulas rather than as a conceptual structure for explaining and predicting observable phenomena. They believe that understanding involves the ability to recall familiar facts and principles rather than the ability to use such principles to solve diverse or novel problems. They think that learning is predominantly achieved by the rather passive processes of reading, listening, and routine practice rather than by active thought processes needed to restructure one's own knowledge and to use it flexibly. They view mistakes as inadvertent slips or as sources of embarrassment and correspondingly try to dismiss them as quickly as possible. (By contrast, more sophisticated learners would study their mistakes to gain valuable insight about their thought processes. They would also deliberately try to develop methods for preventing, detecting, diagnosing, and correcting any possible mistakes.)

Such naive student conceptions about science and thinking can markedly affect their learning goals and methods and can thus have far-reaching consequences on what students ultimately learn. Yet, these naive conceptions are often not explicitly addressed in most science courses. Indeed, they are sometimes even reinforced by prevailing teaching practices. (For example, students' mistakes are more often penalized than viewed as interesting objects of study and learning.)

DESIRED FINAL STATE OF THE STUDENT

A specification of desired instructional outcomes requires consideration of the following two basic questions: (a) What are the observable student capabilities desired as the goal of instruction?, and (b) What kinds of underlying knowledge and thought processes can lead to such desired capabilities?

Specification of Instructional Goals

All too often teachers or curriculum developers strive to "cover" adequately the important topics of a scientific discipline. Rapid advances in the sciences also lead to an overwhelming temptation to squeeze increasingly more factual knowledge into students' heads, or to introduce some of the graduate-course material of yesteryear into the introductory science courses of today. Unfortunately, many topics can be covered without any significant student learning. Furthermore, excessive preoccupation with adequate coverage of scientific subject matter often impedes a careful analysis of desirable instructional goals.

The relevant issues can be clarified by accepting the premise that education should ultimately prepare students to cope effectively with the world in which they will have to live and work. This world is today deeply affected by scientific and technological knowledge. This knowledge is increasingly voluminous (leading to a "knowledge explosion"); increasingly abstract and highly symbolic (e.g., concerned with electromagnetic fields and wave functions, and not just with visible levers and gears); and rapidly changing (threatening many people with premature obsolescence in their careers).

The mere acquisition of factual knowledge is no longer sufficient to deal with such a complex and changing world. Instead, effective education must increasingly strive to prepare students to apply abstract and symbolic knowledge, to solve novel problems, and to learn independently in order to extend their knowledge and adapt to new situations. Needless to say, such intellectual skills are much more difficult to teach than mere factual knowledge. The goal of preparing students effectively for our modern world thus forces us to address increasingly ambitious instructional goals.

Underlying Knowledge and Thought Processes

Prevailing teaching practices, which focus predominantly on the content and logical structure of scientific subject matter, are insufficient to attain such instructional goals. Indeed, a systematic pursuit of such goals requires a sufficiently detailed understanding of the underlying thought processes whereby scientific knowledge can be flexibly used. Such an understanding should then provide a sound basis for designing instruction to teach scientific problem-solving and learning skills.

However, it is a major challenge to understand the cognitive processes leading to effective scientific work and problem solving. For example, expert scientists usually cannot articulate much of the underlying knowledge responsible for their own good performance. This underlying knowledge is largely "tacit," outside the range of their conscious awareness, and can only be made more explicit with considerable difficulty. (The situation is analogous to that of native speakers of English who speak very well, but are quite unable to articulate the underlying rules of grammar and sentence construction which they implicitly use.) The extent of this tacit knowledge becomes amply apparent in attempts to program computers to solve scientific problems without human intervention; for the computer requires explicit instructions about thinking strategies of which human scientists are often quite unaware.

But even success in explicating the tacit knowledge of expert scientists is insufficient for purposes of instructional design. For instance, expert scientists do not always behave optimally. Furthermore, even if they do, they often rely on tacit knowledge and intuitive recognition processes acquired by dint of years of experience. Instruction lasting for some days or weeks cannot possibly aim to have students simulate the behavior of such experts. Instead, it must often identify and teach explicit thought processes whereby students can achieve good intellectual performance in less intuitive ways.

The following paragraphs illustrate some of the salient issues that arise in attempts to understand and teach scientific thinking and problem-solving skills.

Form of Underlying Knowledge

Nature of Concept Knowledge. Scientific concepts such as *area, acceleration,* or *energy* are the basic building blocks of scientific conceptual structures and are essential ingredients for any scientific problem solving.

Many observations, such as those reported in the previously cited investigations, indicate that students are often unable to interpret adequately the scientific concepts that they were taught. For example, in one of our studies students could answer correctly only about 40% of the qualitative questions about the concept acceleration which they had studied for more than two months in an introductory college physics class (Reif, 1987a).

Closer examination reveals the following main characteristics of students' underlying knowledge about a scientific concept: (a) This knowledge is often quite incoherent, consisting of little more than loosely connected knowledge fragments about the concept, and (b) Even when students can state the definition of a concept, they are often unable to interpret it in particular instances.

By contrast, the effective and efficient interpretation of a scientific concept requires underlying knowledge with the following main features (Reif, 1985, 1987b): (a) The concept must be accompanied by procedural knowledge specifying explicitly what one must actually *do* to interpret the concept in any particular instance; (b) The entire knowledge about the concept must be coherent so as to allow inferences between various knowledge elements. Such coherence facilitates remembering the concept, regenerating it if partially forgotten, detecting and remedying inconsistencies, and extending the concept to unfamiliar situations; (c) The concept must be integrated in a larger conceptual structure (including preexisting knowledge) which is globally coherent and thus free of inconsistencies; and (d) Efficient concept interpretation is facilitated by a compiled repertoire of "intuitively" recognized knowledge about the concept. However, it is essential that such intuitive scientific knowledge be consistent with formal scientific knowledge, be testable against it, and be carefully discriminated from everyday intuitive knowledge.

Prevailing teaching methods often fail to ensure that the concept knowledge acquired by students has the preceding characteristics. In particular, scientific concepts are often specified implicitly by examples, by vague analogies, or by descriptions without procedural specifications of how they are to be interpreted. Furthermore, students are rarely actively engaged in exploring a concept to make inferences or detect inconsistencies; hence their resulting conceptual knowledge remains fragmented. (As indicated later in this chapter, alternative teaching methods can remedy such deficiencies.)

Organization of Students' Knowledge.   The organization of knowledge is of crucial importance in determining how readily the knowledge can be used. For example, if file folders are randomly arranged in a large file cabinet, the valuable information in these folders becomes so difficult to retrieve that it is almost useless. Yet, there are many indications that the scientific knowledge learned by students often remains poorly organized, and that this lack of good organization contributes to students' difficulties.

A cognitively interesting—and ultimately pedagogically important—question is the following: How can knowledge be effectively organized so as to facilitate remembering, retrieval, detection of inconsistencies, modification, generalization, and effective problem solving? For example, cognitive research suggests that some hierarchical forms of organization may be particularly useful. As an illustration, geographical information is commonly organized in such hierarchical form: A coarse map of the United States indicates only major geographical information, but is elaborated into regional maps showing more detailed information, each of which is further elaborated into maps of states showing even more detailed information, etc.

In a study carried out by Eylon and me (1984) we investigated such organizational issues in the simple case of scientific arguments. Such an argument (e.g., a proof or problem solution) consists of a sequence of steps leading from a premise to a desired conclusion. However, it may be organized in different ways. For example, it may be organized purely linearly as a sequence of some 15 detailed steps leading from premise to conclusion. Alternatively, it may be organized hierarchically as a sequence of some four grossly described major steps, each of which is further decomposed into three or four subsidiary detailed steps. Our study showed that students, who had learned an argument organized in such a hierarchical form, were substantially better able to remember the argument, to find mistakes in it, or to modify it than students who had learned the same argument organized in a linear form.

The implication of these remarks is that instruction must pay at least as much attention to the organization as to the content of the scientific knowledge acquired by students. Note that it is not sufficient that instructional materials and presentations be well-organized; it is also necessary that the knowledge actually incorporated in students' heads be well-organized.

Prevailing teaching practices often fail to pay adequate attention to the effective organization of the knowledge acquired by students. Indeed, they are sometimes even deleterious because they provide students with poor models of knowledge organization. For example, arguments and illustrative problem solutions in textbooks are almost always presented in purely linear sequential form. Although such presentations may be logically impeccable, they are psychologically far from optimal. As already pointed out, there is good evidence that a hierarchical organization of such arguments would make them easier to learn and to use. As another example, some physics textbooks (e.g., Ohanian, 1985) include "formula sheets" which are merely long lists of miscellaneous physics formulas. Such lists reinforce students' erroneous notions that physics knowledge consists basically of a bunch of formulas. Furthermore, formulas summarizing important principles would be more useful organized hierarchically than in "laundry list" form.

Problem-Solving Methods

Problem solving is centrally important in science. It provides the essential means to attain the scientific goal of making numerous inferences and predictions.

Most students find problem solving difficult. Furthermore, many emerge from their science courses with poor problem-solving skills and without problem-solving methods appreciably more systematic than random search.

This situation is actually not too surprising. Problem solving methods are rarely explicitly taught. Indeed, most problem-solving instruction focuses on problem *solutions* (i.e., on the products of problem solving), rather than on the *processes* whereby such solutions can be generated. For example, most textbooks and teachers present sample solutions to typical problems. But such a solution provides very little information about the process whereby it was generated. For example, it does not reveal how decisions were made to choose one particular principle rather than another, how to avoid solution paths which lead nowhere, or how to recover from an impasse. In other words, a sample solution can be admired and judged for correctness, but it does little to help the student make all the judicious decisions needed to find a solution to a novel problem.

To overcome such difficulties, one needs to identify and then teach explicitly some useful general problem-solving methods. These must include methods for initially describing or redescribing a problem so as to bring it into a form that facilitates its subsequent solution; judicious search and decision methods that help to find a solution; and methods to assess the merits of the resulting solution and to revise it if necessary (Reif, 1983; Reif & Heller, 1982). The following brief comments point out some of the pertinent issues.

Initial Problem Description. The initial description of a problem is very important since it may crucially determine the difficulty of solving the problem or whether it can be solved at all. For example, in one of our investigations (Heller & Reif, 1984) we found that students, who had successfully completed an introductory physics course with grades of B or better, failed to correctly solve about 50% of a set of mechanics problems because of their faulty initial descriptions of these problems.

The knowledge structure about a particular scientific domain specifies the entities of interest in this domain, the special concepts useful for describing these entities, and the properties of such concepts. This knowledge structure thereby suggests implicitly how to describe situations in that domain. It should then also be possible to formulate, and ultimately to teach, explicit procedures for describing any problem in this domain.

For example, we were able to formulate such explicit procedures for describing any problem in elementary mechanics. (These procedures specify how to describe the motion of any particle by its position, velocity, and acceleration; and how to describe the interaction of any such particle by the long-range and short-range forces on it.) When students were induced to describe problems according to these general procedures, they generated almost perfect descriptions and vastly improved their success in finding the solutions of the problems thus described. (Heller & Reif, 1984).

By contrast, ordinary science courses rarely teach students how to describe problems effectively—at least not with sufficient explicitness. For example, textbooks in mechanics attempt to teach how to describe interactions of particles by forces drawn in free-body diagrams, but they do not point out the need to describe equally carefully the motion of particles (a curious omission since mechanics is centrally concerned with the relation between motion and interaction). They stress the need to list all forces, but often do not specify how to enumerate all such forces. Furthermore, they do not sufficiently specify the properties of some relevant forces (e.g., friction forces), and sometimes they even specify these properties incorrectly.

Decision Methods Facilitating Search for Solutions. The essential difficulty of problem solving is the need to make judicious decisions so as to select among many possible solution paths, most of which lead to nowhere, the one (or very few) which lead to the desired goal. Yet, such decision methods are rarely taught explicitly in prevailing science courses.

Although there are no methods that guarantee good decisions, there are methods which substantially enhance the probability of making useful decisions. Some such methods could be taught and would certainly be better than the haphazard approaches used by many students. Merely identifying available options can be helpful. (For example, the mere question, "What principle could you apply to what system?" can help students identify available options and get them thinking along useful directions.) To decide among such options the student may look one step ahead to predict their anticipated consequences and then choose the option with the most desirable consequences. Another useful decision method, *means-ends analysis,* involves diagnosing the currently existing difficulties of a problem and then trying to solve a subsidiary problem that aims to reduce some of these difficulties. As another example, problems are rarely solved most efficiently by immediately reaching for equations, as many students are wont to do. Instead decisions are usually made more easily by progressive refinements, first considering the main features of a problem described qualitatively in words or pictures, and only then focusing on quantitative details.

Such decision methods are rarely taught in science courses. Indeed, prevailing instruction is sometimes even counter-productive by overemphasizing mathematical formalism and symbol manipulations at the expense of qualitative descriptions and methods of progressive refinement.

Assessment of Solutions.   Once a problem solution has been obtained, it needs to be assessed and revised if necessary. Some general assessment criteria, useful for testing any solution, can readily be formulated. For example, is the solution clearly specified (i.e., free of ambiguities or undefined symbols)? Is the solution internally consistent (e.g., do equations contain quantities with consistent units)? Is the solution correct by criteria of consistency with external information (e.g., does the answer agree with expected qualitative functional dependencies or with the answers for simple special cases)? Is the solution optimal (e.g., as simple as possible)?

Such systematic assessment methods are relatively simple to implement and teach. Nevertheless, quite a few of them are not explicitly taught in most science courses.

DESIGNING THE INSTRUCTIONAL PROCESS

The consideration of the preceding issues, dealing with underlying knowledge and thought processes that facilitate good intellectual performance, provides a basis for systematically designing instruction whereby such intellectual performance may be attained.

Such systematic design is ordinarily not incorporated in most science courses. For example, teaching is most often done by explaining relevant scientific topics, giving examples, and providing practice; the presumption is that such instruction is sufficient to make students capable of using their newly acquired knowledge effectively. Sometimes special environments (e.g., laboratories or computational environments) are provided where students can explore and learn by discovery; the presumption here is that students will acquire useful knowledge and even transfer their learning to other domains.

Some learning does occur under the preceding conditions. However, important issues are neglected in such simple instructional approaches. Available evidence also indicates that the actual learning outcomes are often rather disappointing. This is scarcely surprising because the kinds of knowledge and thought processes required for science are quite complex and not easily acquired. Thus, more systematically designed instruction and explicit teaching are often required to help students learn scientific knowledge that they can use reliably and flexibly.

Such systematic teaching needs to analyze in adequate detail the initial knowledge and thought processes of a student, and those required for the desired intellectual capabilities. Such an analysis then permits the design of explicit instructional methods to ensure that the student learns the desired kinds of knowledge and thought processes.

Some recent investigations show that such a systematic approach to instruction can be quite effective. They also indicate that explicit instructional methods are by no means restricted to the training of mundane skills; properly designed, they can also be valuable in teaching fairly sophisticated cognitive strategies.

A good example of this kind of systematic instructional approach is provided by the work of Brown and Palincsar (Palincsar & Brown, 1984; Brown & Palincsar, in press) who aimed to teach reading with good comprehension. Their examination of the initial state of children who read poorly revealed that the children's difficulties were not primarily caused by inabilities to decode words and sentences, but by failures to extract meaning. Brown and Palincsar then formulated a model of good reading as an active problem-solving process whereby a person seeks to extract meaning from text by repeatedly using a few specific strategies: (a) asking questions about the text, (b) seeking to clarify what has been read, (c) summarizing to organize the information in one's mind, and (d) making predictions to test one's understanding.

The preceding analysis led Brown and Palincsar to devise an instructional method designed to induce students to read with explicit use of these strategies. In this method, called *reciprocal teaching,* the instructor and student alternately assume the teaching role to coach each other in using these strategies while reading. In this way the strategies are explicitly articulated, demonstrated, and actively practiced in a context where the instructor can closely monitor the students' performance and gradually get them to use the reading strategies independently.

This instructional approach proved highly effective. It improved students' reading comprehension test scores from 15% to 85% after about 20 training sessions and also had enduring effects in students' performance outside of English classes. Furthermore, this instructional approach, which originally started as a psychological experiment, is now used throughout the school system of Springfield, Illinois.

Another example of systematic instruction, more directly relevant to science, is provided by some work done by my collaborators and me (Labudde, Reif, & Quinn, 1988). The starting point was an investigation of students' initial knowledge about scientific concepts (such as the physics concept acceleration) which they had been taught in ordinary science courses. This investigation revealed that many students were unable to answer even simple qualitative questions about such concepts. As already mentioned in a previous section, students' difficulties could be traced to the fragmentation and uninterpretability of their acquired knowledge. In a later section, we also presented an analysis specifying the characteristics of the underlying knowledge leading to effective and efficient ways of interpreting scientific concepts. (Reif, 1987b).

Building on this analysis, we tried to devise an instructional method that would teach students to interpret effectively the concept acceleration. This method aimed to make students' knowledge about this concept interpretable and coherent by teaching them an explicit procedure for interpreting the concept and letting them consistently apply this single procedure to a variety of special cases. Furthermore, the method was aimed at making students' entire knowledge (including preexisting everyday knowledge) coherent by letting them use this same procedure to detect, diagnose, and correct the errors committed by themselves or by others.

This instructional method was tested in an experiment where students were initially able to answer only about 40% of questions about acceleration. After the instruction, which lasted about half an hour, they were able to answer about 95% of such questions and no longer invoked erroneous knowledge fragments about the concept. The results of the experiment were thus quite encouraging and suggest ways of extending such teaching methods into actual teaching practice.

Other examples of systematic and explicit instructional approaches are summarized in a paper by Collins, Brown, & Newman (in press). For example, such approaches have been successfully implemented to teach writing (Bereiter & Scardamalia, 1987; Scardamalia, Bereiter, & Steinbach, 1984), problem solving in mathematics (Schoenfeld, 1985), and recursive programming in LISP (Pirolli, 1986).

IMPLEMENTING EDUCATIONAL IMPROVEMENT

Systematic principled instructional design is an essential prerequisite for substantial improvements in science education. Unfortunately, it is not sufficient. Even highly successful instructional approaches, well-validated on a small scale, are difficult to translate into forms producing widespread educational impact. Indeed, any hope of achieving significant educational improvements requires that one address complex intellectual and social issues not amenable to simplistic solutions.

For example, additional monies devoted to education would certainly be helpful, but even this is not enough to overcome existing obstacles to substantial educational progress. One of the most critical of these obstacles is a shortage of appropriate talent. At the level of educational research and curriculum development, there is a great shortage of first-rate talent with good analytic skills and the relevant expertise in science, cognitive science, computation, artificial intelligence, and education. It is rare when one individual possesses these different kinds of expertise; even working groups of individuals collectively possessing these complementary kinds of expertise are very rare and usually below critical size. At the level of classroom science teachers the shortage is very well known. Low pay and attractive opportunities elsewhere keep good talent away from teaching positions. Moreover, there are few institutional mechanisms for effectively training teachers in scientific disciplines as well as in modern educational approaches. All such talent shortages cannot be overcome by mere money, but also require wise policies and adequate time.

As another example, computers and other modern information technologies have great potentialities as useful instruments for research, powerful media of expression and communication, and effective means of educational distribution. However, such information technology, like any other kind of technology, is merely a tool that can be used or misused. Its educational value cannot be realized without good educational design based on an adequate understanding of relevant cognitive and social processes, or without adequately extensive investments in talent, time, testing, and financial resources. Such investments needed to exploit technology are commonplace in modern industries which, unlike medieval "cottage industries," are highly capital-intensive (amortizing large initial capital investments by widespread and repetitive use of their products). But how can information technologies be effectively exploited in the social and economic context of our present educational system which, like a cottage industry, is largely labor-intensive (content to pay many teachers repeatedly, but not prepared to make substantial initial investments)?

Some Suggestions

Despite such difficulties, I believe it is possible to make some modest suggestions that could be implemented and would have major long-range educational impact. In particular, the suggestions in the following paragraph seek to address fundamental current problems: (a) the lack of superior talent devoted to education, (b) the lack of effective deployment of limited existing talent resources, and (c) the lack of persuasive examples demonstrating the efficacy of more scientific and technological approaches to education.

1. Special fellowships established by the National Science Foundation could help appreciably in attracting good students, well trained in the natural or engineering sciences, to graduate programs leading to innovative educational careers. Such fellowships, with stipends extending over three years or so, would provide students with financial inducements to pursue such graduate work. Even more importantly, the existence of such fellowships, well-publicized and spon-

sored by a scientifically well-respected national foundation, would legitimize innovative work in modern education—designating it scientifically important and worthy of pursuit by good analytic minds.

2. The National Science Foundation has recently advocated the establishment of science and technology centers in various scientific or engineering fields. The establishment of one or two such national centers for educational science and technology could substantially advance promising new educational approaches and have major beneficial impact on science education in this country. Recommendations to establish such educational centers have been made in several recent national reports (National Academy of Sciences, 1984; Linn, 1986). Such educational science and technology centers could: (a) bring together critical-size interdisciplinary groups of individuals jointly possessing the requisite kinds of expertise in science, cognition, computation, and education; (b) provide essential computational and other physical resources; (c) be important training centers for graduate students and postdoctoral fellows; (d) provide beneficial foci of interaction with classroom teachers; and (e) become exemplary exporters of good new educational ideas and artifacts.

3. Prototype educational projects, that are designed on the basis of cognitive analyses and that judiciously exploit new information technologies, would be very valuable to test new ideas and techniques on a modest scale facilitating careful development and assessment. They would foster beneficial interaction between educational research and development, and would provide good bases for extension to larger-scale practical instruction. Visibly successful prototype projects, especially when accompanied by exportable instructional programs, would also be powerful sources of educational dissemination.

Need for Changed Attitudes

The preceding suggestions are not visionary and could be implemented. The main obstacles are due more to prevailing attitudes than to purely scientific or technical factors. For example, education is today predominantly regarded as an art, craft, profession, or field of philosophical argumentation—just as medicine or agriculture were regarded in the last century. Educational wisdom is supposed to derive more from the experience of practitioners than from any scientific insights. Universities narrowly view education as simply teaching, that is, as a service function, rather than as a more encompassing intellectual discipline. By contrast, universities do not view medicine only as clinical practice, but as a far more encompassing scientific discipline. Consequently, education is not approached using the same criteria and is not considered worthy of the same investments as medicine or other applied sciences.

However, some of these attitudes are becoming increasingly inappropriate (Reif, 1986; Resnick, 1983). Recent years have seen substantive advances in the cognitive sciences (information-processing psychology, artificial intelligence, linguistics) and dramatic advances in computers and other information technologies (Gardner, 1985). These scientific and technological advances have the potential of markedly transforming and improving education just as other scientific and technological advances historically transformed fields like medicine or agriculture into applied sciences with far-reaching social benefits. Indeed, beginnings of this transformation are becoming noticeable as some new kinds of individuals, well trained in scientific or technological fields, have recently come to devote their talents to educational problems and to approach these in more analytic ways.

Hence, it is appropriate to regard education today as an emerging applied science with promising, but yet largely unrealized, potentialities. Educational progress then hinges crucially on the willingness to treat education by the same criteria as those used to foster other emerging scientific or engineering fields (e.g., like medicine or agriculture some decades ago, or like high-temperature superconductivity today). It would then be useful to undertake efforts to attract, train, and support superior talent that could further education by more analytic and technological approaches; to establish some centers where interdisciplinary talent of this kind could be nucleated and productively engaged; and to encourage prototype projects to explore the new educational approaches and translate them into significant improvements in science education.

ACKNOWLEDGMENTS

Some of the work reported in this chapter was partially supported by the National Science Foundation under grant No. MDR-8550332.

REFERENCES

Bereiter, C., & Scardamalia, M. (1987). *The psychology of written composition*. Hillsdale, NJ: Lawrence Erlbaum Associates.

Brown, A. L., & Palincsar, A. S. (in press). Reciprocal teaching of comprehension strategies: A natural history of one program for enhancing learning. In J. Borkowski & J. D. Day (Eds.), *Intelligence and cognition in special children: Comparative studies of giftedness, mental retardation, and learning disabilities*. New York: Ablex Publishing.

Caramazza, A., McCloskey, M., & Green, B. (1981). Naive beliefs in "sophisticated" subjects: Misconceptions about trajectories of objects. *Cognition, 9,* 117–123.

Clement, J. (1982). Students' preconceptions in elementary mechanics. *American Journal of Physics, 50,* 66–71.

Cohen, R., Eylon, B., & Ganiel, U. (1983). Potential difference and current in simple circuits: A study of students' concepts. *American Journal of Physics, 51,* 407–412.

Collins, A., Brown, J. S., & Newman, S. E. (1989). Cognitive apprenticeship: Teaching the craft of reading, writing, and mathematics. In L. B. Resnick (Ed.), *Knowing, learning, and instruction: Essays in honor of Robert Glaser*. Hillsdale, NJ: Lawrence Erlbaum Associates.

diSessa, A. (1983). Phenomenology and the evolution of intuition. In D. Gentner & A. Stevens (Eds.), *Mental models*. Hillsdale, NJ: Lawrence Erlbaum Associates.

diSessa, A. (1987). Knowledge in pieces. In G. Forman & P. Pufall (Eds.), *Constructivism in the computer age*. Hillsdale, NJ: Lawrence Erlbaum Associates.

Driver, R., Guesne, E., & Tiberghien, A. (Eds.). (1985). *Children's ideas about the physical world*. Milton Keynes, England: Open University Press (Taylor & Francis.)

Eylon, B., & Reif, F. (1984). Effects of knowledge organization on task performance. *Cognition and Instruction, 1,* 5–44.

Gardner, H. (1985). *The mind's new science: A history of the cognitive revolution*. New York: Basic Books.

Halloun, I. A., & Hestenes, D. (1985a). The initial knowledge state of college students. *American Journal of Physics, 53,* 1043–1055.

Halloun, I. A., & Hestenes, D. (1985b). Common sense concepts about motion. *American Journal of Physics, 53,* 1056–1065.

Heller, J. I., & Reif, F. (1984). Prescribing effective human problem-solving processes: Problem description in physics. *Cognition and Instruction, 1,* 177–216.

Helm, H., & Novak, J. D. (Eds.). (1983). *Proceedings of the International Seminar on Misconceptions in Science and Mathematics*. Ithaca, NY: Cornell University.

Kahnemann, D., & Tversky, A. (1982). On the study of statistical intuitions. In D. Kahnemann, A. Tversky, & P. Slovic (Eds.), *Judgment under uncertainty: Heuristics and biases* (pp. 493–508). Cambridge, England: Cambridge University Press.

Labudde, P., Reif, F., & Quinn, L. (1988). Facilitation of scientific concept learning by interpretation procedures and diagnosis. *International Journal of Science Education, 10,* 81–98.

Linn, M. C. (Ed.). (1986). *Establishing a research base for science education: Challenges, trends, and recommendations* (Report of a NSF-sponsored conference.) Berkeley: University of California.

McCloskey, M., Caramazza, A., & Green, B. (1980). Curvilinear motion in the absence of external forces: Naive beliefs about the motion of objects. *Science, 210,* 1139–1141.

McDermott, L. C. (1984, July). Research on conceptual understanding in mechanics. *Physics Today, 37,* 24–32.

National Academy of Sciences (1984). *Report of the research briefing panel on information technology in precollege education* (Research Briefings 1984). Washington, DC: National Academy Press.

Ohanian, H. C. (1985). *Physics*. New York: W. W. Norton.

Palincsar, A. L., & Brown, A. L. (1984). Reciprocal teaching of comprehension-fostering and comprehension-monitoring activities. *Cognition and Instruction, 1,* 117–175.

Pirolli, P. (1986). A cognitive model and computer tutor for programming recursion. *Human-Computer Interaction, 2,* 319–355.

Reif, F. (1983). Understanding and teaching problem solving in physics. In *Research on physics education: Proceedings of the first international workshop* (pp. 15–53). Paris: Centre National de la Recherche Scientifique.

Reif, F. (1985). Acquiring an effective understanding of scientific concepts. In L. H. T. West & A. L. Pines (Eds.), *Cognitive structure and conceptual change* (pp. 133–151). New York: Academic Press.

Reif, F. (1986, November). Scientific approaches to science education. *Physics Today, 39,* 48–54.

Reif, F. (1987a). Instructional design, cognition, and technology: Applications to the teaching of scientific concepts. *Journal of Research in Science Teaching, 24,* 309–324.

Reif, F. (1987b). Interpretation of scientific or mathematical concepts: Cognitive issues and instructional implications. *Cognitive Science, 11,* 395–416.

Reif, F., & Heller, J. I. (1982). Knowledge structure and problem solving in physics. *Educational Psychologist, 17,* 102–127.

Resnick, L. B. (1983). Mathematics and science learning: A new conception. *Science, 220,* 477–478.

Scardamalia, M., Bereiter, C., & Steinbach, R. (1984). Teachability of reflective processes in written composition. *Cognitive Science, 8,* 173–190.

Schoenfeld, A. H. (1983). Beyond the purely cognitive: Belief systems, social cognitions, and metacognitions as driving forces in intellectual performance. *Cognitive Science, 7,* 329–363.

Schoenfeld, A. H. (1985). *Mathematical problem solving.* New York: Academic Press.

Trowbridge, D. E., & McDermott, L. C. (1980). Investigation of student understanding of the concept of velocity in one dimension. *American Journal of Physics, 48,* 1020–1028.

Trowbridge, D. E., & McDermott, L. C. (1981). Investigation of student understanding of the concept of acceleration in one dimension. *American Journal of Physics, 49,* 242–253.

Viennot, L. (1979). Spontaneous reasoning in elementary mechanics. *European Journal of Science Education, 1,* 205–221.

6

Interactive Learning Environments and the Teaching of Science and Mathematics

Ann L. Brown and Joseph C. Campione
University of California at Berkeley

There is considerable converging evidence that by the end of grade school, children achieve success at mastering basic skills of reading, writing, and arithmetic. There are disturbing signs, however, that many students lack a firm conceptual grasp of the goal of the activities in which they engage in school. It seems that they can perform the necessary subskills or algorithms on demand, but they do not grasp their significance. We have argued elsewhere (Campione, Brown, & Connell, 1988)[1] that several features of grade school education could be responsible for this state of affairs. First, there is a clear emphasis on direct instruction with strong teacher control. Lower-level skills are taught before higher-level understanding, causing predictable problems of metacognition (Brown, 1975, 1978). Students fundamentally misunderstand the goal of early education; they come to believe that reading *is* decoding and that math consists *only* of quickly running off well-practiced algorithms without error. This emphasis on skill training is stressed to an even greater degree for low-achieving students, those for whom explicit instruction in understanding is particularly necessary. Strategies are rarely taught. When practice in understanding is provided, frequently it, too, is treated as consisting of decomposable skills. Such activities are presented as ends in themselves, rather than as a means to a more meaningful end. Little attention is paid to the flexible or opportunistic use of strategies in appropriate contexts. This common approach leads to the acquisition of "inert knowledge" (Whitehead, 1916) that cannot be applied broadly or flexibly. We argue that one major reason for this state of affairs is that traditional educational

---

[1]Portions of the introduction of this chapter are similar to sections of a previous chapter by Campione, Brown, & Connell (1988).

practice rarely incorporates metacognitive and contextual factors in learning (Campione, et al., 1988). In this chapter we will consider traditional methods of teaching science and mathematics in grade school and then look at some innovative alternatives that we are currently developing.

LIMITATIONS OF STANDARD INSTRUCTIONAL PRACTICES

A major problem with mathematics education is that although the majority of school aged children acquire the ability to run off algorithms, significantly fewer come to understand what they are doing; they do not understand the logic of, for example, place value notation. Below we list some of the reasons why.

Emphasis on Direct Instruction

The typical mode of instruction is one in which the teacher lectures and the students listen. Children assume the role of passive, rather than active, participants. It is as if the knowledge the teacher has can be transmitted directly to the students; the metaphor is that of pouring information from one container (the teacher's head) to another (the student's head). In mathematics class, the typical pattern is for teachers to work out problems on the board and for students to practice examples of the same type as seatwork. There is much less discussion and argument in mathematics classes than in social studies classes, even when the same teacher is working with the same students (Stodolsky, 1986).

Lack of On-Line Diagnosis

Teachers do not expend much effort in making on-line diagnoses of individual students' processing capabilities. The instruction proceeds at a predetermined rate and sequence that is dictated by the curriculum employed in the particular classroom. Teachers appear to follow a "curriculum script" (Putnam, 1987) prescribing the sequence of skills and activities through which students are expected to progress at the same rate. Although the teachers in Putnam's study did engage in some evaluation of student responses, their major aim was to ensure communal progress through the curriculum-driven agenda.

Basic Skills Before Understanding

In the grade school curriculum there is a powerful tendency to concentrate on "basic skills" rather than so-called "higher-level understanding." For example, in reading instruction in the early grades the emphasis is on decoding before comprehension. Similarly, in mathematics, the emphasis is on automaticity in the application of algorithms. In general, these skills are emphasized at the expense of understanding. The obvious reason for this approach is that young learners, lacking basic skills, need to practice them to the point of automatization, but it is possible to teach such basic skills in an atmosphere conducive to better understanding. Extrapolating from experimental studies of instruction, we would describe the instruction students often receive in school as "blind" in that students are rarely told about why they practice the activities they do. This contrasts with informed, self control instruction, in which students are informed of the reasons why the skills are taught, and are further encouraged to monitor, regulate, and control their own learning (Brown, Bransford, Ferrara, & Campione, 1983; Brown, Campione, & Day, 1981). One clear consequence of blind instruction is a lack of ensuing transfer: Students taught skills blindly do not apply them to related problems. If flexible use of instructed skills is the goal, students need to be informed of the purposes of the skills they are taught and given instruction in the monitoring and regulation of those resources.

Absence of Explicit Strategy Instruction

This emphasis on decontextualized learning of subskills is accompanied by a lack of explicit instruction regarding the *strategies* or *heuristics* that experts flexibly deploy when attempting to learn. However, there is by now considerable evidence that less capable students do not acquire a variety of such cognitive strategies unless they are given detailed and explicit instruction in their use (Brown, et al., 1983; Campione, Brown, & Ferrara, 1982). It is also true that the more complex the strategy in question, the more explicit the instruction needed, even for more capable students (Brown, et al., 1983; Brown, et al., 1981; Day, 1986; Rohwer, 1973). An excellent example of this point with highly selected students is Schoenfeld's (1985) finding that in order to induce college math students to make effective use of Polya-like (1973) heuristics, such as examining special cases and establishing subgoals, it was necessary for him to present a detailed description of the operation of the heuristics, together with considerable guided practice in their use. The idea that complex problem-solving strategies will emerge from instruction aimed at instilling their constituent subskills is an illusion.

Differential Treatment Effect

This general emphasis on basic skills and subskills is even more extended in time and pronounced in the case of weaker students. Those perceived by their teachers to be poor readers get extensive decoding practice at the expense of instruction aimed at comprehension (Brophy & Good, 1969; Collins, 1986). So too in mathematics teachers introducing new topics (for example, remainder division) structure their feedback and emphases differently with groups of higher-and-

lower-ability students (Petitto, 1985). The weaker the students, the more pronounced the tendency to repeat instruction in the basic algorithm. Shortcuts and
strategies are suggested only to the "brighter" students. But the weaker the
students, the less likely they are to acquire those higher-level strategies without
explicit instruction.

ALTERNATIVE LEARNING ENVIRONMENTS

Given these traditional emphases in grade school education, it is not surprising
that children have problems understanding what they are doing. Two aspects of
their metacognitive knowledge (Brown, 1975, 1978; Flavell & Wellman, 1977)
are faulty. First, children's conscious and reportable information about the learning process is inadequate and, as a result, their ability to self-regulate learning,
that is, to monitor and control their own learning processes, is inadequate. This is
not only a problem for young learners, however; even college students tend to
overestimate their level of understanding, suffering from an "illusion of comprehension" (Glenberg & Epstein, 1985; Glenberg, Wilkinson, & Epstein, 1982;
Maki & Berry, 1984) that results in little effort to understand deeply. But the
emphasis in grade school on skills learned in isolation produces particular difficulties for young learners who tend to acquire encapsulated and "inert" skills
(Whitehead, 1916), available when clearly marked by context (as they are, for
example, on many standardized tests) but not serviceable as general tools for
learning. Although the skills may be learned eventually, the control structure
needed to apply them flexibly and appropriately can be notably absent from the
repertoires of even talented students (Davis, 1984; Schoenfeld, 1983, 1985).

Because of these problems we have suggested that instruction aimed at specific skills, or even strategies, is less effective than creating learning environments
in which the goal is to enhance students' conceptual understanding of any skills
or strategies they might adopt. We believe that it is essential that learners reflect
on the goal of their learning activity, rather than engage in blind drill and
practice, even when that drill and practice is devoted to appropriate procedures
(Brown & Campione, 1986).

The question is how can we get students to reflect on their own level of
understanding. One method is to have them engage in group problem-solving
activities of some kind. Understanding is most likely to occur when students are
required to explain, elaborate, or defend their positions to others; the burden of
explanation is often the push needed to make students evaluate, integrate and
elaborate knowledge in new ways. Hatano and Inagaki (1987) argue that comprehension is enhanced if students are asked to explain their views and clarify
their position, or, better still, if they are directly challenged by others.

> First, in the process of trying to convince or teach other students, one has to
> verbalize, or make explicit that which is known only implicitly. One must examine

one's own comprehension in detail and thus become aware of any inadequacies, thus far unnoticed, in the coordination among those pieces of knowledge. Second, since persuasion or teaching requires the orderly presentation of ideas, one has to better organize intra-individually what one knows. Third, effective argumentation or teaching must incorporate opposing ideas, in other words, coordinate different points of view inter-individually between proponents and opponents or between tutors and learners.

Peer interaction, or dialogical interaction in general, such as discussion, controversy, and reciprocal teaching, tends to induce persistent comprehension activity directed to the target. It creates and amplifies surprise and perplexity, produces discoordination, and relates the target to one's domains of expertise and interest. It also invites students to "commit" themselves to some ideas, by asking them to state their ideas to others, thereby placing the issue in question in their domains of interest. In addition, the social setting makes the enterprise of comprehension more meaningful. (Hatano & Inagaki, 1987, p. 40)

Hypothesis–Experiment–Instruction

Hatano and Inagaki came to these conclusions after studying a Japanese science-education technique known as Hypothesis–Experiment–Instruction (Itakura, 1962) that has been successfully applied to mathematics and social studies. Students are given a problem with three or four alternative answers. The alternatives include some choices based on common misconceptions or "bugs" (Brown & Burton, 1978). Students make independent choices from these alternatives, which are then tabulated on the blackboard. Students are encouraged to defend their choices, and consider alternatives in whole class discussions. They may change their choices. Finally, students' choices are tested by observing an experiment: After presenting the problem, the teacher's role is to act as a neutral chairperson; the students take control of the discussion and experimentation.

Consider the following example of a science problem.

Suppose that you have a clay ball on the end of a spring. You hold the other end of the spring and put half of the clay ball into the water. Will the spring become (a) shorter, (b) longer, or (c) retain its length?

Before discussion, the class is divided with 35% who predict the spring will get shorter (the correct answer), 24% who believe the spring will get longer, and 41% who think it will stay the same length. Typical of the students' responses are the justifications given in Table 6.1. After expressing their opinion, the children are encouraged to argue against competing opinions. For example, arguing against the "stay the same" position, Kurokawa states:

Your opinion is strange to me. You said, "The weight of the clay ball won't change because it is only half immersed in water." But you know, when the person's head is above the water, his weight is lighter in water.

TABLE 6.1[2]
Student Response to the Clay Ball and Spring Problem

---

a. Predict spring will get shorter (35%)

Kurokawa: The iron sinks into the water, but I think the clay ball won't. The water has a power to make things float, so the spring will become shorter.
Sano: The clay ball won't float, but in general all the things become lighter in water, so the clay ball will also be lighter in water than in air.
Imai: I feel lighter in water. I have this experience when I take a bath.
Shimazaki: The water has a power to make things float. Therefore, I think the water will make the clay ball float to some extent.

b. Predict spring will get longer (24%)

Odagiri: The clay ball consists of tiny particles—atoms. When we put the clay ball into the water, the water will be absorbed into the space among tiny particles. This makes the spring longer.
Ide: The spring will be longer because the clay ball will sink.
Fukuda: Because the water will be absorbed into tiny particles which the clay ball consists of.

c. Predict spring will stay the same (41%)

Kimura: The nature of the clay will not change when we put it into the water.
Chiyo: The weight of the clay will remain the same when we put it into the water.
Ehara: The clay ball is as heavy in water as in air.
Kurosaka: The water has a power to make completely immersed things float, but that power will not operate if it is only half immersed.

---

[2]The data in this table are from "Facilitation of Knowledge Integration through Classroom Discussion" by K. Inagaki (1981) *The Quarterly Newsletter of the Laboratory for Comparative Human Cognition, 3,* p. 26. Copyright 1981 by LCHC. Reprinted by permission.

And arguing about the correct answer that the spring will get shorter we have the following exchange:

Nouchi: I don't agree with the idea that the clay ball is as heavy in water as in air. I suppose the water has a power to make things float.

Kurosaka: Even if the water has a power to make things float, the clay ball will not float, I suppose.

Shiga: If the clay is a very small lump, I suppose, the water can make it float.

Momozaki: A thing will float when its weight is lighter than the weight of the water with the same volume. I wonder if the weight of the clay ball is lighter than that of the water.

Defending the correct answer we see:

Shimazaki: You said a few minutes ago, "A person's weight is lighter than that of water." However, a person won't float in the air. This means that a person weighs less in water by the power to make things float which the water possesses. This power also operates on clay balls. So the clay ball will become lighter to some degree in water. Thus, I think, the spring will be a little shorter.

Sano: Let's imagine the scene that we pick up a pebble in a pool. When we throw the pebble into the water, it goes down slowly into the water. This means that the pebble weighs less to some degree in water, I think.

After discussion, the majority of children believe the spring will get shorter (62%) and this number increases after experimentation. Indeed, the general results from these procedures is that experimental subjects show greater interest, offer better explanations, and generalize the information more broadly to new but related problems (Hatano & Inagaki, 1987; Inagaki, 1981; Inagaki & Hatano, 1977).

Reciprocal Teaching

Our own experiments with interactive learning environments have focused on *reciprocal teaching* (Brown & Palincsar, in press; Palincsar & Brown, 1984). This procedure was developed initially to improve reading comprehension. Groups of students and an adult teacher work together in discussion groups consisting of, ideally, six students. Each member of the group takes a turn serving as both learning leader and supportive critic. The learning leader is responsible for orchestrating the discussion, and the supportive critic encourages the learning leader to explain the content and help resolve misunderstandings. In order to give a comfortable repetitive structure to the discussions, the learning leader is asked to practice four strategies. He begins the discussion by *asking a question* on the main content and ends by *summarizing*. If there is disagreement, the group rereads and discusses potential candidates for question and summary statements until they reach consensus. Summarizing provides a means by which the group can monitor its progress, noting points of agreement and disagreement. Particularly valuable is the fact that summarizing at the end of a period of discussion helps students establish where they are in preparation for tackling a new segment of text. Opportunistic attempts to *clarify* any comprehension problems that might arise are also an integral part of the discussions. And finally, the leader asks the group to *prediction* about future content. These particular strategies both promote comprehension and provide the student with concrete methods of monitoring their understanding. They are concrete heuristics for getting the discussion going.

Reciprocal teaching is a form of guided cooperative learning because of the provision of expect scaffolding by the adult teacher. Not only does the teacher serve as a model of expert performance, but throughout provides guidance and feedback tailored to the need of the current discussion leader. It is the teacher's job to see that responsibility for the comprehension activities of the group is transferred to the students as soon as possible. As students master one level of involvement, the teacher increases the demands so that students are gradually called upon to function at a more challenging level, finally adopting the leader/critic role fully and independently. The teacher then fades into the background and acts as a sympathetic coach, allowing the students to take charge of their own learning.

The reciprocal nature of the procedure guarantees student engagement and motivation. Students are forced to explain and justify their position to others and in so doing come to understand better themselves. An example of this explanatory process can be seen in the dialogue in Table 6.2 that takes place among a group of low achieving, third grade, minority children. The learning leader, Subject 1 (S_1), does not fully understand the meaning of the term camouflage because she unduly influenced by the inconsiderate text comment that "a chameleon can take on the color of its background." This she takes to mean any color. The ensuing discussion forces the learning leader to reevaluate her understanding and come to terms with the constraints placed on the mechanisms. Note Subject 1's confusion and persistence. She attempts to understand by asking about special cases. She appears to really want to know, and eventually, in her final summary, she appears satisfied. It is also Subject 1 who sets up the motivation to find out *how* the mechanism works. Several days later, when discussing another instance of camouflage, the walking stick insect that disguises itself by changing its shape, Subject 1 shows she has mastered this idea.

> OK, it goes invisible because it looks like a twig. It's frozen and twig-like on the tree and on the ground. Some things change color, to nature colors, green and brown; some things hold still, some things pretend to be twigs, I get it.

Seeking adequate explanations for events seems to characterize good problem solving in adults (Chi, Bassok, Lewis, Reimann, & Glaser, 1989) and children (Brown & Kane, 1988). The reciprocal teaching procedure encourages explanation and the search for deeper levels of understanding (Hatano & Inagaki, 1987).

The reciprocal teaching procedure has been used successfully as a reading and listening comprehension intervention by average classroom teachers with academically at-risk grade and middle school children. Students typically begin the intervention scoring approximately 30% correct on independent tests of comprehension and continue in the program for at least 20 instructional days. We count as successful any student who achieves an independent score of 75% correct on five successive days (Palincsar & Brown, 1984). With this as the criterion, approximately 80% of the students are judged to be successful. Further-

TABLE 6.2
The Meaning of Camouflage

S_1:	(Question) What does camouflage mean?
S_2:	It means you invisible like G.I. Joe (spontaneous analogy)
S_1:	(Question) What color can it (the chameleon) be?
S_3:	Brown
S_1:	No
S_2:	Green?
S_1:	No
All:	What? (confusion)
S_1:	(Question) What color would it be if it was on a fire engine—red, right? And if it's on a car—black, right? And on a cab—yellow, right?
All:	No! No!
S_4:	It can only be greenish or brownish, like in nature.
S_1:	(Indignant) It says (reading text), "A chameleon can take on the color of its background"—so it can be *any* color, right?
S_4:	No, No! It can only be colors like brown earth and green trees and yucky color like mud—like G.I. Joe.
S_6:	So he matches.
T:	(Scaffolding weaker student: repeating text line) Yes he matches, he's almost invisible.
S_6:	He changes colors so his enemy won't get him.
T:	Good. That's right.
S_1:	(Question—confused) But what if he's not in the grass, what if he's in a whitish kinda color?
S_6:	He would turn white?
All:	No.
S_1:	(Question—still confused) But what if he doesn't have that color? What if he can't turn that color?
S_5:	He just moves on down to another spot that has his color.
S_4:	No, only green and brown and yucky.
S_1:	(Question, persistent) Can it be blue like water in the forest?
S_3:	Water in the forest isn't blue, it's yucky colored, so chameleons can be yucky colored too.
S_1:	(Summary) OK. OK. I summarize, it change color.
T:	What does?
S_1:	(Sigh) Chameleon change color to hide from its enemies. It can be green—umh brown—yucky color (pause) so it has to stay in the forest 'cause those colors are there.
S_1:	(Question) I have another question. How does it change color?
T:	Good question. Anyone know?
S_6:	It doesn't say.
T:	Do we really know how it changes?
S_1:	No.
T:	Any predictions? No?
T:	(Prediction) I think it might be a chemical in its body, let's read to find out.

more, students maintain their independent mastery anywhere from six months to a year after instruction ceases (Brown & Palincsar, 1982; Brown, Palincsar, Ryan, & Slattery, work in progress; Palincsar & Brown, 1984); they generalize to other classroom activities—notably science and social studies—and they improve approximately two years on standardized tests of reading comprehension (Palincsar & Brown, 1984).

We will give the details of just one recent study in which at-risk minority third grade children were trying to learn a coherent body of knowledge about animal defense mechanisms such as camouflage, mimicry, protection from the elements, extinction, parasites, and natural pest control. The themes repeated during the discussions and were also taken up in the daily independent tests of comprehension.

Twenty days of such discussions led to dramatic improvement in both comprehension processes and theme understanding. Figure 6.1 shows the daily independent assessment results of four groups. Students were assigned to reading or listening procedures depending on their decoding competence. RTE refers to reciprocal teaching groups where the teacher mentioned the theme explicitly and RTI refers to groups where the theme is implicit and the teacher did not intervene. However, as most of these groups concentrated on the repeated themes anyway, this proved to be an insignificant variable. The two control groups are C (control), untreated except for taking the pre- and posttest, and P (practice), a group that took all the daily assessments but had no instruction. As can be seen in Fig. 6.1, the reciprocal teaching groups' improvement is large and reliable. Even 12 months later the effect of instruction was apparent.

Each day the discussion and test passages were directly analogous. For example, under the natural pest control theme, the children might discuss the Manatees (large sea mammals) that, for their own protection (from sharks), were moved inland where they took to eating the water hyacinths that had previously clogged Florida's inland waterways. The Manatees were thus welcomed by the residents because they provided a biological (rather than chemical) solution to an environmental problem. Immediately after discussing this example of a biological deterrent, or natural pest controller, the students read and answered questions on a passage that contained the analogous problem of how to rid a garden of mosquitoes, where they are told that (a) purple martins eat mosquitoes and (b) purple martins like to live in man-made bird houses. This crucial information is buried under other facts about the life style of these birds. When asked, in the questioning, how the gardener might rid himself of mosquitoes, an observant child would respond like Jeremy:

> the house-owner could build a home for purple martins at the bottom of the garden . . . but I think Raid is best—but it's just like the Manatees we talked about . . . and the ladybugs eating the farmer's a- a- (Teacher—aphids) right aphids—we talked about that last week.

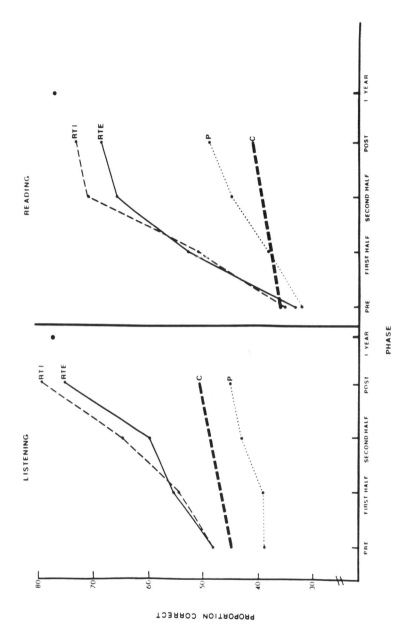

FIG. 6.1. Proportion correct on daily independent tests of comprehension.

121

Regular practice greatly improves the ability to use analogous information to solve problems; that is, guided practice creates a mind-set to reason by analogy (Brown & Kane, 1988). The children began by noting few of the analogies, but after several days they were able to solve the analogies with an 80% success rate and this reaches 100% when the animals physically resemble one another (Brown, 1989; Gentner & Landers, 1985).

The students duly note the repetitive themes in their discussions. In Table 6.3, a group of six children are discussing the critical paragraphs in passages about natural pest controllers. Note that the children readily hone in on the usefulness of ladybugs, remember ladybugs 12 days later when discussing the Manatees, and finally, even a year later, they discuss the analogy between ladybugs and lacewings. Also important to note is the fact that the children, although scarcely eloquent, are in charge of the discussion; the teacher provides encouragement, support, and occasional individual scaffolding.

Not only did the children remember how to conduct the reciprocal teaching dialogues and continue to score well on independent tests, they also remembered the content. When asked to sort pictures of animals into the six themes, they scored 85% correct immediately after the study and 82% correct one year later. Responses were scored as correct when the child could name the theme and justify why the animal in question was an exemplar of that theme. Finally, on both the long- and short-term tests the children were able to classify novel exemplars of the themes and place them in appropriate habitats. Reciprocal teaching experience enables the children both to learn a body of coherent, usable, knowledge and to develop a repertoire of strategies that will enable them to learn new content on their own.

RECIPROCAL TEACHING AND MATHEMATICS UNDERSTANDING

Reciprocal teaching of text comprehension strategies was originally designed in response to the overwhelming evidence that many children fail to develop such skills on their own (Baker & Brown, 1984a, 1984b; Brown, 1980; Paris & Myers, 1981). Similarly, there is considerable agreement that the way in which mathematics is taught in school leads to students' failure to understand what they are being taught, with a consequent inability to make flexible use of the knowledge they have at their disposal (Lesh, 1982). Many children master basic algorithms if provided with enough drill and practice, but they have considerable difficulty achieving a robust understanding of the conceptual basis of these algorithms (Gelman & Greeno, 1989; Resnick, 1988). Students need practice connecting their fragmentary knowledge into systems of "meaningful mathematics" (Davis, 1984; Noddings, 1985). When this opportunity is provided by an expert classroom teacher, the results are quite dramatic (Lampert, 1986). Lampert's children argue about the meaning of mathematical expressions and attempt

TABLE 6.3
Repetitive Themes in Reciprocal Teaching Dialogues:
Natural Pest Control Third Grade High Risk Children (N = 6)

Day 5 Ladybugs

Student 1:	(*Question*) What do they eat?
Teacher:	What do what eat?
Student 1:	(*Question*) The ladybugs. What do ladybugs eat?
Student 2:	Aphids, little white bugs.
Student 1:	(*Question*) Right. Why do farmers like them?
Student 3:	Because they eat the little bugs off the farmers plants.
Student 1:	That's the answer I want. (pause)
Student 1:	(*Question*) I have another question, where do they live?

(overlapping discussion of potential places to live not accepted because they are not mentioned in the text)

Student 4:	I know, they crawl on leaves and rosebuds as in the grass.
Student 1:	(*Summary*) Okay it's about the ladybugs that crawl in the grass and help the farmer by eating bad little insects.
Teacher:	Good summary.

Day 17 Manatees

Student 4:	(*Question*) How does the Manatee clean up the river?
Student 3:	By eating water plants?
Student 4:	No you missed one word
Student 3:	By eating water hedge-whatever
Teacher:	Hyacinth
Student 4:	Yea that's right
Student 4:	(*Question*) How many years ago people moved some of the Manatees from the sea in the inland river? Amelia
Student 6:	A few years
Teacher:	(*Scaffolding*) Anne (S$_4$)—another way you could have asked that question would be—*when* did the people move the Manatees.
Student 4:	(*Question*) What did the people want the Manatees to eat?
Student 2:	The plants
Student 4:	I want the whole sentence.
Student 2:	The people wanted the Manatees to eat the water hyacinths that grow in the river
Student 4:	(*Summary*) Okay, that's it; it tells where the people moved them and what they wanted them to eat, and why.
Teacher:	Good summary.
Student 1:	(*noting analogy*) The Manatees went through and ate all the plants so that's helping like the ladybugs because they eat all the aphids, bad bugs.

Year Follow-up Day 1 Lacewings

Student 6:	(*Question*) Okay, what do lacewings eat?
Teacher:	Good question.
Student 5:	Bugs and insects
Student 3:	Aphids and other bugs

(*continued*)

123

TABLE 6.3 (*continued*)

Year Follow-up Day 1 Lacewings

Student 2:	And, they eat the farmers' crops.
Students:	No.
Student 6:	(*Clarification and question*) No the bad bugs eat the farmers' crops. Okay let me see (pause). Are insects nice to crops?
Student 1:	Not always, some are.
Student 6:	(*Question, clarify*) What's the story about, any one need clarifying?
Student 2:	It's about how the lacewing is destroying the insects that are in the fields trying to eat crops.
Student 6:	Okay, people often think of insects as just good for nothing bugs.
Students:	You told the answer. That's not a question.
Student 6:	(*Question*) Okay, let me see (pause). Why does the farmer like the (pause), oh yea, we already know this. I'll ask it anyway, Why does the farmer like the lacewing?
Student 1:	(*Analogy*) Because they can stop other things from eating crops. I remember we read about farmers that don't want animals on their crops because they kill 'em.
Student 2:	They put spray on it?
Student 1:	No ladybugs.
Student 4:	The ladybugs ate them all up so they don't hurt no crops.
Student 6:	So they quit messing up crops.

to convince each other of the appropriateness of the algorithms they invent. The children engage in lively discussions about the meaning of what they are doing, and it is these reflective processes that are largely absent from traditional mathematics classes. In reality, children rarely talk about mathematical entities and, as a result, they have difficulty talking about mathematics (Collins, Brown, & Newman, 1989; Resnick, 1988). Obviously, there is a pressing need to develop procedures for improving understanding in mathematics learning (Davis, 1984; Resnick & Omanson, 1987).

There is considerable agreement that some form of cooperative learning environment where students and teachers work and talk together can facilitate this level of analysis (Brown & Palincsar, 1989). A variety of such interventions have been dubbed cognitive apprenticeships by Collins, Brown, and Newman (1989). They emphasized the domain general nature of interactive learning environments by comparing three "so called" success stories: (a) reciprocal teaching of reading comprehension, conducted primarily with low-achieving grade and junior high students (Palincsar & Brown, 1984); (b) Bereiter and Scardamalia's (1987) work on writing composition with gifted sixth graders; and (c) Schoenfeld's (1985) teaching of mathematics problem solving to selected undergraduates at major universities. When comparing these different domains and learners, Collins et al. point out the surprising similarity between successful interventions across domains and learners.

By concentrating on the similarities, Collins et al. de-emphasize the notable differences between these programs, the two major differences being: (a) the domains and (b) the developmental status of the learners. Obviously, the actual heuristics will differ across domains, a point we will illustrate here by describing how we set about modifying reciprocal teaching to help promote mathematics learning. When contrasting reciprocal teaching of reading and mathematics in this way, we highlight the domain general philosophy of cognitive apprenticeships and the domain-specific procedures necessary to enhance understanding.

The other crucial difference is the developmental one; obviously the developmental status of the learner must influence what is taught. To make this point even more obvious, consider this example. Two of the important features of cognitive apprenticeships are the notion of *reflection*—students are asked to consider the problem-solving trace of an expert—and the very similar notion of *abstracted replay*—where students are asked to reflect on their own prior problem-solving traces. Schoenfeld (1985) uses these procedures to good effect when working with selected college undergraduates, but we do not think it would help younger children to reflect on Schoenfeld's problem-solving efforts; they simply would not have the requisite background knowledge. Similarly, if left to muddle through and reflect on their own traces, young children may not benefit much either. Implicit in these notions is the idea of developmental level; the difference between the expert and the novice must be small enough so that there can be communication. We would like to argue that we need to consider not only task analysis, but also diagnosis of entering levels and successive developmental models of competence; there are clearly intermediate stages on the road to expertise. Indeed, the expert may only be the right model for other experts. An essentially developmental concept such as the "zone of proximal development" (Vygotsky, 1978) or the "region of sensitivity to instruction" (Wood, Bruner, & Ross, 1976) needs to be added to the developmentally neutral idea of expert scaffolding favored by Collins et al., a Skinnerian notion that was intended to be age, domain, and species independent (Brown et al., 1983). Not anyone can be scaffolded by anyone. There must exist a bandwidth of competence (Brown & Reeve, 1987) through which a particular learner can navigate at any point in time. In short, when designing learning environments, one must deal with the constraints set by the particular domain in question *and* with the learner's cognitive status.

We would like to argue that the biggest engineering feat in designing these environments is providing a natural way of externalizing mental activities that are usually covert (Brown, 1980). Unlike apprentice weavers (Greenfield, 1984), the cognitive apprentice cannot learn from observation alone. One can, perhaps, learn to knit without instruction, but cognitive events are not usually observable; the trick is to make them so. We first need to decide just what cognitive events we wish to externalize. We believe that one should select the cognitive activities that need to be externalized and reflected upon through an analysis of the error patterns, or reflection failures, of the target learners.

Consider how this was done for reciprocal teaching of reading comprehension. In reciprocal teaching, questioning, summarizing, clarifying, and predicting were not randomly selected activities. We settled on them only after diagnosing that they were comprehension-monitoring activities rarely engaged in by weak students. They were also selected because: (a) They did not evolve spontaneously; (b) They could be readily taught—the student could produce some questions or summaries right from the start, thereby assuring that some type of discussion got going; (c) The activities were readily engaged in by inexperienced learners when instructions to "monitor your understanding" or "be strategic" provoked blank stares; and most important, (d) They were selected because a byproduct of summarizing what one has just read, asking for clarification, etc. is to force comprehension monitoring. If you cannot summarize, you need to reread, seek help, etc. The four activities became rituals that ensured that a discussion took place and forced comprehension monitoring of oneself and others.

The question then became how to do this in mathematics. What would the comprehension-fostering activities be in mathematics problem solving that would serve the function met by summarizing, questioning, etc.? Obviously these strategies, appropriate though they were for reading, could not be transported unchanged to the new domain. To get some idea of how to proceed, we again observed groups of children engaged in the target task, in this case solving algebraic word problems, and diagnosed six characteristic monitoring errors. The students had trouble: (a) extracting the relevant facts from the story, (b) estimating an approximate answer, (c) drawing visual representations, (d) matching equations to the drawing, (e) checking arithmetic facts, and (e) sense-making (to which we shall return).

Given this diagnosis, what ritual activities comparable to questioning, summarizing, etc. could we invent that would also: (a) scaffold a discussion and (b) force reflection? Note that the choice of the activities was determined not by how experts would attack the problems but by what students were failing to do on their own.

To take the place of summarizing, etc., we introduced three problem-solving boards for the students to work on. All other aspects of the interactive learning environment remained the same. The students worked together with an instructor in small cooperative groups. The students and adult teacher again took turns acting as learning leaders and supportive critics responsible for leading a discussion aimed at understanding algebra word problems. The procedure embodied expert modeling, scaffolding, and coaching on the part of the teacher; the method forced externalization of strategies, monitoring of progress, and attempts to impose meaning. What differed is that the strategies selected to scaffold the discussion were tailored to the domain.

In our first mathematics modification (Campione, et al., 1988) learning leaders guide the group in working on three successive blackboards designed to help students proceed systematically. In addition, these procedures generate an external record of the group's problem solving, which can then be monitored, evaluated, and reflected upon. The three boards were: (a) the *Planning Board,* where the group extracts the relevant facts embodied in the word problem; (b) the *Representation (drawing) Board,* where the students draw diagrams illustrating the problem; and (c) the *Doing Board,* where they translate the drawings into the appropriate equation(s) and compute the answer. Students discuss their approaches and help each other reflect upon the visual trace of their joint work (Collins, et al., 1989).

The subjects were above average students attending a summer school prior to entering seventh grade. They could do the arithmetic required, out of the context of word problems, but were experiencing metacognitive difficulties of the type mentioned above.

Figure 6.2 shows some intermediate states in the problem-solving process that illustrate the use of the boards. In the top panel, the students have extracted the relevant information and have begun to develop a representation of the problem. They have also translated it into an equation (on the Doing Board). However, the equation does not have a unique solution, so they are forced to go back to the drawing board and elaborate their sketch. In the next panel, we see additional work that leads to an appropriate equation, and the computational stage is begun (they solve for the number of calories in the milk). Later in the sequence they finish the problem and engage in some checking activities to make sure that the result makes sense and is consistent with the problem statement.

The group worked on three different types of problems over the course of 20 days of instruction: (a) single variable linear equations, (b) two variable linear equations, and (c) monomial by binomial equations. The design included several pretests evaluating computational skill as well as entering ability to deal with these kinds of problems. Essentially, the students could not solve these problems independently when the intervention was begun. A posttest assessed improvement on the target problems, as well as performance on some very open-ended transfer problems. A control group received the same introduction to the problem types as the experimental group and samples of worked out examples. They also took the same tests as did the experimental group.

The main results are shown in Fig. 6.3, where it can be seen that the experimental group outperformed the control group on both the target and transfer problems. This pattern of results is encouraging. The discussions did proceed smoothly, student engagement did result, and the Planning, Drawing, and Doing routine became established. The students were enthusiastic and willing to spend considerable periods of time working on individual problems, both in class and at home. Their performance improved significantly more than a control group given equivalent practice and feedback on the main problem types but no discussions. We should point out, however, that this success was achieved under

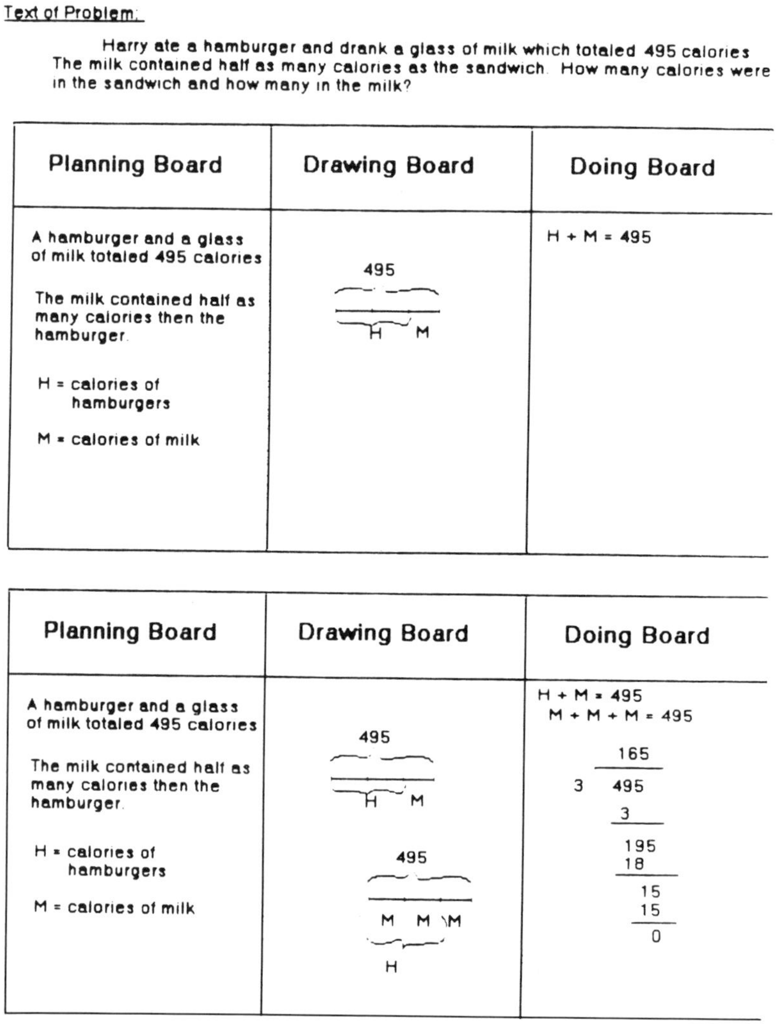

<u>Text of Problem:</u>

Harry ate a hamburger and drank a glass of milk which totaled 495 calories
The milk contained half as many calories as the sandwich. How many calories were
in the sandwich and how many in the milk?

Planning Board	Drawing Board	Doing Board
A hamburger and a glass of milk totaled 495 calories The milk contained half as many calories then the hamburger. H = calories of hamburgers M = calories of milk	495 ⌒ — ⌒ H M	H + M = 495

Planning Board	Drawing Board	Doing Board

A hamburger and a glass of milk totaled 495 calories
 	
The milk contained half as many calories then the hamburger.
 	
H = calories of hamburgers
 	
M = calories of milk
	495 ⌒—⌒ H M
	
	495 ⌒—⌒ M M M ⌣ H
		H + M = 495
		M + M + M = 495
		
		   165
		3 495
		    3
		
		   195
		    18
		    15
		    15
		     0

FIG. 6.2. Examples of the reflection boards used in the initial math study.

optimal circumstances where the teacher was an expert teacher of mathematics. In addition, the group was ideal. It was small in size (six students). It consisted of students who were especially selected because they had good computational skills, but experienced difficulties of a metacognitive nature. And the students were well-motivated volunteers. Furthermore, the range of problems was severely restricted to three repetitive cases for which the students had been taught visual representations. Therefore, they did not need to invent a novel representa-

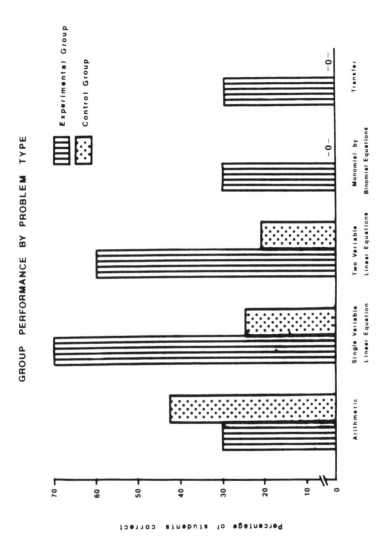

FIG. 6.3. Performance on the three problem types in the experimental and control groups.

129

tion, they merely selected the appropriate model from a choice of three familiar ones.

We turn now to the ensuing failure from which we can learn a great deal about the difficulty of creating interactive learning environments in the classroom settings.

The failure experiment was intended to be a replication of the preceding study in the classroom, but it went awry. The main problem was the nature of the group and the teacher's reaction to it. The students this time were not preselected for their mathematics skill and indeed it turned out that several of them could not do some of the arithmetic computations, nor could they solve the relevant equations. As a result, they could not serve as effective learning leaders. In addition, the group size was increased to 12, including four who could not speak English very well. After some negotiation, we did manage to persuade the classroom teacher to reduce the group to eight, but eight sixth graders with varying math and language skills continued to prove unwieldy.

In addition to the suboptimal group composition, a seemingly simple change in the procedure was (unbeknownst to us) introduced. In order to increase the quality of the audio recording, the teacher decided to replace the large public boards with smaller ones placed on the table between the children and the teacher. This simple move made the boards less available for group inspection. Indeed, it effectively neutralized the boards as an object of group reflection. Now approximately one third of the children did not look at the boards, and another third had to look at them upside down most of the time. As the teacher took over the actual writing, the children appeared to give up ownership of the boards altogether. The boards became the teacher's, occasionally shared with a child or two. The intended interactive cooperative nature of the group was never realized.

The changed social situation fundamentally affected the teacher's behavior in predictable ways. His role became more typical of the traditional classroom teaching reviewed in the first part of this chapter. His scaffolding attempts became more intrusive and authoritarian obviously in response to the students' inattention and faulty math. He took on most of the planning and monitoring functions, relegating the lower level mechanics to those children who could carry them out. Those who could not, were rarely called on. One child appeared more competent in mathematics, although her English was faulty. Seeing this, the teacher called on her most often. This is a variant of the differential treatment effect due to perceived competence described earlier (Petitto, 1985).

This example also illustrated a problem with the scaffolding notion borrowed from apprenticeship systems in the workplace, where scaffolding is often taken to mean that the adult (expert or mastercraftsman) takes responsibility for the work that novices cannot do. But if the teacher does this, how will the students ever learn? We need some notion, such as that implied by the zone of proximal development, of a negotiated arena of activity in which the child can interact productively with the minimum amount of support, either in the form of an adult teacher, more capable peers, computers, or even physical devices like the reflection boards. Students need practice in what they currently cannot do, not in areas in which they are competent.

The most notable change in the group activity was the loss of turn-taking and cooperation. The teacher decreased his wait time dramatically, and occasionally was unable to pick up on a promising lead from a student because it was not the path he was looking for. An unfortunate casualty of the loss of turn-taking was student engagement and practice. Some students clearly ceased to be involved. And this made it much more difficult for the teacher to engage in on-line diagnosis of the students' levels of understanding. Instead of a cooperative social situation featuring shared expertise and co-construction of meaning, we ended up with a fairly typical mathematics class with a teacher teaching and some students engaged in individual practice.

An additional dramatic consequence of this breakdown was that it was the mechanics of that problem that were delegated to cooperating students. As such, the discussion also focused on mechanical manipulations and the students lost contact with the semantics of the problem. As an example, consider a case where the students were trying to solve a problem in which they were told that someone had some one cent and three cent stamps, there were 17 more three cent than one cent stamps, and that the total value of the stamps was $1.11. They were to find out how many three cent stamps the person had. At one point they generated the equation $1(x) + 3(x + 17) = 111$, and could identify the quantities involved, that is, they realized that the 3 symbolized the value, in cents, of a three cent stamp, that $(x + 17)$ represented the number of three cent stamps, and that the product stood for the value, in cents, of the three cent stamps. In trying to solve for x, the product was expanded to $3x + 51$, an expression that caused the students to lose sight of the actual quantities involved. The discussion bogged down while attention was focused on solving the equation, an extremely difficult one for these students. The teacher then took control, organized the effort toward generating a solution, and allowed the students to do only simple arithmetic when it was needed. A value of x was eventually obtained, but by then the students were so attuned to the mechanics that they clearly did not appreciate the nature of the "solution" they had uncovered. Not surprisingly, they could not solve a subsequent problem involving only numeric changes.

This tale of woe illustrates just what is expected of the ideal teacher (Lampert, 1986). The teacher needs domain knowledge and must completely understand the mathematics or science. In addition, the teacher needs knowledge of the processes of learning in general, and of developmental trajectories in the domain in question. Diagnosis of students' understanding and, finally, the ability to apply wisdom opportunistically under the pressures of a classroom setting are also required. This is no small feat!

As we prefer to end on a more optimistic note, we turn now to our current attempts to come up with a more robust procedure. Even if our replication had been a success, we were not totally satisfied with the procedure. Intuitively, the planning, drawing, and doing boards did not seem to serve exactly the same function as summarizing, questioning, etc. Experts doubted that they would solve problems this way, whereas most skilled studiers report using the equivalent of self-testing via summarization or clarification-seeking when learning from texts. In addition, the use of standard diagrams was appealing to some of our colleagues, whereas others said they would never use them. Finally, the metacognitive or control processes that we wanted to externalize did not emerge as the main focus of discussion.

In collaboration with Allan Collins, we developed the alternative reflection board illustrated in Figure 6.4. It was designed to be directly responsive to the six characteristic monitoring errors of the target students (difficulty extracting goals, estimating, drawing diagrams, matching equations to diagrams, checking arithmetic and sense-making). The board was designed to support a problem-solving emphasis. The solution path was parsed into four general categories: goal setting, planning, problem solution, and sense-making/checking.

Goal Setting involves stating clearly the exact problem that is to be solved, along with identifying any unknown quantities that are relevant to the problem.

The Planning phase is the most elaborate and consists of two general sets of activities, one "required" and one "optional." The required component involves the student identifying the quantities given in the problem along with the relations among those quantities. It is emphasized that whenever a quantity is selected, a description of that quantity must be appended so that students do not lose touch with the semantics of the problem. If the students wish to do so, they can also draw a sketch representing the problem statement and use that to facilitate their planning. It is emphasized that the sketch may help identify the quantities needed to satisfy the goal. Estimation is another optional activity that is incorporated into the Planning phase. If desired, students can try to predict what the answer might look like, for example, specify a range of possible solutions or constraints on the solution.

When the group is satisfied that they have extracted all the necessary information, the students proceed to the Problem-Solution phase, where they attempt to generate their specific answer. This can be accomplished either by going directly to some arithmetic activities (adding, multiplying, etc.) or by taking the intermediate step of generating a more abstract algebraic representation of the problem (an equation).

Once an answer has been generated, the Sense-making/Checking phase begins. This phase involves several related activities, including checking the actual computation for arithmetic errors; checking the answer against the problem statement, or the estimation done during the Planning phase, to see if the answer makes sense in the context of the actual problem; and then reflecting on the overall problem solving process.

During the final phase, the students reflect upon the steps taken in solving the

FIGURE 6.4

Problem: John has 4 marbles more than Karen,
who has twice as many marbles as Linda.
The 3 together have 24. How many does John have?

Goal: State Goal(s)
State Unknown(s)

Number of marbles that John has = X
Y = number of marbles Karen has
Z = number of marbles Linda has

Planning

STATE GIVENS AND RELATIONSHIPS BETWEEN THEM
$Z = 2 Y$ (Karen has twice as many as Linda)
$X = Y + 4$ (John has 4 more than Karen)
$X + Y + Z = 24$ (all 3 together = 24)
  (Error not noticed)

Estimating, Sketching etc.

If all had the same, John would have 8 ($3 \times 8 = 24$), but John has 4 more than Karen, and Karen twice as many as Linda, so Linda is small, Karen middle and John has the most. Estimate 12 approximately.

Problem Solution

Tutor: You want to find one
Tutee: Unknown (prompt) I know more about Y so let's get rid of the Z, can I?

Equation:
$(Y + 4) + Y + \quad Z = 24$
$\quad Y + 4 \ + Y + 2Y = 24$
$\qquad\qquad 4Y + \quad 4 = 24$
$\qquad\qquad\qquad 4Y = 20$
$\qquad\qquad\qquad\quad Y = \ \ 5 \ = \text{Karen}$
$\qquad\qquad\qquad\quad X = \ \ 9 \ = \text{John}$
$\qquad\qquad\qquad\quad Z = 10 \ = \text{Linda}$
  (Error still not noticed)

Sensemaking & Checking

---

1) ANSWER MAKE SENSE?
2) COMPUTATIONS CORRECT?
3) REVIEW METHOD(S) OF SOLUTION

Checking:
1) $5 + 9 + 10 = 24$
2) John $= 9$
   Karen $= 5$, $\ 9 - 5 = 4$
3) Error accepted...:
   Accept the answer without validating relation between Karen and Linda

---

problem, discuss choice points that were encountered, talk about alternatives that were or might have been considered, etc. Particular attention is given to any fix-up strategies that might have been brought into play—methods of identifying unproductive paths and ways of getting the process back on track.

The final activity involves what we term problem extension. Here the idea is to pose a variety of related problems, either by modifying some of the quantities included in the original problem, or by exchanging some of the givens for unknowns. The goal is to make clear to the students that they have been working on not only a single problem, but also one of a family of related problems. To the extent that they understand the quantities involved in the original problem and the relations among those quantities, they should be able to "play" with their specific solution somewhat and solve these related problems. Thus, extension serves to accentuate the need for understanding the problem, and provide a way of checking to see if the students do in fact understand the nature of the solution they have generated. This is training for transfer, which we strongly endorse (Brown & Campione, 1978, 1984).

This modified procedure does result in more discussion of a metacognitive nature, but as can be seen from the trace shown in Figure 6.4, it does not always lead to the correct answer. Here Allan Collins is tutoring the first author by leading her through the stages of the reflection board. Because the problem is simple, they have no problem stating the goal—the number of marbles John has, which they label X because the tutee believes algebra has to have things like Xs and Ys. They then go on to map out the givens and the relations among them, still connecting X, Y, and Z with the quantities. Unfortunately, they confuse Karen and Linda and write $Z = 2Y$, which means that Linda has twice as many as Karen rather than the correct $Y = 2Z$. Supremely confident in their joint effort, they go to the problem solution, lose track of the goal and quantities and come up with an answer that satisfies one of the constraints of the problem: $5 + 9 + 10$ does add up to 24. They check this and are duly satisfied with it, but overlook the fact that Karen has half as many as Linda and ignore the constraint generated via estimation, that is, that John has the most marbles. Figure 6.5 shows the second attempt where they redo the problem and get a different set of answers that also obey the problem constraints. Noting the differences prompts a discussion of which set of numbers obey *all* the constraints and are therefore the correct ones. Unfortunately, these interactions were not tape recorded, we only have the record of the trace.

This procedure is being streamlined and used with eighth graders with Allan Collins and Joe Campione serving as tutors together with practiced classroom math teachers. The initial results are encouraging. The new reflection board serves the intended purpose of provoking group discussion and argumentation. It remains to be seen if such experiences result in better independent learning than traditional practice, as it did in the initial math study (Campione, et al., 1988)

FIGURE 6.5

Problem: John has 4 marbles more than Karen,
who has twice as many marbles as Linda.
The 3 together have 24. How many does John have?

Goal: State Goal(s)
State Unknown(s)

Number of marbles John has = X = Goal
Y = Karen's marbles
Z = Linda's marbles

Planning

STATE GIVENS AND RELATIONSHIPS BETWEEN THEM
Unknowns:
X = number of marbles John has = Goal
Y = number of marbles Karen has
Z = number of marbles Linda has
Relations:
$Y = 2Z$ Karen = Linda × 2 ($2Z$)
$X = Y + 4$ John = Karen + 4 ($Y + 4$)
$X + Y + Z = 24$

Estimating, Sketching etc.

John has the most = 12

Problem Solution

Equation:
$$(Y + 4) + Y + Z = 24$$
$$(Y + 4) + Y + (1/2)Y = 24$$
$$(2\text{-}1/2)Y = 20$$
$$Y = 20 \div (2\text{-}1/2) = 8$$

$$\boxed{\begin{array}{l} 2.5 \mid 20\ = \\ 25 \mid 200\ = 8 \end{array}}$$

$Z = 4$
$X = 12$
$$X + Y + Z = 24$$
John has 12 marbles = Goal

Sensemaking & Checking

---

1) ANSWER MAKE SENSE?
2) COMPUTATIONS CORRECT?
3) REVIEW METHOD(S) OF SOLUTION

Checking:
1) $12 + 8 + 4 = 24$
2) John = Karen plus 4
    $Y + 4 = 12$
    $Y = 8$
    $8 + 4 = 12$
3) Karen = 8
    Linda = 4 $4 \times 2 = 8$
4) Answer fits with estimated number.

---

and in the reciprocal teaching of simple biological themes (Brown, Palincsar, Ryan, & Slattery, work in progress).

SUMMARY

We have reviewed our research program aimed at developing supportive contexts for learning in math and science. Based on our previous work in the reciprocal teaching of reading comprehension, we have attempted to transfer the philosophy behind the reading work to the new domains. In describing the similarities and differences between reading, science, and math, we illustrate the domain-specific nature of the activities that promote discussion, argumentation, explanation, and reflection, as well as the domain-independent philosophy of the interactive learning environments.

Common to the learning environments is the key notion of *supportive contexts* for learning. Four main principles are involved: (a) Fostering conceptual understanding of procedures (Gelman & Greeno, 1989; Resnick, 1988) rather than just speed and accuracy should be the aim of assessment and instruction; (b) Expert guidance is used to reveal as well as promote independent competence; (c) Proleptic teaching aims at one stage beyond current performance, in anticipation of levels of competence not yet achieved individually, but possible within supportive learning environments; and (d) Microgenetic analysis permits estimates of learning as it actually occurs (Brown, et al., 1983; Brown & Campione, 1986).

In conclusion, the main thrust of our research program is to attack the time-honored problem of transfer (Brown, 1989; Brown & Campione, 1984). Although we may not expect transfer of knowledge or domain-specific procedures (learning biological principles may not influence mathematical insights), we do expect transfer at the level of control structures, at the level of the weak but general methods (Newell, 1979) that enable access to new learning. By (a) making the control processes explicit, (b) making reflecting and checking part of learning in a variety of domains, and (c) using the dialogue approach generally, we hope to create an environment in which novices can learn effectively in a new domain, because they are experts at seeking, evaluating, monitoring, and upgrading information in collaborative and individual settings.

ACKNOWLEDGMENTS

The preparation of this manuscript and the research reported in it were supported by a grant from the McDonnell Foundation and by Grant HD 05951 from the National Institute of Health. The authors thank Allan Collins and Annemarie Palincsar for their continuing advice on this project. We would also like to thank

Don Holste and the students at the Urbana Schools who participated in the math studies and Jim Casey and the students of Franklin School, Champaign, who participated in the biological themes study.

REFERENCES

Baker, L., & Brown, A. L. (1984a). Cognitive monitoring in reading. In J. Flood (Ed.), *Understanding reading comprehension* (pp. 21–44). Newark, DE: International Reading Association.

Baker, L., & Brown, A. L. (1984b). Metacognitive skills and reading. In D. Pearson, M. L. Kamil, R. Barr, & P. Mosenthal (Eds.), *Handbook of reading research* (pp. 353–394). New York: Longman.

Bereiter, C., & Scardamalia, M. (1987). *The psychology of written composition.* Hillsdale, NJ: Lawrence Erlbaum Associates.

Brophy, J. E., & Good, T. (1969). *Teacher-child dyadic interaction: A manual for coding classroom behavior.* Austin: University of Texas, The Research and Development Center for Teacher Education.

Brown, A. L. (1975). The development of memory: Knowing, knowing about knowing, and knowing how to know. In H. W. Reese (Ed.), *Advances in child development and behavior* (Vol. 10, pp. 103–152). New York: Academic Press.

Brown, A. L. (1978). Knowing when, where, and how to remember: A problem of metacognition. In R. Glaser (Ed.), *Advances in instructional psychology* (Vol. 1, pp. 77–165). Hillsdale, NJ: Lawrence Erlbaum Associates.

Brown, A. L. (1980). Metacognitive development and reading. In R. J. Spiro, B. C. Bruce, & W. F. Brewer (Eds.), *Theoretical issues in reading comprehension* (pp. 453–481). Hillsdale, NJ: Lawrence Erlbaum Associates.

Brown, A. L. (in press). Analogical transfer and learning in children. What develops? In S. Vosniadou & A. Ortony (Eds.), *Similarity metaphor and analogy.* Hillsdale, NJ: Lawrence Erlbaum Associates.

Brown, A. L., Bransford, J. D., Ferrara, R. A., & Campione, J. C. (1983). Learning, remembering, and understanding. In J. H. Flavell & E. M. Markman (Eds.), *Handbook of child psychology* (Vol. 3, pp. 77–166). New York: Wiley.

Brown, A. L., & Campione, J. C. (1978). Permissible inferences from cognitive training studies in developmental research. In W. S. Hall & M. Cole (Eds.), *Quarterly Newsletter of the Institute for Comparative Human Behavior, 2*(3), 46–53.

Brown, A. L., & Campione, J. C. (1984). Three faces of transfer: Implications for early competence, individual differences, and instruction. In M. Lamb, A. Brown, & B. Rogoff (Eds.), *Advances in developmental psychology* (Vol. 3, pp. 143–192). Hillsdale, NJ: Lawrence Erlbaum Associates.

Brown, A. L., & Campione, J. C. (1986). Psychological theory and the study of learning disabilities. *American Psychologist, 41*(10) 1059–1068.

Brown, A. L., Campione, J. C., & Day, J. D. (1981). Learning to learn: On training students to learn from texts. *Educational Researcher, 10,* 14–21.

Brown, A. L., & Kane, M. J. (1988). Preschool children can learn to transfer: Learning to learn and learning from example. *Cognitive Psychology, 20,* 493–523.

Brown, A. L., & Palincsar, A. S. (1982). Inducing strategic learning from texts by means of informed, self-control training. *Topics in Learning and Learning Disabilities, 2*(1), 1–17.

Brown, A. L., & Palincsar, A. S. (in press). Guided cooperative learning and individual knowledge acquisition. In L. B. Resnick (Ed.), *Cognition and instruction: Issues and agendas.* Hillsdale, NJ: Lawrence Erlbaum Associates.

Brown, A. L., & Reeve, R. A. (1987). Bandwidths of competence: The role of supportive contexts in learning and development. In L. S. Liben (Ed.), *Development and learning: Conflict or congruence?* (pp. 173–223). Hillsdale, NJ: Lawrence Erlbaum Associates.

Brown, J. S., & Burton, R. R. (1978). Diagnostic models for procedural bugs in basic mathematics skills. *Cognitive Science, 2,* 155–192.

Campione, J. C., Brown, A. L., & Connell, M. L. (1988). Metacognition: On the importance of understanding what you are doing. In R. I. Charles & E. A. Silver (Eds.), *Research agenda for mathematics education: The teaching and assessing of mathematical problem solving* (pp. 93–114). Hillsdale, NJ: Lawrence Erlbaum Associates.

Campione, J. C., Brown, A. L., & Ferrara, R. A. (1982). Mental retardation and intelligence. In R. J. Sternberg (Ed.), *Handbook of human intelligence* (pp. 392–490). New York: Cambridge University Press.

Chi, M. T. H., Bassok, M., Lewis, M. W., Reimann, P., & Glaser, R. (1989). Self-explanations: How students study and use examples in learning to solve problems. *Cognitive Science, 13,* 145–182.

Collins, A., Brown, J. S., & Newman, S. E. (in press). Cognitive apprenticeship: Teaching the craft of reading, writing, and mathematics. In L. B. Resnick (Ed.), *Cognition and instruction: Issues and agendas.* Hillsdale, NJ: Lawrence Erlbaum Associates.

Collins, J. (1980). Differential treatment in reading groups. In J. Cook-Gumperz (Ed.), *Educational discourse.* London: Heinemann.

Davis, R. R. (1984). *Learning mathematics: The cognitive science approach to mathematics education.* Norwood, NJ: Ablex.

Day, J. D. (1986). Teaching summarization skills: Influences of student ability level and strategy difficulty. *Cognition and Instruction, 3*(3), 193–210.

Flavell, J. H., & Wellman, H. M. (1977). Metamemory. In R. V. Kail, Jr. & J. W. Hagen (Eds.), *Perspectives on the development of memory and cognition* (pp. 3–33). Hillsdale, NJ: Lawrence Erlbaum Associates.

Gelman, R., & Greeno, J. G. (in press). On the nature of competence: Principles for understanding in a domain. In L. B. Resnick (Ed.), *Knowing and learning: Issues for a cognitive psychology of instruction.* Hillsdale, NJ: Lawrence Erlbaum Associates.

Gentner, D., & Landers, R. (1985). Analogical reminding: A good match is hard to find. *Proceedings of the International Conference on Systems, Man and Cybernetics.* Tucson, AZ.

Glenberg, A. M., & Epstein, W. (1985). Calibration of comprehension. *Journal of Experimental Psychology: Learning, Memory and Cognition, 11,* 702–718.

Glenberg, A. M., Wilkinson, A. C., & Epstein, W. (1982). The illusion of knowing: Failure in the self-assessment of comprehension. *Memory and Cognition, 10,* 597–602.

Greenfield, P. M. (1984). A theory of the teacher in the learning activities of everyday life. In B. Rogoff & J. Lave (Eds.), *Everyday cognition: Its development in social context* (pp. 117–138). Cambridge, MA: Harvard University Press.

Hatano, G., & Inagaki, K. (1987). A theory of motivation for comprehension and its application to mathematics instruction. In T. A. Romberg & D. M. Steward (Eds.), *The monitoring of school mathematics: Background papers. Vol. 2. Implications from psychology, outcomes of instruction* (Program Report 87-2, pp. 27–66). Madison: Wisconsin Center for Educational Research.

Inagaki, K. (1981). Facilitation of knowledge integration through classroom discussion. *The Quarterly Newsletter of the Laboratory for Comparative Human Cognition, 3,* 26–28.

Inagaki, K., & Hatano, G. (1977). Amplification of cognitive motivation and its effects on epistemic observation. *American Educational Research Journal, 14,* 485–491.

Itakura, K. (1962). Instruction and learning of the concept force. Kasetsu—Jikken—Jugyo [hypothesis—experiment—instruction]: A new method of science teaching. *Bulletin of the National Institute for Educational Research, Vol. 52.*

Lampert, M. (1986). Knowing, doing, and teaching multiplication. *Cognition and Instruction, 3*(4), 305–342.

Lesh, R. (1982). *Metacognition in mathematical problem solving.* Unpublished manuscript, Northwestern University, Chicago.

Maki, R. H., & Berry, S. L. (1984). Metacomprehension of text material. *Journal of Experimental Psychology: Learning, Memory and Cognition, 10,* 663–679.

Newell, A. (1979). One final word. In D. T. Tuma & F. Reif (Eds.), *Problem solving and education: Issues in teaching and research.* Hillsdale, NJ: Lawrence Erlbaum Associates.

Noddings, N. (1985). Formal models of knowing. In E. Eisner (Ed.), *Learning and teaching the ways of knowing: Eighty-four yearbook of the National Society for the Study of Education: Part II.* Chicago: Chicago University Press.

Palincsar, A. S., & Brown, A. L. (1984). Reciprocal teaching of comprehension-fostering and monitoring activities. *Cognition and Instruction, 1*(2), 117–175.

Paris, S. G., & Myers, M. (1981). Comprehension monitoring and study strategies of good and poor readers. *Journal of Reading Behavior, 13,* 5–22.

Petitto, A. L. (1985). Division of labor: Procedural learning in teacher-led small groups. *Cognitive and Instruction, 2,* 233–270.

Polya, G. (1973). *How to solve it: A new aspect of mathematical method* (2nd ed.). Princeton, NJ: Princeton University Press.

Putnam, R. T. (1987). Structuring and adjusting content for students: A study of live and simulated tutoring of addition. *American Educational Research Journal, 24,* 13–48.

Resnick, L. (1988). Treating mathematics as an ill-structured discipline. In R. I. Charles & E. A. Silver (Eds.), *Research agenda for mathematics education: The teaching and assessing of mathematical problem solving* (pp. 32–60). Hillsdale, NJ: Lawrence Erlbaum Associates.

Resnick, L., & Omanson, S. F. (1987). Learning to understand arithmetic. In R. Glaser (Ed.), *Advances in instructional psychology* (Vol. 3, pp. 41–95). Hillsdale, NJ: Lawrence Erlbaum Associates.

Rohwer, W. D., Jr. (1973). Elaboration and learning in childhood and adolescence. In H. W. Reese (Ed.), *Advances in child development and behavior* (Vol. 8, pp. 1–57). New York: Academic Press.

Schoenfeld, A. H. (1983). Beyond the purely cognitive: Belief systems, social cognitions, and metacognitions as driving forces in intellectual performance. *Cognitive Science, 7,* 329–364.

Schoenfeld, A. H. (1985). *Mathematical problem solving.* New York: Academic Press.

Stodolsky, S. (1988). *The subject matters: Classroom activity in math and social studies.* Chicago: University of Chicago Press.

Vygotsky, L. S. (1978). *Mind in society: The development of higher psychological processes.* (M. Cole, V. John-Steiner, S. Scribner, & E. Souberman, Eds. and Trans.). Cambridge, MA: Harvard University Press.

Whitehead, A. N. (1916). *The aims of education.* Address to the British Mathematical Society. Manchester, England.

Wood, D., Bruner, J., & Ross, G. (1976). The role of tutoring in problem solving. *Journal of Child Psychology and Psychiatry, 17,* 89–100.

7 A Practical Authoring Shell for Apprenticeship Learning

William J. Clancey
Kurt Joerger
Institute for Research on Learning
Cimflex Teknowledge, Inc.

ABSTRACT

A novel practical authoring and tutoring program is described that enables a teacher to adapt an existing expert system for use in teaching. The expert system is run in a normal manner by a student, typically using predetermined cases selected by the teacher. The expert system stops at *breakpoints* (to be described later) set during the authoring phase, and invokes a general tutoring program that probes the student's understanding of relevant data and intermediate conclusions. The *authoring program* analyzes the knowledge base and helps the teacher determine which breakpoints will produce meaningful interactions. Breakpoints can be varied for different cases, and a library of cases can be built into a sequenced lesson curriculum. The authoring and tutoring systems are both rule-based programs written in the same language as the knowledge base, making them easy to change and easy to interface with the expert system.

INTRODUCTION

One of the greatest instructional advances in the past decade has been the development of knowledge-based models of expert reasoning. For the first time, we are able to capture and replicate part of the process by which an experienced problem solver gathers information about a problem; forms intermediate abstractions summarizing the problem situation (e.g., diagnoses, design specifications); and relates this understanding to possible courses of action (e.g., repair plans, process-control plans). In *knowledge-based tutoring,* a general instructional program can be used to interact with students using different knowledge bases (Clancey, 1987). Separating the domain knowledge base from the teaching knowledge provides tremendous engineering efficiency, allowing the teaching knowledge to be more clearly formalized, evaluated, and reused.

However, the knowledge-based tutoring paradigm has several practical difficulties. Among the most important is the requirement that the knowledge base be well-formed in order to interface properly with the general instructional program. Although knowledge acquisition programs might one day semiautomate this process, the reality today is that the endeavor is one that only experienced programmers (or experts trained to be programmers) can accomplish after many months of tedious design and debugging. Indeed, research of the past decade has only further increased our standards of knowledge representation desirable for teaching, while tools for constructing such programs lag far behind or are not generally available.

This leaves us with an immense practical problem for meeting the potential of knowledge-based tutoring. Are teachers simply to wait until knowledge acquisition programs are more advanced? Can anything be done to provide useful capability on commercially available and supported tools, encouraging a more rapid application of our research, particularly in the universities? Is there a practical method for exploiting Artificial Intelligence (AI) research in a simple way, perhaps compromising some of the current research concerns focusing on knowledge representation and explanation capability, but nevertheless providing something useful for teachers in classrooms today?

With these questions in mind, we sought to develop a novel means for using expert systems for teaching and gain some of the benefits immediately for teachers with minimal resources or knowledge engineering capability. On the basis of other reported research (Bahill and Ferrell, 1986) that demonstrated that teachers were ready to develop programs for their classes, we decided to develop a shell for simple, rule-based expert systems. In particular, we sought to avoid the problems encountered in GUIDON, which required such a rule base to be well-formed, and to put aside the ideals of NEOMYCIN (Clancey, 1986), which imposed epistemological distinctions that go beyond most teachers' experience and programming capabilities. We wanted an authoring program that could cope with messy knowledge bases—containing arbitrary program code in places—yet still carry on a meaningful interaction with a student.

Furthermore, in response to requests from industrial users of expert systems, we sought to provide a mode for running an expert system that would enable the user to learn on the job, in the course of his routine operation of the program. The goal is to convey to the user some of the routine problem-solving knowledge in the program in order to use the program responsibly as a tool—better understanding how it operates—and build on the program's methods through the user's own experience. This is one form of *apprenticeship learning,* an instructional approach that emphasizes the importance of learning in the context of actual problems (Collins, Brown, and Newman, 1986). As will be seen, the environment we have designed provides other features of apprenticeship learning, including modeling the process of expert reasoning, helping the student through difficult areas (scaffolding), and sequencing problems and support over time (fading).

This method of teaching—with rules of practice encoded in a knowledge base—is suitable for many professions, including such diverse problem areas as loan-risk analysis, medical diagnosis, geological deposit analysis, engineering analysis or design, and manufacturing process control. The main examples in this paper are drawn from a simple structural engineering application involving determination of how a structure such as an airplane wing will behave when it is loaded in a certain way and stressed over time. Notice that apprenticeship learning, as reflected by the domain-specific nature of a knowledge base, does not involve learning scientific or mathematical principles in isolation or abstractly; rather it involves applying principles in the context of real-world problems.

Our approach is to have a running expert system invoke the tutoring system at breakpoints set by a human teacher in a way that can be customized for individual cases (i.e., any particular problem, such as a Boeing-747 airplane wing to be analyzed) and students. That is, we transfer some of the burden of adapting the expert system to the teacher, who selects and orders the concepts (i.e., any domain-specific term, such as whether a structure is thin-walled or solid) and example cases that will be discussed with the students. On the other hand, the authoring program handles the technical details of determining which rules are compatible with the teaching program and preparing the necessary code that enables an interruption to occur during the expert system's operation. Furthermore, the teaching program handles all details concerning the form of probing that is appropriate for the situation, adapting to the role of the chosen concepts in the current case, and the student's response to its questions. Questioning itself follows a simple Socratic format involving more detailed probing for incorrect responses, such as that applied in SCHOLAR and GUIDON. With the instructional design separated from the knowledge base in this way, it is possible to experiment systematically with different interactive styles. Thus, standard instructional designs can be collected, standardized, and shared, rather than redundantly and implicitly encoded for each problem or domain (Clancey, 1987).

This design is implemented in a working program called TRAINING EXPRESS.[1] The program is implemented in the M.1[2] knowledge base language and provides a teaching interaction with domain knowledge encoded in arbitrary M.1

---

[1]TRAINING EXPRESS is a trademark of Cimflex Teknowledge, Inc. The program runs on personal computers with a color monitor in 640K of resident memory. Interfaces are coded in C.

[2]M.1 is a trademark of Cimflex Teknowledge, Inc. M.1 is a rule-based language used for developing expert systems, distinguished from similar languages by how variables can be used to write general rules.

knowledge bases, subject to some limits discussed below. TRAINING EX-PRESS has been integrated with a database system, DBASE III +, which the authoring program uses to store a "curriculum" or library of cases to be run by a student as well as records of student performance. An important part of the design is that the teacher can vary breakpoints from case-to-case, allowing probing to vary with problem difficulty. Furthermore, cases need only be partially specified, so a student can vary the input and explore how the expert system behaves in different situations, while still being probed in a manner the teacher finds compatible with the prespecified fixed inputs.

Sections that follow provide scenarios of teacher and student interaction, a discussion of the novelties, advantages, and limitations of the approach, a discussion of the status and future development of the product, and conclusions about the practical development of AI research.

SCENARIOS: TEACHER AND STUDENT INTERACTIONS

In this section we describe the basic operation of TRAINING EXPRESS. For simplicity, we will continue to refer to the authoring program and the teaching program, which are in fact completely separate pieces of code, implemented as two M.1 knowledge bases (see Fig. 7.1). In this section, we first establish some basic terminology, describe the flow of control, and provide working examples of the program.

Basic Operation

When a student using TRAINING EXPRESS runs a new or "canned case," the expert system invokes the teaching program at what we call breakpoints. Each breakpoint corresponds to a concept in the knowledge base (e.g., "the name of the disease requiring therapy," "the shape of the substructure being analyzed"). Breakpoints are set by the human teacher in an interactive program called the authoring system. (M.1 uses an expression-value language. Therefore, an expression such as "shape(SUBSTRUCTURE)" bundles an object [e.g., the substructure being analyzed] with an attribute [e.g., "shape"]. This simplifies the knowledge representation task. In this paper "concept" and "expression" are used synonymously. We use "expression" when referring to the M.1 language in order to be precise.) This program reads the domain knowledge base and prepares a *secondary knowledge base,* which contains cross-index information (e.g., "rule-56 mentions the geometry of the structure") and declarations written in the M.1 language that cause breakpoints to occur (e.g., "when the expert system has determined the substructure-deflection, invoke the tutor"). When the teaching program is invoked, it asks the student questions, which we call *probes:*

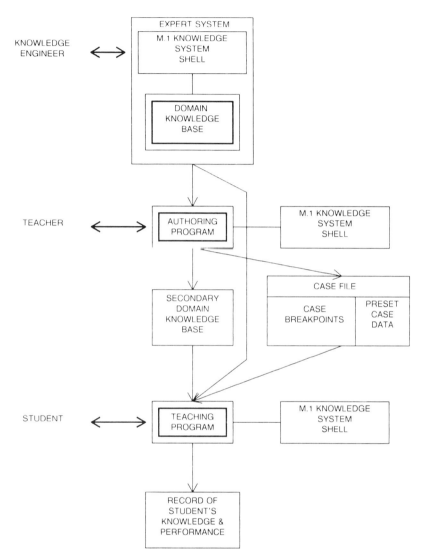

FIG. 7.1. Basic flow of control: A knowledge engineer or teacher interacting with the M.1 shell creates a domain knowledge base; this
knowledge base is loaded and analyzed by the authoring system,
which interacts with the human teacher to set breakpoints, define
cases, and provide additional text explanations; these files are then
available to the teaching system, which interacts with a student while
the expert system is in the normal course of solving a problem.

What conclusions can be made now? What factors support these conclusions? What value for a given factor is consistent with the conclusions? The teaching program indicates which student answers are correct, wrong, or missing, and probes in more detail when mistakes are made. Domain rules are referenced by name, so the student can examine the knowledge base, using already available M.1 commands like "list" and new commands supplied with the tutor, such as "evidence."

In summary, with more detail, the two M.1 programs constituting TRAINING EXPRESS do the following:

The Authoring System

- interacts with a human teacher to determine which expressions should be discussed with the student (breakpoints);
- analyzes rules *concluding* these expressions (a rule *concludes* an expression if its right-hand side or action part states that the expression has a particular value; thus rule-56 concludes that the shape of the substructure is a shell; "shape(SUBSTRUCTURE)" is the expression, "shell" is the value);
- creates a secondary knowledge base with breakpoints marked;
- enables the teacher to edit and store text that introduces individual cases, explains the significance of individual breakpoints, justifies individual rules, provides a synonym for referring to a knowledge-base concept, and defines the meaning of the knowledge-base concept;
- enables the teacher to store a library of cases to be sequenced in a particular way (curriculum) for presentation to a student;
- enables the teacher to examine and maintain a database of student records pertaining to their performance and knowledge exhibited when using the teaching program.

The Teaching Program

- loads the domain knowledge base and secondary knowledge base created through interaction with the teacher;
- interacts with the student to select an appropriate case;
- interacts with the student during the expert system consultation to probe his understanding and provide definitions and explanations upon request;
- stores a record of the student's performance and knowledge.

It is important to realize that *an M.1 domain knowledge base can be used without changes*. Furthermore, in the authoring system, the teacher simply selects breakpoints from a single multiple-choice question. Everything else is handled automatically. In addition to the secondary knowledge base, which provides cross-index information making the teaching program more efficient, the interface between the domain knowledge base and tutor consists of a single M.1 statement for each breakpoint. For example,

whenfound(shape(S) = X) = whenfoundtutor(shape(S)).

This specifies that when the M.1 expression shape(S) has been determined by the expert system in the context of a particular case, with any value (indicated by the variable X but the conclusion cannot be "unknown"), the M.1 expression "whenfoundtutor(shape(S))" should be sought as the next goal. Thus an interruption occurs during the normal processing of the expert system. Rules concluding "whenfoundtutor(ANY)" match the new goal ("whenfoundtutor-(shape(S))"). These rules interact with the student, probing him about the expression shape(S). Thus there are really two M.1 knowledge bases running together, the domain expert system and the teaching program. Before going into more detail, we present an example.

AUTHORING SYSTEM EXAMPLE DIALOGUE

The following examples demonstrate TRAINING EXPRESS using a simplified version of SACON (Bennett & Engelmore, 1979), an expert system developed to aid civil engineers in setting up a complex software package that performs structural analysis (stress and deflection analysis of structure behaviors under particular loads). The task of SACON is to abstract the structure being analyzed, use heuristic rules to approximate the structure's behavior, and select one or more computer programs that will provide more detailed analysis of the expected behaviors.

In the following transcript, the teacher types *author sacon* after the M.1 > prompt. The following lines are generated by the authoring system. The teacher selects the breakpoints from a menu of expressions that the authoring system extracted from the knowledge base. The program then reads the rules concluding expressions selected by the teacher and compiles information to be placed in the secondary knowledge base.

M.1 > author sacon

Loading the sacon knowledge base . . . done.

Analyzing the knowledge base for expressions concluded by rules . . . done.

Which expressions would you like the tutor to discuss with the student?
1. alpha-gamma(SS)
2. analysis-class(STRUCTURE)
3. analysis-recs(STRUCTURE)
 . . . {section omitted}
20. nd-deflection(SS)
21. nd-stress(SS)
22. nonlinearity(STRUCTURE)
23. rms(NUMLIST)
24. shape(SUBSTRUCTURE)
25. sm-and-dm-parms(LC)
26. ss-deflection(SS)
27. ssnonlinearity(SS)
28. ss-stress(SS)
29. stress(STRUCTURE)
30. stress-bound(LOADING)
31. stress-criterion(SUBSTRUCTURE)
32. sum([FIRST | REST])
33. sumsquares([FIRST | REST])
34. youngs-modulus(SS)

> > **2,24,26,27,28**

Reading the rules for analysis-class(STRUCTURE)...

rule-10...rule-11...rule-12...Error: Ignoring
    listof(stress(STRUCTURE))
rule-13...rule-64...rule-27...done.

Reading the rules for shape(STRUCTURE)...
rule-54...rule-55...rule-56...done.

Reading the rules for ss-deflection(SS)...
rule-58...rule-59...rule-60...
rule-50...Error: Ignoring listof(DB,loading(SS) = L and
    deflection-bound(L) = DB)
rule-53...Error: Ignoring SQ/L
    Error: Unable to find values for variable on right hand side of
    rule-53

{section omitted}

Save secondary knowledge base?

> > **y**

Saving the secondary knowledge base...done.

To run the tutor in a fresh M.1, load TREX.KB, and give the command, "teach sacon."

Notice that the authoring program prints error messages, indicating which domain rules cannot be successfully analyzed and why. With this information, the teacher can decide to omit the breakpoint or, if it is something relevant to teaching, rewrite or seek help for rewriting the offending rules. In practice, the offending rules or rule clauses concern a computation which is an artifact of the implementation such as collection of expressions in list structures. Such computations can often be hidden by defining an intermediate expression with a value that summarizes the computation (e.g., the ratio "SQ/L" in rule-53 might be replaced by an expression "the square-footage divided by the loading" or some more descriptive domain term such as "loading distribution," which would then be concluded by a separate rule).

Indeed, we find an obvious correspondence: The clean, easily analyzed parts of the knowledge base tend to contain the basic associations we wish to convey to a student. The procedural and computational parts can be ignored by the teaching program. This strategy works especially well, given the approach of interrupting the working expert system. That is, breakpoints can be deliberately defined to bypass difficulties to explain computations, and to focus on the essential domain facts and associations.

The specific limitations of the authoring system change over time as we add new capabilities to the analysis rules and as M.1 itself becomes more complex with new constructs that make analysis more difficult. The program has fairly extensive capability to deal with numeric computations involving variables such as those in SACON. In the current version, the chief restrictions involve list manipulation. For example, rules concluding a value that is a list are flagged by the analysis program. A break will still occur, but the teaching program states conclusions rather than probing.

The authoring program is menu-driven with many other capabilities as summarized in the preceding paragraphs. Each case is conceived to be a lesson, in the sense that it brings out a certain set of concepts in a particular problem context. The teacher defines a case by running the domain expert system within TRAINING EXPRESS, and using the menus to define problem data (called *presets*), define terms, provide text to appear when a breakpoint occurs, explanatory text for specific rules, etc. Other menus are used to sequence lessons and provide introductory and concluding text.

TEACHING PROGRAM EXAMPLE DIALOGUE

Here again we show TRAINING EXPRESS output for the SACON knowledge base using typescript format rather than the window-menu system provided for the student. In this example, the selection of the case file and other text provided by the teacher, which would normally be seen by a student, is not shown. Student input is in **bold face.** Teaching program output is in *italics*. Other text is output by the expert system. Notice that the program refers to SACON rules by number, which the student can later use to print the rule or get other information about it, such as an explanation or literature reference.

M.1 > **teach sacon**

Welcome to Training Express, the tutor for M.1 knowledge bases.

Loading the knowledge base...done.
Loading the secondary knowledge base...done.

To run the tutor, use the "go" command.

M.1 > **go**
What is the name of the structure you wish to analyze?
> > **747-wing**
How much error (in percent) is tolerable for the analysis?
> > **5**
Do you want to examine the responses of the 747-wing, evaluate its instability, or both?
> > **examine**
Does the 747-wing have any time dependent terms in its equations of equilibrium?
> > **no**
Is heating or cooling of interest in determining the structural responses of the 747-wing?
> > **no**
What are the names of the substructures of the 747-wing?
> > **total-wing**
What is the material composing most of the total-wing?
> > **high-strength aluminum**

Enter the average longest dimension of the total-wing in inches.
$>>$ **1420**
Enter the weight of the total-wing in pounds.
$>>$ **46000**
Is the total-wing thin-walled or solid?
$>>$ **thin-walled**
What is the configuration of support for the total-wing?
$>>$ **one-side**
Is the total-wing to be modelled as a 1, 2, or 3 dimensional system?
$>>$ **2**
Would you describe the construction of the total-wing to be a
* network. (truss and/or frame elements)
* continuum. (sheets and/or blocks of material), or
* network and continuum. (a combination)?
$>>$ **continuum**
Is the overall geometry of the total-wing curved or planar?
$>>$ **curved**

———— *Tutoring Interruption* ————

What do you conclude about the shape of the total wing?
  1. beam
  2. plate
  3. shell
$>>$ **2**
No, the expert didn't conclude that the shape of the total wing is plate.
You should have concluded that the shape of the total wing is shell.
Which factors support the choice of shell?
  1. construction of the total wing
  2. modelling-dimensionality of the total wing
  3. geometry of the total wing
$>>$ **2**
Correct.

rule-56 mentions construction of the total wing and geometry of the total wing.
What geometry of the total wing is consistent with shell?
$>>$ **options**
Your response must be chosen from the following:
curved, planar.

What geometry of the total wing is consistent with shell?
*> > **planar***
Incorrect: planar.
rule-56 mentions curved.

——— *Returning to Consultation* ———

{section omitted}

Do the supports of the 747-wing involve Coulomb friction, non-linear springs, and/or gapping?
*> > **no***

——— *Tutoring Interruption* ———

What do you conclude about the ss-deflection of the total wing?
  1. kinematic-collapse
  2. excessive-deflection
  3. incremental-strain-failure
*> > **3***
Good, the expert concluded that the ss-deflection of the total wing is incremental-strain-failure.
You should also have concluded that the ss-deflection of the total wing is excessive-deflection.

Which factors support the choice of excessive-deflection?
  1. sname
  2. integrity-goal(STRUCTURE)
  3. non-dimensional deflection of the total wing
  4. errorclass
  5. non-dimensional stress of the total wing
  6. loading of the total wing
  7. cycles(L)
*> > **2,5,6***
Incorrect: integrity-goal(STRUCTURE), nd-stress of the total wing and loading of the total wing. See rule-59.
rule-59 mentions errorclass and non-dimensional deflection of the total wing.

What errorclass is consistent with excessive-deflection?
*> > **list rule-59***

rule-59:
  if errorclass = moderate and
   nd-deflection(SS) = ND and
   ND > = 0.0111111
  then ss-deflection(SS) = excessive-deflection.

What errorclass is consistent with excessive-deflection?
> > **moderate**
Good, that is the conclusion considered by the expert. See:
rule-59.

{section omitted}

The following analysis classes are relevant to the analysis of your structure: general-inelastic

* *

The following are specific analysis recommendations you should follow when performing the structure analysis:

- Activate incremental stress-incremental strain analysis.
- Model nonlinear stress-strain relations of the material.
- Solution will be based on a mix of gradient and Newton methods.
- Special code should be written to scan peak stress at each step and to evaluate fatigue.

{section omitted}

Discussion of the Tutoring Example

As can be seen, SACON runs in the normal way, using either a case preset by the teacher or a case chosen by the student. The "whenfound" declarations inserted by the authoring system cause the tutoring interruptions to occur. The teaching program is passed the name of a concept (M.1 expression). Under rule-based control, the teaching program then examines the conclusion made for the current expression, probes the student, provides feedback, and probes at deeper levels when the student makes mistakes. For example, when the student states a value concluded by the program, the tutor says, "Good, the expert concluded. . . . " When a value is missing, the tutor says, "You should also have concluded . . . "

It is not the aim of this research to argue for particular probing strategies, but rather to demonstrate what is easily possible and plausibly useful. As indicated, these probes are similar to what have appeared in several other programs, particularly GUIDON.

The current version of TRAINING EXPRESS illustrates three types of probes:

1. Ask student to state a value for an expression:
   What do you conclude about EXPRESSION?
   For example, "What do you conclude about the substructure deflection?"
   This probe occurs when the expression is determined ("whenfound"); this is a primary breakpoint set by the teacher.
2. Ask student to state expressions that support a value:
   What factors support the choice of VALUE?
   For example, "What factors support the choice of excessive deflection?"
   This probe occurs when the student fails to mention a value with a certainty greater than 50 in the first probe. The question is asked for each "strongly believed missing value."
3. Ask the student what values for an expression are consistent with the conclusion made:
   What EXPRESSION is consistent with VALUE?
   For example, "What errorclass is consistent with excessive deflection?"
   This probe occurs when the student fails to mention a factor in the second probe.
   The question is asked for each "missing factor."

The pedagogical strategy used here prompts the student to use existing knowledge to draw conclusions. When the student leaves something out, the program gives the answer and checks if the student can now remember why it is correct. A Type 2 probe requests the concepts that are related to a conclusion. A Type 3 probe more specifically asks the student to recollect the rule that relates two concepts by requesting the relevant value. As a whole, this approach engages the student in the ongoing consideration of the case, offering practice in applying general knowledge (a set of rules) to a particular problem (the case being discussed). It would also be possible to ask the student to list relevant case data before the program requests it, thus going much further in drawing the student into the problem-solving activity.

SOME IMPLEMENTATION DETAILS

M.1 has several distinctive features that make it particularly suitable for instructional application:

- Both the knowledge base and cache of conclusions about the case are accessible and modifiable under rule-based control, facilitating communication between the expert system and teaching program.
- Variables can be used freely, allowing rules to be written in general form. In particular, domain principles can be stated in general form, and teaching rules use variables to refer to domain concepts.
- External functions can be used to access a database and control the window-menu display.

The reasoning done by the teaching system is often complex. For example, consider the domain rule from SACON shown in Fig. 7.2.

The author shell extracts the "factors" from this rule:

```
rulefactors(rule-56) =
    [construction(SUBSTRUCTURE),
    modelling-dimensionality(SUBSTRUCTURE),
    geometry(SUBSTRUCTURE)].
```

The program knows that MD is a variable and knows which clause binds it. It knows that the rule concludes about an expression using a variable. Therefore, before discussing the rule with the student, the teaching program silently reapplies the rule and extracts the bindings for all of the variables. Thus, in the program's output we see "geometry of the total wing." This has been generated from geometry(total-wing) from a template supplied by the teacher during the authoring phase.

Much more complicated analysis is possible. For example, the program can determine that a rule clause is irrelevant because it sets a variable that is not needed for matching a domain fact. For example, if the fact wine(red,ANY,dry) = zinfandel is in the knowledge base, a clause setting a variable that matches ANY will be irrelevant to the successful application of the rule. The most difficult analysis involves determining the possible values a rule might conclude

---

rule-56:

```
if   construction(SUBSTRUCTURE) is unique and
     modelling-dimensionality(SUBSTRUCTURE) = MD and
     MD > = 2 and
     geometry(SUBSTRUCTURE) = curved
then shape(SUBSTRUCTURE) = shell.
```

FIG. 7.2. Domain rule from SACON encoded in M.1

---

if listof(EXP, kbentry(RULE:if PREM then ACT) and
              conjuncts(PREM) = PROPS and
              member(PROPS) = TERM and
              expression(TERM) = EXP) = LST and
    dremdupls(LST) = SIMPLELST
then rulefactors(RULE) = SIMPLELST

FIG. 7.3.   Typical TRAINING EXPRESS rule written in M.1

when a variable appears on the right-hand side, for example, if . . . = W then
wine = W.

As indicated, all analysis and teaching operations are carried out by M.1
rules. This is particularly advantageous for generating questions for the student
and parsing his answers. In effect, the teaching program is carrying on a kind of
"consultation," in which the probes are questions and the student's responses are
data. All records of the student-teacher interaction are thus stored in M.1's cache
and available for further reasoning by M.1 rules. In comparison, GUIDON was
implemented in a stylized version of Lisp, requiring time-consuming and complicated translation between EMYCIN's data structures and the list structures of
GUIDON. Records of GUIDON's reasoning were stored in a completely different representation than MYCIN's consultation results, making the tutoring code
much more complex and more difficult to maintain.

Figure 7.3 is a typical rule from TRAINING EXPRESS. It retrieves a rule
from the knowledge base (using the kbentry primitive), determines the conjuncts
of the premise (using other analysis rules) and maps over them, extracting the
"expressions" (also determined by other analysis rules). Duplicates are removed
from the resulting list. In all fairness, list manipulation is important in the tutor
and such rules could not be successfully analyzed by the authoring program
itself. So, although the syntax is compatible, we don't quite have a teaching
program that can teach about itself.

ADVANTAGES AND LIMITATIONS

The general pedagogical approach used is that of a case-method tutor. That is,
the program teaches by discussing the application of general knowledge in specific situations. In many respects, this knowledge-based tutor has one of the
simplest designs conceivable for a tutor of this type. The human teacher selects
the concepts that will be discussed with the student. The case is prepared by the
teacher or the student. In contrast, GUIDON only works with canned cases it has
solved before. Students solve the problems on their own, which requires the
entire knowledge base to be well-structured, so each concept and rule can be

discussed explicitly with the students, depending on their needs. In TRAINING EXPRESS only the concepts the human teacher wishes to focus on need be expressed in well-written rules. That is, we implement a form of *apprenticeship scaffolding,* skipping over the difficult areas that can fruitfully be deferred.

Other advantages are apparent by comparison to other intelligent tutoring systems (Sleeman & Brown, 1982):

As in SCHOLAR, TRAINING EXPRESS contains teaching interaction knowledge that is separate from the domain knowledge base. However, TRAIN-ING EXPRESS reasons about an expert system knowledge base which can solve problems, not just a static semantic net of facts.

Like WHY, TRAINING EXPRESS probes a student's understanding of causal associations (in so far as they are encoded in the domain rules). However, the domain relations it interprets are applied in the context of some problem-solving task (e.g., diagnosing a particular device); they constitute more than a static causal model by describing how physical systems behave normally.

TRAINING EXPRESS is similar to SOPHIE-1 in that it evaluates a student's hypotheses for consistency with available data. However, it probes the student's understanding of the general rules, going beyond just indicating correctness and providing a model of expert behavior (as in SOPHIE-III).

Thus, TRAINING EXPRESS has the advantage of any knowledge-based tutor—conveying knowledge about categories of problem situations and solutions as well as facts—with the pragmatic advantage of not requiring a perfectly well-formed knowledge representation. TRAINING EXPRESS works with any operational M.1 expert system without modifications. The teacher can design an arbitrary M.1 knowledge base and enter rules for the students to learn, pairing example cases that illustrate typical problem situations with breakpoints that illustrate key points in the reasoning. The tutor is conceptually simple, since it is not required to be in control of the interaction at all times, and this helps the teacher predict how the knowledge-base design will affect the program's behavior.

Indirectly, the design of the program is also intended to provide directed motivation for cleaning up knowledge bases. We believe that it is unrealistic to expect classroom teachers, with their limited discretionary time for planning course material, to structure subject matter along the dimensions used in our research programs (e.g., NEOMYCIN, PROUST). Instead, we want a teaching program to run without errors and produce a useful dialogue, even if it can't discuss every expression or rule with the student. This design focuses knowledge-base improvements. The output of TRAINING EXPRESS makes clear the kind of teaching interaction clean rules allow, motivating and guiding the teacher (or a programmer assistant) in the process of rewriting the rules for the concepts that the program is to discuss. The authoring system also provides suggestions about how rules can be rewritten using intermediate expressions that hide procedural computations.

Our experience also indicates that knowledge bases built specifically for teaching are stand-alone, small, and relatively easy to reconfigure. In practice a teacher is only likely to create a knowledge base good enough for the cases to be taught. In fact, the number of rules is not as important in many respects as the value of being able to show the student how actual problem cases are solved. Even a 20–rule system can reveal many subtle issues. The limitations of the system provide an excellent starting part for classroom discussion, and finding such limitations can be an important activity in learning from a knowledge-based tutor.

Thus, having the student adopt the attitude that the program is an object of study to be evaluated rather than vice versa, may mitigate some of the limitations of the approach. In particular, the apprenticeship method we follow doesn't challenge the student to step through the entire problem himself. Interruptions occur while the expert system, not the student, is solving the problem.

However, the teacher can encourage the student to vary the supplied cases and probe the program to print the rules or unwind the reasoning using M.1's WHY command. Thus, following the curriculum might be just a means of providing background for the more important activity of defeating the program and explaining why it fails.

The basic idea introduced here of having an authoring program help a teacher configure a knowledge-based teaching program can be applied to other forms of knowledge-based tutoring. For example, a simulation knowledge base, embodying a function-structure model of some process, might be analyzed in a similar way to point out difficulties the general teaching program will encounter or prompt the teacher to supply textual explanations for the student.

A teacher using a shell like TRAINING EXPRESS must cope with a teaching strategy that he or she did not design as well as a knowledge representation language that might be difficult to master. We believe that we have established important design principles for future systems by having the authoring program guide the teacher in revising the knowledge representation and (within some bounds) requiring the teaching program to cope with knowledge-base structures that it can't completely interpret.

In one sense, the analysis program is acting as a student, looking over the knowledge base and complaining about constructs that are difficult to understand. Like other computer assisted instruction (CAI) authoring systems, the analysis program performs some of the tasks of a knowledge acquisition program. Indeed, TRAINING EXPRESS could be used directly as a means of familiarizing experts with the capabilities of an existing expert system. The knowledge engineer could set breakpoints that will query the expert to fill in gaps in the knowledge base. Then, rather than comparing the expert's conclusions, factors, and values to the (nonexisting) rules in the program, an augmented TRAINING EXPRESS could write rules as the expert justifies his conclusions. This demonstrates again the striking promise of knowledge-based programs as mentioned in the opening paragraph of this paper. In particular, it shows the value of programs that can reason about other programs, which is well illustrated by TRAINING EXPRESS' ability to write M.1 code that will interrupt an expert system, determine how variables are set and used, and even rewrite domain rules before they are presented to a student.

STATUS AND FUTURE WORK

TRAINING EXPRESS has been used with approximately a dozen M.1 knowledge bases, in domains ranging from patent law to structural analysis and medicine. The program has been released to selected customers for testing. The perceived value of the program appears to be very sensitive to the familiarity of the audience with the knowledge base and with the complexity of the knowledge base. That is, the program is most suitable for students who are already familiar with the domain of discourse of the expert system. The program is most suitable for knowledge bases with a high proportion of heuristics and facts relative to procedural code (e.g., for conveying text to the user). Industrial R & D researchers were particularly interested in using the program during the test and verification cycle for M.1 knowledge-base developers in the manner described above.

Knowledge engineers also find that TRAINING EXPRESS has considerable value over a simple WHY and HOW explanation facility. In particular, TRAINING EXPRESS allows the developer to trap expressions that are always inferred by the system, but never presented to the user in the expert system's output. In addition, TRAINING EXPRESS can present glossary, concept, and rule explanations when a break occurs. Currently M.1 and most rule-based systems only provide explanations when a question is asked.

Development of TRAINING EXPRESS is continuing with users who wish to construct knowledge bases primarily for teaching, rather than adapting large existing programs, which poses more difficulties. Limitations such as those involving list manipulation will be handled as the need arises.

CONCLUSIONS

TRAINING EXPRESS is a knowledge-based tutoring system that is designed to exploit the essential benefits of expert systems research in a classroom or on-the-job setting. We seek a trade-off, building on the advantages of a simple, rule-based design, and skirting the disadvantages by shifting some of the decision to a human teacher. In particular, we exploit the advantage of an expert system as a model that can be inspected and reasoned about by a teaching program. We separate the teaching program from the domain knowledge so teaching strategies are stated in a general way and easily reused. The breakpoint design focuses the student interaction on the essential reasoning points, skirting messy code or unimportant details. Finally, the teacher is given a convenient facility for entering a library of cases which, combined with the breakpoints, demonstrate the domain heuristics and facts the teacher wishes to convey. We believe that this combination of features—interactive probing, situated (in context) learning, and sequenced examples—represents in large part the essential advantages of knowledge-based tutoring.

We believe that our design is a good example of the 80/20 rule: From 20% of effort involved in designing programs like NEOMYCIN and GUIDON, we believe we have derived 80% of the educational benefit. In many respects, TRAINING EXPRESS is a case study in how academic research ideals can be adapted to application realities. Many trade-offs were made in moving from GUIDON to TRAINING EXPRESS for the following reasons:

- to make the program more reliable,
- to build on well-known advantages of more traditional approaches (e.g., the window-menu system for defining cases and creating a curriculum), and
- to provide a program that non–AI specialists could use today.

Indeed, our collaboration has revealed that there is an often unrealized middle ground between state-of-the-art research and commercial engineering. We didn't just take an existing research design (such as GUIDON) and "apply" it. We radically redefined the nature of the entire interaction between student and program. We brought the human teacher into the process and gave the teacher an essential role. We developed a powerful authoring program that can relate rules and facts, coping with difficult problems involving variables. Then we applied the idea, packaging it with conventional hardware (e.g., personal computers) and software (the "C" programming language), and integrating it to a conventional database and window–management system.

We found that producing a commercial product provides an entirely new orientation to research. In some respects, this is the well-known distinction between basic and applied research. We were faced with the constraints of a target audience and providing obvious value over current techniques. We weren't interested in just developing a theory or speculating on new instructional methods. We believe that as hardware makes delivery of complicated AI-based systems commonplace, many other researchers will want to engage in the same kind of collaboration, producing good research that anticipates use by actual teachers and students.

REFERENCES

Bahill, A. T., & Ferrell, W. R. (1986). *An introductory course in expert systems* (Technical Report). University of Arizona, Systems and Industrial Engineering Dept.

Bennett, J. S., & Engelmore, R. S. (1979). SACON: A knowledge-based consultant for structural analysis. *Proceedings of the Sixth International Joint Conference on Artificial Intelligence* (pp. 47–49). Tokyo, Japan.

Clancey, W. J. (1986). GUIDON to NEOMYCIN and HERACLES in twenty short lessons [ONR Final Report 1979–1985]. *The AI Magazine*, 7(3), 40–60.

Clancey, W. J. (1987). *Knowledge-Based tutoring: The GUIDON program.* Cambridge, MA: MIT Press.

Collins, A., Brown, J. S., & Newman, S. E. (1986). *Cognitive apprenticeship: Teaching the craft of reading, writing, and mathematics* (BBN Technical Report 6459).

Sleeman, D., & Brown, J. S. (Eds.). (1982). *Intelligent tutoring systems.* New York: Academic Press.

8 An Applied Science of Instructional Design

Jill H. Larkin
Carol Scheftic
Ruth W. Chabay
Carnegie Mellon University

The past 20 years have provided the following intellectual and technological advances which could form the basis for an applied or "engineering" science of education—that is, a science of instructional design. These new intellectual and technological tools could make possible substantial advances in the design and delivery of effective instruction at a time when the need for more teachers and better instruction is critical.

1. Advances in cognitive science give us a new and more precise understanding of the knowledge we need to teach, and they give us new and more sophisticated techniques for programming computers. Cognitive science, a relatively new field, has developed primarily from the combined interests of psychologists using computers to model the complex phenomena of human reasoning and computer scientists seeking ways to use a computer to respond flexibly from a collection of stored knowledge, rather like a human does, instead of simply following the steps of a predefined program.

2. Advances in technology give us, at reasonable cost, machines sufficiently powerful to use techniques from the cognitive and computer sciences.

   Machines at costs currently around $1,000 to $3,000 are sufficiently powerful to run programs reflecting the insights of cognitive science, and the costs of powerful computers continue to decrease.

3. Twenty years of experience in using computers as an educational medium gives us enough command of that medium to apply it effectively.

   The computer is a novel instructional communication medium. Every medium requires experience for optimal use. (Television of 40 years ago now appears awkward and naive compared with the elegance of today's best programs.)

4. The desperate need for better school instruction in mathematics and science provides an opportunity for technology to take a more central role in the delivery of instruction.

Particularly in mathematics and in the physical sciences, schools have difficulty attracting and retaining talented teachers, who often have other appealing career possibilities. The current rate at which teachers are being trained is very small. In the foreseeable feature, there will be only a small fraction of the number of qualified teachers needed to educate the technically literate work force that is so important to our nation.

5. There is a need, and an opportunity, to attract and train good scientific talent to apply cognitive science and technology to the solution of educational problems.

Although the field of educational research has always involved fascinating and socially important problems, it has historically had difficulty in attracting top quality research talent. The intellectually-exciting developments listed in the preceding paragraphs are making this field more appealing to many of our best young researchers.

This paper is largely based on a proposal to the National Science Foundation's program for Science and Technology Centers.[1] Its purpose in this volume is to describe the integrated design and development effort needed to capitalize on the research and technological advances previously listed, and thereby begin to have significant impact on the educational problems of our nation. This is instructional design in the "large"—an effort to build a new educational engineering science that links basic research and educational applications, as the physical engineering sciences link the basic physical sciences and practical engineering applications.

The three main sections describe the following in more detail: (a) the current unique opportunities for launching such an applied science; (b) the requirements for doing so; and (c) an agenda for an Educational Science and Technology Center that would provide these requirements for progress.

CURRENT UNIQUE OPPORTUNITIES

The decreasing cost and increasing capabilities of computers have sparked new developments in psychology, technological tools, and educational technology. Current problems in the educational system provide yet another reason to explore

---

[1]This program was initially mandated by Congress in 1987. Proposals are being evaluated, and it is possible that some will be supported under funds made available in 1989.

the potential for using cognitive science and computers in instruction. The following paragraphs elaborate on these opportunities.

The New Discipline of Cognitive Science

The computational power and precision of the computer have allowed psychological researchers to begin to build models that capture some important features of human reasoning. Early models were able to replicate human performance on simple and very tightly-organized reasoning tasks, like puzzles, logical inferences, and theorem-proving. More recent computer-implemented models of thinking describe reasoning in a variety of domains relevant to education, for example, geometry, algebra, physics, and chemistry.

During the same period, computer science has developed increasingly sophisticated techniques for programming powerful computers. In particular, there has developed a field called *Artificial Intelligence* (AI), that is concerned with programming techniques that allow a computer to respond flexibly from a collection of stored knowledge, rather like a human does, instead of simply following the steps of a predefined program.

Computer models of human problem solving and artificial intelligence programs have much in common. Together, they form the foundation of a new discipline concerned with precise models of reasoning, called *cognitive science*.

Because of their desire to understand general reasoning in scientific fields, cognitive scientists have a strong interest in understanding learning and reasoning in subjects taught in schools. Typical textbook problems are constructed with the intent to illustrate, on a small scale, the reasoning required in the discipline. Therefore the reasoning processes used to solve such problems are ideal objects for psychological study. The observations suggest the nature of scientific reasoning, but the amount of reasoning required is small enough to be manageable. As cognitive scientists learns more about scientific reasoning, there begin to be studies of larger-scale problem solving. The substantial literature on problem solving in school subjects accumulated over the last 10 years provides a scientific basis for a new approach to education.

There are a growing number of efforts to use these computer models of human reasoning as the basis for instructional computer programs. Because these programs incorporate the ability to solve a variety of problems in a human-like way, they can, in principle, provide flexible individual guidance to a student trying to work a problem. For example, an instructional program in geometry can detect both whether a student's choice of a theorem is legal and whether it is helpful in producing a solution (Anderson, Boyle, & Reiser, 1985). Because the program need not have prestored solutions, it can, in principle, work with the student, following any proof that is constructed. In short, the goal is that a computer program like this could act as an individual "tutor" for individual students. Since there is appreciable evidence that individual tutoring is by far the most effective and efficient means of instruction (Bloom, 1984), using computers as a cost-effective means of giving many students individual tutors could substantially increase the efficiency and effectiveness of our nation's education.

The precise thinking required by computer models also has supported the development of systematic instruction based on a detailed description of the skill to be taught, though not necessarily implemented on a computer. Among the instructional efforts based on these precise skill descriptions are a number of very promising success stories. For example, Heller and Reif (1984) increased physics students' abilities to solve mechanics problems from 40% to 80%. Freeland and Larkin (1987), who taught students a simplified form of organic chemistry, found that compared with textbook-like instruction, materials based on a computer-implemented model let students maintain their problem-solving skills more effectively and transfer them more reliably to different problems. The differences were dramatic—about 40%. Anderson et al. (1985) found that college students learning the computer language LISP from a computer-based tutor spent only about half the time of students working with conventional materials, and still surpassed their classmates on both a final exam and a demanding programming project. Palincsar and Brown (1984) dramatically increased the reading comprehension of middle-school students who were very poor readers.

In short, in the laboratory, we see the possibility for not just small evolutionary changes in educational effectiveness, but for dramatic revolutionary changes. It is time to exploit this possibility and to explore its potential for addressing our very real national educational problems.

Technology Advances

The original computer-implemented models of human reasoning required large, expensive computers. But computers have increased in power and decreased in cost. There is now a class of computers (e.g., the latest entries in the Macintosh and IBM families of machines) sufficiently powerful to include models of human reasoning, but sufficiently cheap to be within the affordable range for educational institutions during the next few years. For an educational institution, a powerful microcomputer currently costs from $1,000 to $3,000. Trends show that costs decrease by about ½ every two years, so these machines should be available in the near future at prices of only a few hundred dollars.

Experience in Using Computers in Education

The last two decades have seen the emergence of an educational software industry. While not all of educational software is of good quality (neither are all books), during this period an appreciable number of people have cultivated the design and development skills needed to use the computer effectively as an educational communication medium.

Skillful communication using a computer is not easy. Early educational uses of the computer were clumsy, confusing, and boring, like early educational television (e.g., the "talking heads" of Sunrise Semester). Moreover, the computer is an interactive medium, which further increases the complexity of the skills needed to design and deliver top quality instruction through its use. But today there are developers with up to 20 years of actual practice in this medium, people who have the experience and taste to use the medium in reliably attractive and effective ways (Chabay & Sherwood, in press).

Combining the preceding three developments (the growing research base in cognitive science, the availability of suitable computers, and the accumulated experience with the use of computers in education), the conditions are ideal for exploiting these assets for potentially dramatic help in addressing our nation's crucial education problems.

Societal Changes in the Educational System

During roughly the same few decades, there has been considerable change in the educational system, including:

- fewer school-age children and an older population with less motivation to support education financially;
- continuing difficulty in attracting high-quality talent to the low status and underpaid field of teaching—a problem particularly severe in the case of mathematics and science teachers, whose training gives them many other attractive employment opportunities;
- a seniority employment system that encourages transfer of inadequately prepared teachers from other fields into the mathematics and science classes, rather than hiring new individuals trained in these fields;
- a resulting increase in class size, with mathematics and science classes too often taught by underprepared teachers.

This situation means that many school systems currently are simply unable to provide quality instruction in mathematics and science. In this setting the possibility of exploiting computers to support instruction becomes attractive. Essentially, a school system could opt for the same kind of productivity emphasis that has characterized all advanced industry. By investing in computers and effective software, the school could provide science and mathematics education of a quality currently unavailable to most students.

These comments certainly do not imply that computers are likely to provide instruction comparable to that of the best teachers, or that the need for concurrent efforts to attract and retain good human talent to teaching is eliminated. But given the severe difficulties our schools not encounter and the scientific developments previously discussed, it seems possible that a serious effort to exploit cognitive science and technology could improve the quality of mathematics and science education in the many schools that simply cannot provide excellent teachers in these areas.

The Nucleus of a Research Community

In the past few years, a small group of talented researchers from outside fields (largely cognitive psychology, computer science, and the physical sciences) have entered the field of education. This (albeit small) influx of excellent talent provides a current opportunity to encourage better educational research and development and to build a community of researchers to tackle this exciting task.

NEEDS FOR EXPLOITING THESE OPPORTUNITIES

The opportunities outlined in the preceding section are unique and offer our nation a chance at educational innovation that may not appear again. The exploitation of this opportunity, however, requires resources discussed in this section.

An Interdisciplinary Team

Despite both a relevant research base that many agree is ready for exploitation, and a 20–year experience base in practical development, computers have not yet had a major impact on education. This lack of impact should not be surprising. Current educational computing programs are produced by several different groups. There are basic researchers who largely produce laboratory mock–ups, not robust deployable products. There is a cottage industry of individual developers who can produce excellent products, but only at great cost, working with limited tools, and without reliable guidelines. Finally, there are teams of instructional designers and programmers, who are frequently based within large publishing houses and work within a framework adapted from print (or maybe video), but designed to supplement the sale of textbook series. In order to make an impact we need to build on current research and experience to produce the necessary infrastructure for technology transfer—an applied or engineering science and technology of education.

This science and technology of education requires a collaborative team including expertise in the following disciplines:

- Cognitive Psychology
- Education
- Mathematics and Physical Science
- Modern Computer Science, including Artificial Intelligence and Software Engineering

• Practical Development and Deployment of Educational Computing Programs

Sustained Effort over a Period of Years

For two reasons, the stable support provided by center funding is essential for current progress in educational science and technology.

First, it is not enough that the individuals with the disciplinary skills in the preceding list interact occasionally. There are very different mind sets and traditions in different disciplines. For example, researchers produce experimental materials and computer programs that need to work only "well enough" for their experiments. They ordinarily have no experience with the grueling attention to every defect and detail that characterizes successful software developers. More substantively, researchers think "big" and build prototypes; developers must think "small," because every detail must really work. While applying such concentration to the specifics, however, it is very easy to lose sight of advances and alternatives at the macroscopic level.

Establishing a working group, where people from the various disciplines really communicate and contribute to each others' work, is an effort that takes months or years. Only center funding provides enough time to build such a group and then capitalize on its coordinated expertise.

Second, systematic application of cognitive science to develop new technologies is currently limited by the lack of good talent in this field. There are very few first-rate scientists currently working in this area, and therefore much of the talent for even an initial effective effort must be recruited and trained. If this were systematically done by even a few centers, then some of these individuals would move into other settings (universities, schools of education) providing a critical mass of ongoing research nationwide. Only center funding provides sufficient stability to develop new talent in the following ways: (a) by finding mechanisms to encourage good students (with top quality scientific training) to pursue careers devoted to the advancement of modern education, (b) by providing them with the opportunities (presently available in only very few places) to obtain good training in cognitive science, computers, and the other fields needed for such careers, and (c) by providing some opportunities for the training of postdoctoral fellows and some educational practitioners (such as teachers or designers of educational software).

An Integrated Core Group with Complementary Talents

It is essential to have adequate concentrations of individuals with complementary expertise in cognitive science, computers, artificial intelligence, education, and the physical and mathematical sciences; all of these disciplines are needed for an analytic approach to education. It is not necessary that all members be experts in all of the above areas; they should, however, have true expertise in one or two such domains, and sufficiently good understanding and interest in several related fields to permit significant collaboration with their colleagues who have complementary specialties.

Assembling such a group, however, is only part of the task. For a center to be more than a collection of individuals each doing their own work, there must be appreciable interaction and investment in developing a common set of goals and understandings of the entire complex of disciplines. All members must be part of a working group, where people from different disciplines regularly communicate and contribute to each other's work on a variety of projects and over an extended period of time.

Good Facilities and Technical Support

It is essential to support available talent with good facilities, especially with computational and other information technologies that are effectively maintained, reflect the current state of the art, and are updated to keep pace with the rapid changes occurring in such technologies. Furthermore, the financial base of support must be sufficiently stable and long-term to allow people to undertake significantly large research and development projects.

The Need for an Educational Science and Technology Center

Meeting these needs requires time to build an organization capable of making significant contributions toward solving the difficult problems of education. A long-term commitment is essential both to attract first-rate talent, and to build an integrated interdisciplinary team with shared goals. At this time, using significant research results to build usable classroom prototypes is extremely labor-intensive, with a life cycle of many years. In many cases the tools and facilities do not exist, and building them is part of the task.

It is for these reasons that stable, long-term funding for a few centers is crucial to capitalizing on today's unique set of opportunities.

AN AGENDA FOR AN EDUCATIONAL SCIENCE AND TECHNOLOGY CENTER

This paper began with five factors that create a unique opportunity for significantly improving education in the physical sciences. This section describes five corresponding activities that form the core of the agenda for an educational science and technology research group.

Taking optimal advantage of the varied expertise of such a group requires a focus sufficiently limited so that group members all work on interacting problems. Such strong overlap makes communication interesting and valuable and facilitates real collaboration among people with different talents. Too broad an agenda encourages each member to address familiar problems that may not relate to the work of others, and then the center becomes merely a home for individual investigators. What is needed is a team whose combined ability is more than the sum of its individuals.

Build an Applied Cognitive Science Research

The heart of any modern applied science is the development of general guidelines and special-case rules that can guide practical development. The engineering sciences make extensive use of theoretical results from basic sciences, primarily physics and chemistry. But the engineering sciences have extended these results to address special cases and provide useful approximations essential for practical applications. Similarly the basic sciences of psychology and computer science provide results central to informed design of educational computing. But these results must be extended and tested in situations directly relevant to the use of computers to provide instruction. As in the physical engineering sciences, the flow of information is not simply from basic research to applications. By addressing practical needs, the engineering science can identify theoretical problems that may be much less apparent in the controlled conditions of the basic research laboratory. For instance, the applied scientist may develop guidelines or rules of thumb based largely on experience. Elucidating why such guidelines work can be an important challenge to basic research. (For example, practical observations in metallurgy drove the development of the theory of dislocations; similarly, theoretical physicists today work hard to understand the observed high-temperature superconductors.)

Applied cognitive science research is needed to extend basic results to apply to special cases, and to interpret basic results into general guidelines. A number of topics of current interest to researchers (such as problem solving, planning, connecting factual and strategic knowledge, thinking with analogies, reasoning with mental models, understanding diagrams, and the goal setting and strategic planning used by teachers) are directly relevant to the design of educational materials. The following examples of existing work illustrate the kind of cognitive science research that could guide design of educational computer programs.

A consistent set of research results concerning physics students (Chi, Feltovich, & Glaser, 1981; Heller & Reif, 1984; Larkin, 1983; McDermott & Trowbridge, 1980; Trowbridge & McDermott, 1981) indicate that even students who can manipulate mathematics almost always lack important qualitative understanding. Work by Larkin (1983) suggests how these problems might be remedied by teaching directly some of the qualitative reasoning patterns characteristic of skilled scientists.

There are currently two global computer architectures, ACT (Anderson, 1983) and SOAR (Laird, Rosenbloom, & Newell, 1986) that capture important general features of human reasoning. These architectures provide an informative beginning to the tasks of development programs that solve physical-science problems, that learn how to solve them, and that ultimately can teach students how to solve them.

The ACT architecture makes a clear distinction between declarative knowledge (such as facts and formulas) and procedural knowledge (knowledge about how to solve problems). ACT supports an explicit model of how declarative knowledge can be developed into procedural knowledge.

There exist now several computer-based tutors sufficiently well developed to have some classroom use. Anderson et al. (1985) have developed tutors for the programming language LISP, for geometry, and for algebra. His approach gives a high emphasis to reasoning rules, and less emphasis to how a problem is conceptualized or visually represented. Experience with these tutors can furnish a valuable complement to the interests of center members in the latter issues. Soloway and his colleagues (Johnson & Soloway, 1985; Sack & Soloway, 1987) have developed a tutor that can detect real conceptual misunderstandings, not just syntax errors, in students' computer programs. These techniques also can be applied to solving problems in the physical sciences.

Through detailed laboratory observations and modeling, Larkin has elucidated properties of displays, both static and dynamic, that generally aid people in solving problems. Briefly stated, the most striking property of good displays is that they let the user substitute easy perceptual judgments for more complex logical inferences. This and other properties can be used as guidelines for designing computer displays (Larkin & Simon, 1987).

Trowbridge, Scheftic, and Larkin (1987) have developed a computer program called *Sketch,* which coaches a user through the process of sketching a curve by an incremental process. The user must first draw a basic curve (e.g., $y = x^2$ or $y = \log(x)$), then shift or stretch the geometric curve to make it correspond to a more complex equation (e.g., $y = 3 + x^2$ or $y = \log(x)/2$). The program is sufficiently flexible that it can accept equations entered by the student. It can also advise the student when unproductive actions are taken. *Sketch's* teaching strategy is based on a cognitive analysis of the verbal statements of expert teachers describing their teaching methods. The techniques developed with *Sketch* can be refined and used with other programs.

In the longer term, continued applied cognitive research should aid researchers in producing answers to the questions with which they struggle. For example, there are several psychological models that can provide some guidance on following students' steps in problem solving; however, there are no models that offer good guidance on what kind of hint, help, or advice to offer a student who has made an error (Lepper & Chabay, in press).

In contrast to the approach of most current research-based efforts to build computer-based instruction, practical experience from the development of top quality instructional programs suggests that the first and most central problem to address is the nature of the interface. A research center should focus on practical and theoretical aspects of interface design for instructional applications with support from facilities in the programming environment. This focus could suggest important new techniques for artificial intelligence and for human-computer interaction.

Illustrations from several sources suggest that large LISP machines and large development efforts are not necessarily required to exploit many of the insights from research on artificial intelligence and human-computer interaction. Programs using artificial intelligence techniques have been written in standard programming languages. Experience in transporting educational applications between large advanced-function workstations and smaller, more commonly available yet increasingly powerful, microcomputers indicates that instructional design issues, not hardware characteristics, can be the basis for most interaction decisions.

Build Tools and Working Environments

A major function of any modern applied science is building tools that make it possible to produce complex applications with reasonable effort. In educational computing, an effective set of development tools are needed to deal with the following problems:

1. *Sophisticated interfaces:* Modern computer applications make use of windows, mouse interactions, attractive multifont text, and high quality graphics–features which are extremely difficult to program. End users find such applications directly appealing and easy to use. It is very different for the programmers who create these attractive applications. Working with conventional programming tools, they face a bewilderingly complex and daunting programming task, far more difficult than writing programs for earlier computing environments. Not only is the task vastly more complex, but production-level programs normally require lengthy batch compilations, making the programmer's environment much less interactive than that of the end user. These factors have created a situation where only exceptionally skilled programmers can produce modern interactive programs.

2. *Portability:* Few such modern programs are transportable among currently available computer systems, and almost none have realistic hope of ready

transport to new or future systems. Therefore, machines define subgroups of developers who are relatively isolated from each other. Moreover, by reducing the potential market for developer's product(s), subgroups also serve to reduce the incentive for creating those products.

3. *Unification of effort:* Educational computing efforts are so fragmented that it is very difficult to build on the work of others. Developers of "conventional" programs and of those based on techniques of cognitive science know little about each other's work.

If these problems were solved, it would become possible, for at least some teachers, to use the computer as a flexible educational tool, producing their own simple lessons. Good tools would also enhance the productivity of expert and professional developers, letting their products run on multiple machines without reprogramming. A uniform and broadly used set of tools would facilitate easier communication between developers and researchers who currently use vastly different programming languages and techniques ranging from Apple II assembly language to sophisticated derivatives of LISP.

To solve these problems well enough to make effective progress requires a coherent set of development tools—an educational software development environment with the following features:

1. a full programming language with features that make possible the use of modern computer science approaches;
2. special support for nonexpert and educational programmers;
3. sophisticated tools for interpreting both mouse and text input
4. easy control of modern displays including windows, menus, variable fonts, and graphics; and
5. portability among current machines and to machines of the future.

These features are a minimal set required for reasonably efficient educational programming (or any programming where display and interaction are important, and the users are not always expert programmers).

The cT programming language (formerly CMU Tutor), developed at Carnegie Mellon, provides a start for an environment meeting these criteria. cT combines the computational power of a general programming language with features designed to facilitate the construction of dynamic interactive programs.

cT provides a highly productive programming environment. It is easy to learn, with on-line documentation that includes examples of features, in context, that can be immediately executed. It is easy to use, with a tightly integrated graphics editor and incremental compiling that drastically shorten the design cycle. It encourages the development of good user interfaces, for instructional and other purposes, because it is so easy to test and revise until it works effectively. It is also possible to use cT to construct visual interactive interfaces to programs written in other languages, for example LISP, OPS5, and SOAR. This provides one method of utilizing the power of cognitive computer models in a visual interactive medium.

Produce Exemplary Products

Although there currently exist a variety of computer instructional materials that are attractive and useful to students, many authors themselves believe that these materials fall short of realizing the real promise and potential of computer-based learning. It isn't just that specific failings need remedied. There is a feeling that major new insights and new paradigms are required to make a significant difference. However, there are no clear models demonstrating how to combine the results of basic research in cognitive science with practical expertise in educational software design in order to produce artifacts which both reflect new research insights and use the computer medium creatively and effectively.

The output of an educational science and technology center should include, therefore, concrete, demonstrably effective, working examples of educational programs strongly tied to the applied research in learning and teaching carried out at the center and at other educational research institutions. In order to serve as exemplars to the wider community of potential developers of educational computer programs, the instructional programs must be accessible to teachers and researchers who do not have access to expensive, powerful computer workstations. The programs, therefore, must run well on common microcomputers of several varieties and must receive wide distribution, perhaps commercially.

A single program, however, impressive, can have little impact in a real educational setting. To produce materials that can be used with students as an integral component of a curriculum, a center must produce a number of modular packages, each composed of several educational applications which, together, adequately cover a particular topic. Several such modules dealing with important topics in physical science could then constitute an important component of a semester high school or college course.

In addition to exemplary educational materials, a center can offer important principles and guidance to others who wish to develop educational software. Through systematic observation of the development, testing, and revision process, developers can produce documentation that details the steps and problems of the process, and that articulates explicitly some of the tacit principles and development strategies employed by expert developers of educational computer materials. Such practical principles, combined with the principles from applied cognitive science research on which the educational strategies of the programs are based, can help to raise significantly the quality of educational computer applications produced by beginning developers, as well as to advance the state of the art.

The process of applying cognitive principles to the development of computer instructional programs is inherently a cyclical one. In attempting to create concrete, visual, interactive applications of cognitive principles, a developer is inevitably forced to formulate concepts much more precisely and to make myriad other microscopic instructional decisions, which themselves generate further questions about the learning and teaching process. Thus, the development process itself leads to reformulation of existing principles and to the generation of further research issues.

In testing educational programs, students' unanticipated difficulties with concepts, pedagogical strategies, displays, and interfaces lead to revision of programs and to reconsideration of the underlying theoretical and practical principles. In the "formulate, develop, test, revise" cycle, both products and principles are redesigned and reformulated. The output of each instructional computing project in a center, therefore, should include both usable educational products and reformulated principles (theoretical and practical).

Although an educational science and technology center should not be a software factory, the major way to illustrate the effectiveness of an applied educational science is ultimately to produce a selection of usable products that demonstrate this effectiveness and are widely available through the conventional marketplace. Because of the commitment to developing under use, these materials should always be in usable form, but this form may range from something usable by a few knowledgeable colleagues to, ultimately, final products that can be commercially marketed.

Build a Research Community

Currently, the members of the field of cognitive-science-based education are both too few in number and too scattered in location to achieve critical mass anywhere. The benefits of a center can be magnified if it serves as an attractor of new talent to the field and an intellectual focus for the current group of widely separated individuals. Three illustrations of this approach follow:

1. An educational science and technology center could welcome graduate students from computer science, psychology, education, mathematics, and the physical and biological sciences, by providing facilities for postdoctoral fellows and for visiting faculty, and by offering occasional summer workshops for individuals with appropriate scientific background and an interest in educational computing.

2. The InterUniversity Consortium for Educational Computing (ICEC) ran one-week workshops for faculty and support staff from its member colleges and universities. Rather than merely experiencing existing applications, the participants went more deeply into the new technology by writing their own interactive graphics-oriented programs using the tools developed at Carnegie Mellon. In less

than a week, most of the participants produced significant programs that exploited the modern interactive graphics capabilities of the workstations—an unprecedented achievement. ICEC then provided the necessary software to the participants' home institutions, so that the faculty could continue their individual projects.

3. There are, at present, two disparate communities working on the development of educational computer applications. Developers of Intelligent Tutoring Systems (ITS) employ artificial intelligence techniques to create sophisticated programs that are capable of solving complex problems and of diagnosing "bugs" in students' problem-solving procedures. Developers of more traditional Computer Assisted Instruction (CAI) have great practical experience in producing programs that are robust, attractive to students, have good interfaces, and exploit the graphical capabilities of the computer as an instructional medium. There is potentially much power to be gained from a synthesis of these two approaches. A central goal of the educational program development efforts of a center would be to create such a synthesis, drawing on the expertise of its staff in cognitive science, artificial intelligence, and in the production of large bodies of high quality traditional instructional programs. One conference addressing these issues was sponsored by the James S. McDonnell Foundation in the Spring of 1988, and a resulting book will be available in late 1990 (Larkin & Chabay, in press). The momentum from that initial sharing of issues and concerns could be continued and expanded through a center such as the one proposed in this paper.

SUMMARY

The goal of an Educational Science and Technology Center should be to develop an applied science and technology of education, with the following major products:

1. applied research to extend the results of relevant basic sciences (cognitive and computer science) to produce results directly applicable to the problems of education;

2. tools and working environments to make effective educational products substantially more efficient to produce;

3. exemplary instructional computing products that demonstrate the value of the applied science, illustrate the nature of its principles and guidelines, and show how systematic development can provide dramatically improved instruction;

4. continual wide availability of tools and educational materials during their development in order to enrich interactions with colleagues and provide formative feedback from a broad range of users;

5. a research community, built through the visibility of a center, both attracting visitors for training at the center and illustrating an approach to education that can be used elsewhere.

All recent national reports and briefings to the National Science Foundation stress that the promise of using computers to provide widely available, cost-effective instruction should be exploited (Pea & Soloway, 1987; Melmed & Burnham, 1987; National Academy of Sciences, 1985). This opportunity is particularly promising in the area of science and technology because these areas are economically and socially important to the nation's welfare for the following reasons: (a) teachers with good training in these areas are often attracted out of the educational system by higher paying jobs, (b) there exists for these domains a research and experience base ready for exploitation, and (c) the tight organization of these topics makes them comparatively easy to address with systematic instruction.

The research base comes from cognitive science. This research includes psychological models of how both learners and experts execute instructionally relevant processes. It also includes use of modern computer science techniques to build these models and to extend them to instructional programs. The experience base comes from 20 years of building instructional programs. It includes some excellent examples that are very much in accord with the results of cognitive science research, are currently available commercially, and that are used in schools.

The laboratory results on educational computing products suggest that an approach of this kind can make dramatic differences in learning—savings of a factor of 2 in time and increases in learning of 20% to 40%. It is time to get these opportunities out of the laboratory and to develop the capability to place them at the service of our national educational needs.

The problem addressed by an Educational Science and Technology Center is to exploit the recent progress in cognitive science to produce computer-implemented instruction with significant impact on the nation's effectiveness in teaching mathematics and physical science. The approach is to work from both basic research and practical experience to build an engineering or applied science of education that can systematically be used and improved with continued experimentation and development.

REFERENCES

Anderson, J. R. (1983). *The architecture of cognition*. Cambridge, MA: Harvard University Press.
Anderson, J. R., Boyle, F., & Reiser, B. (1985). Intelligent tutoring systems. *Science, 228*, 456–467.
Bloom, B. S. (1984). The 2-sigma problem: The search for methods of instruction as effective as one-to-one tutoring. *Educational Researcher, 13*, 3–16.

Brownston, L., Farrell, R., Kant, E., & Martin, N. (1985). *Programming expert systems in OPS5: An introduction to rule-based programming.* Reading, MA: Addison-Wesley.

Chabay, R. W., & Sherwood, B. A. (in press). A practical guide for the creation of educational software. In J. H. Larkin & R. W. Chabay (Eds.), *Computer-assisted instruction and intelligent tutoring systems: Shared issues and complementary approaches.* Hillsdale, NJ: Lawrence Erlbaum Associates.

Chi, M. T. H., Feltovich, P. J., & Glaser, R. (1981). Categorization and representation of physics problems by experts and novices. *Cognitive Science, 5,* 121–152.

Freeland, R. J., & Larkin, J. H. (1987). *Model-based instructional design for problem solving in organic chemistry* (Technical Report). Claremont, CA: Harvey Mudd College.

Heller, J. I., & Reif, F. (1984). Prescribing effective human problem-solving processes: Problem description in physics. *Cognition and Instruction, 1*(2), 171–216.

Johnson, W. L., & Soloway, E. (1985). PROUST: Knowledge-based program understanding. *IEEE Transactions on Software Engineering.*

Laird, J. E., Rosenbloom, P. S., & Newell, A. (1986). Chunking in SOAR: The anatomy of a general learning mechanism. *Maching Learning, 1.*

Larkin, J. H. (1983). The role of problem representation in physics. In D. Gentner & A. L. Stevens (Eds.), *Mental models* (pp. 75–98). Hillsdale, NJ: Lawrence Erlbaum Associates.

Larkin, J. H. (In press, 1990). *Expert systems and instruction.* Paper prepared for the National Academy of Sciences, National Research Council, Committee on Human Factors.

Larkin, J. H., & Chabay, R. W., (Eds.). (in press). *Computer-assisted instruction and intelligent tutoring systems: Shared issues and complementary approaches.* Hillsdale, NJ: Lawrence Erlbaum Associates.

Larkin, J. H., & Simon, H. A. (1987). Why a diagram is (sometimes) worth 10,000 words. *Cognitive Science, 11*(1), 65–100.

Lepper, M. R., & Chabay, R. W. (in press). Socializing the intelligent tutor: Bringing empathy to computer tutors. In H. Mandl & A. Lesgold (Eds.), *Learning issues for intelligent tutoring systems.* New York: Springer.

McDermott, L. C., & Trowbridge, D. E. (1980). An investigation of student understanding of the concept of velocity in one dimension. *American Journal of Physics, 48*(12), 1020–1028.

Melmed, A. S., & Burnham, R. A. (1987). *New information technology directions for American education: Improving science and mathematics education* (Final Report No. MDR–8652287). National Science Foundation.

Palincsar, A. S., & Brown, A. L. (1984). Reciprocal teaching of comprehension-fostering and comprehension-monitoring strategies. *Cognition and Instruction, 2,* 117–175.

Pea, R. B., & Soloway, E. (1987). Mechanisms for facilitating a vital and dynamic education system: Fundamental roles for education science and technology. Final report for the Office of Technology Assessment, U.S. Congress.

Sack, W., & Soloway, E. (1987). *Assessing the educational benefits of PROUST* (Technical Report). New Haven: Yale University, Department of Computer Science.

Trowbridge, D., Larkin, J., & Scheftic, C. (1987). Computer-based tutor on graphing equations. *Proceedings of the National Educational Computer Conference.*

Trowbridge, D. E., & McDermott, L. C. (1981). An investigation of student understanding of the concept of acceleration in one dimension. *American Journal of Physics, 49*(3), 242–253.

III SOCIAL CONTEXTS OF LEARNING SCIENCE

James G. Greeno
Stanford University and Institute for Research on Learning

The papers in this section open some new windows on factors that are relevant and important for science education, but have not been emphasized in most science-education or cognitive-science research. Most research on science education treats learning in science as something that an individual student does. The papers in this section point out that learning in science is, in fact, a social phenomenon. Most science learning occurs in classrooms, where the activities of learning are carried out. These classes are part of the institution of schooling, to which students' relations differ in ways that fundamentally affect their learning of science. And students live in a society, where their reasoning and understanding of events may or may not be influenced by concepts and principles that they have learned in school.

Researchers concerned with science education could take the position that social factors are not in our department. The papers in this section argue—in my opinion, persuasively—that we researchers of science education need to learn how to study the social phenomena of learning and understanding scientific concepts. Indeed, the research findings presented by Denis Newman, Penelope Eckert, and Geoffrey Saxe, along with research reviewed by Vincent Lunetta, provide important examples of how such research can be carried out productively.

Newman presents a view that challenges the individualistic epistemology that underlies much current pedagogy and research in science education. He invites us to consider cognitive systems that go beyond the individual's mind to include the social and physical contexts of thinking and learning. This view has a major implication for the way we think about meaning. Meanings are results of interactive social processes in situations that frame communication; they are not inherent in concepts. Newman establishes this idea with an example of negotiated turn-taking in a seed-planting activity in nursery school.

Newman characterizes a key aspect of teaching in social interaction with the term *appropriation,* which refers to a teacher's process of using ideas or products given by a student as a basis for discussion and extension of the student's understanding. This is illustrated with lessons about combinations and estimation of divisors, including the point that the situation has to enable students to provide appropriate products to enable productive reflective adjustments or elaborations. Appropriation at another level is illustrated by a lesson on chemical indicators in which results obtained by pairs of students were tallied and discussed by the whole class thereby enabling the pairs of students to see and reflect on their results in a large context. Finally, Newman discusses the design of Earth Lab, a computational network system that enables students to work in small groups as well as to interact at the macro level where the whole class engages in activity that extends over several months.

Penelope Eckert challenges current views of school learning in another way. We usually consider motivation for learning as a property of individuals, and efforts to increase students' motivation are often focused on including material that relates more directly to students' interests, or on providing examples of careers in which the material of instruction is used. Eckert's analysis implies that this individualistic view may keep us from understanding and addressing the major determinants of students' motives and interests relevant to their enrollment in and engagement with science and mathematics in school.

We all know that some students are engaged by school activities and other students are not, both in a general way and particularly regarding science and mathematics. Eckert presents a fascinating and, to my mind, persuasive argument that much of this difference is produced by a feature of the school's social organization. She found that high school students are polarized, with two main groups that currently call themselves "jocks" and "burnouts," differing dramatically in their relations of identification with the school as an institution. For jocks, school activity is a source of personal esteem and contributes to their goals of future advancement through continued education beyond high school. For burnouts, school activity interferes with most things that are important in their lives and provides little relevant experience or information that could be useful for advancing the careers that they anticipate in blue-collar employment. Jocks and burnouts differ fundamentally in their attitudes and processes for seeking and using information; jocks accept the role of information in the hierarchical organization of the school, with information a kind of commodity that is traded for standing and esteem, while burnouts value and share information as a means of maintaining the integrity of their social commitments. These distinctions bear heavily on students' perceptions of science and mathematics, which are perceived (quite accurately) as prototypical cases of information that is transmitted hierarchically, with students expected to receive the material that is sanctioned by the institution. It seems unlikely that significant changes in the appeal of science and mathematics can succeed based only on changes in the content of the material; significant changes in the activities that determine relations of students with the subject matter must also take place.

Geoffrey Saxe presents a contribution to the growing body of research findings about everyday cognition. There has been considerable research interest in the "problem" of transfer and inert knowledge: How is it that students can become proficient in school tasks but not use the concepts of those tasks outside their classrooms? Recent researchers have looked at this question from a different perspective—by examining the capabilities of individuals to reason and solve problems in everyday settings, whether or not they have learned relevant concepts in school. Most previous research along this line has emphasized successful reasoning, especially involving quantities. These findings challenge the epistemological assumption that knowledge of symbolic expressions and procedures is general if the expressions are general. The alternative view is that knowledge of symbolic procedures can be highly specific, applying mainly to physical symbols, rather than to the quantitative properties and relations that symbolic expressions are meant to denote. A different analysis of knowing focused on relations between cognitive agents and situations may be needed to understand fundamental properties of learning and reasoning.

Saxe presents findings that complicate the picture in an interesting and important way. He reports results of a natural experiment that allowed comparison of children with different amounts of schooling reasoning about prices and profits of the candy that they sold as street vendors and comparison of children with equal amounts of schooling, some of whom were experienced candy sellers and others who lacked commercial experience. On some problems that were solved successfully by candy sellers with different amounts of schooling, sellers with more schooling made more use of procedures that involved symbolic notation of arithmetic, and sellers with less schooling seemed to depend more strongly on the specific features of currency, rather than numerical symbols indicating denominations. On the other hand, in solving arithmetic problems, experienced sellers made more use of regrouping strategies, compared to equally schooled but commercially inexperienced students who relied more on standard algorithmic procedures. Saxe's findings show that the qualitative reasoning of everyday cognition and the symbolic procedures of school mathematics are blended in children's cognitive functioning—they don't reside in insulated compartments. The interactions between school learning and everyday cognition are subtle, indeed, and the research task of clarifying their interactions will be considerable and fascinating.

Vincent Lunetta approaches these issues with the perspective of a science educator. His chapter focuses on cooperative activities that can occur in science learning in classroom and laboratory settings. He reviews research that is beginning to clarify ways in which cooperative settings can be beneficial to learning.

He points out that in science education, laboratory settings could be used in ways that would particularly encourage cooperative learning, but that they are rarely used in that way. Lunetta also considers a distinction between "people's science" and "organized science," emphasizing the need to increase the intersection between the conceptualizations and attitudes of science that are presented in school and those that students bring to their instructional experience.

Lunetta's chapter presents a set of questions for research relevant to science education. One set of questions involves processes of group problem solving and discussion that enhance science learning, such as finding ways to make science problem solving more interesting and engaging for students and identifying characteristics of teaching and the organization of groups that result in productive cooperative learning. A second set of questions involves design of instructional activities and computational resources that facilitate learning through sharing and interpretation of data. A third set of questions involves identifying characteristics of excellent teachers and the social and organizational factors of exemplary schools and school systems, and sociological and anthropological studies of the social and cultural contexts in which students are immersed, both in school and in the society.

In her commentary on the conference papers, Jean Lave, in direct terms, presents a challenge to the individualistic view of cognition. In Lave's words, "It seems important from a relational, contextual, theoretical perspective, to lay out the socio-logic with which social studies of learning and sociological studies of schooling can be brought together to illuminate research on math and science learning. The main point . . . is that a synthesis of primarily psychological with sociological analyses of learning is not just just crucial, but is also feasible." Lave's synthesis would be an assimilation of the psychological to the sociological. "I will argue that processes of knowing and learning and the structure of knowledge itself are *made* in socially organized activity."

Looking at schools as social institutions, and drawing on Eckert's findings, Lave brings in a strong indictment. "Our public school system is organized as a social institution, in short, in which sustained learning stands in conflict with and opposition to the major categories and intentions and activities with which all members of the institution operate. Schools are basically anti-learning institutions. The authoritarian character of the school personnel's management of access to information is a major source of learning rejection by burnouts. And the math and science curricula play a special role, intensifying these problems." Lave offers a research agenda, framed by her characterization of sociological factors, including study of ways to engage students in learning by altering the hierarchical asymmetry of institution; examination of instances in which ethnic minority students have achieved high levels of academic excellence; study of contradictions between beliefs and realities about the role of schooling in the society; inquiry into sociological factors that prevent changes in instructional methods from having major effects; and study of the effects of removing administrative constraints that enforce didactic methods and providing teachers with a clear vision of goals for valuable student learning.

Lave's and Lunetta's proposals for research directions have significant common elements, but they are quite differently focused. In combination, they present an exciting prospect. We need to continue to study processes and systems that work and to understand how they work in order to provide models of better instruction and more powerful resources for teaching and learning. We also need to understand how structural features of our current systems of science education work against the goals of productive and meaningful learning and understanding for fundamental reform. The reports of research and discussions in this section portend a lively and productive period of research as we move toward increased attention to the social contexts of science learning.

9 Using Social Context for Science Teaching[1]

Denis Newman
BBN Systems and Technologies Corporation

The social context of a classroom is actively created by teachers and learners. Instead of looking at the context as the *background* for cognitive work, we might look at it as an *externalization* of cognitive work. The teacher creates a social system in the classroom that supports certain kinds of discourse and activities; students collaborate within the system, contributing observations, answers, and concrete products such as text, projects, and data. The cognitive system includes the externalized tools, texts, data, and discourse all of which is produced by and for the activities. This takes us a step away from traditional psychological approaches that look at the individual mind as the basic unit of analysis. Studying the interactional system, reveals unique properties that support cognitive change. This research perspective then provides another way to look at the design of technology to support interactional systems that are productive of change.

In this chapter, I will make a case for using the socially situated cognitive system as the unit of analysis. This perspective has consequences for studying cognitive change, for organizing science instruction, and for using computers to support social processes. We find in instructional interactions that the data students produce (answers, texts, etc.) can be reinterpreted in terms of the organizational system that is the teacher's frame of reference. We also find that teachers have ways of organizing the work of the students so that their products become

[1]Research for this chapter was supported in part by the National Science Foundation under grant No. MDR-8550449. Any opinions, findings, conclusions or recommendations expressed in this paper are those of the author and do not necessarily reflect the views of the foundation. Author's address: BBN Systems and Technologies Corp. 10 Moulton St. Cambridge, MA 02138. (617) 873-4277. Internet: dnewman@bbn.com.

part of a larger cognitive task distributed among the group. The relationship to the larger group provides other ways of looking at the individual student's contribution. The larger interpretive frame in these cases is not just the context for the student's thoughts, it is also the organization by which the work is getting done.

I will present a series of examples of instructional interactions that support two related propositions:

> What is outside the head is just as much a part of the cognitive system as what is inside the head.

> Group instruction can be more than just a poor approximation of an individual tutorial.

The propositions derive from the sociohistorical theory proposed by Vygotsky (1978, 1986) and elaborated in recent work (Newman, Griffin, & Cole, 1989). This theoretical approach draws attention to the fact that tasks and understandings are observed first in social interaction before being internalized as an individual's capacity. Fundamental to the first proposition is the notion that meaning is actively constructed in interaction and that an utterance or action can come to mean something different by virtue of utterances or actions that follow it. This indeterminacy of the meaning of actions in their context of use is the basis for a process of cognitive change. Fundamental to the second proposition is the notion that activities can be distributed among a group of students, such that distinctions that might be hard for an individual student to maintain can be encoded in the organization. The organization becomes an interpretive frame that provides the basis for a change in understanding.

The first example, from a nursery-school science lesson, illustrates the basis for both propositions. Two examples from a 4th-grade classroom illustrate a basic process in the construction of new understandings between teachers and students. Another example from the same classroom illustrates the way that lesson content can be usefully distributed among students in a class. A final example, from a 6th-grade class, illustrates how technology can assist in the distribution of a task and with setting the stage for constructive teacher-student interactions.

PLANTING SEEDS AND MAKING CLAIMS

In nursery schools, science instruction can occur in organized settings at a very young age. A science lesson with a group of 3-year-olds illustrates the two propositions. First, we see that the meaning of a child's utterance results from what is made of it in the subsequent interaction. Second, we see that the teacher uses the social organization of the lesson, which is constructed in the interactions with the children, as the interpretive frame for the individual utterances. The lesson organization also uses each child's turn as a means for group instruction. In so doing, the children contribute to a lesson that is more complex than any of their individual contributions.

In a videotaped lesson, 7 children were gathered at the table while the teacher conducted a springtime activity involving planting seeds in a planter that was to be placed on the window ledge. First the planter was passed around the table so that each child had a turn mixing the peat moss and soil with their hands, and then each child had a turn poking a hole with his or her finger, choosing a seed from the tray, placing it in the hole, and covering it up. The activity was organized as sets of turns. But 3-year-olds cannot always be counted on to be patient while 6 other children get to play around in dirt.

In the context of this lesson, one little boy made the following assertions:

Teacher: When you think you've mixed it, you can wipe your hands off and then give the next person a turn.

Jeff: I am the next person.

(As Don is patting down the soil on his seed)

Jeff: I'm gonna- now I wanna. *I'm* gwanna plant a bean seed.

In both cases he is making what we call a *claim,* a type of speech act quite familiar in the nursery school (Dore, Gearhart, & Newman, 1978; Newman & Gearhart, 1978). Claims generally take the form of a statement of fact ("I'm first," "It's my turn," etc.) but are never statements of preexisting facts. When they are successful, the statement becomes true as a result, partially, of being asserted. The conditions for success appear to be that there is something like a turn that is currently claimable and that the claim is the first one since the appearance of the claimable item. Although the second condition is usually easy even for 3-year-olds to decide, the first condition can be somewhat less clear. The assertion of the claim itself may lead to mutual recognition that a turn is up for grabs and to the success of the claim and the truth of the assertion.

After Jeff's first claim attempt, the teacher replied, "I think it's going around the table. Who would be the next person?" After the second one, she replied, "Okay, Jeff, walk around over here." There are possibly several reasons for the difference in uptake. In the first case, which occurred as children were taking turns mixing the dirt, the around-the-table order was being well maintained by the teacher. In the second case, while children were planting seeds, the tidy order had been interrupted by the need to reorient the planter. It is clear, however, that if Jeff had not made his second claim, the teacher would probably not have invited him to take the next turn. At the same time, it was far from clear that the next turn was up for grabs because up to that point, the teacher had assigned all the turns. The success of the claim was established by the discourse move that followed it, not by any clear precondition.

In social interaction, the meaning of an action can be constructed or reconstructed by the actions that follow it. In a recent article, Fox (1987) documents in records of naturally occurring adult conversation a similar phenomenon which she calls "interactional reconstruction." The fact that meaning can be changed retrospectively requires that we continue to revise the constructive theories of communication. Further motivation for an interest in the retrospective reinterpretation comes from analysis of instructional interactions. As the student's utterances and actions enter the discourse being structured by the teacher, they can take on new meanings (Newman, Griffin, & Cole, 1989). A process we call "appropriation" requires this type of backward meaning construction. As I document below, a teacher may appropriate a student action into a larger frame of reference giving it new meaning.

In instructional interactions, the frame of reference is very often the classroom organization that the teacher creates for accomplishing the lesson. The case of the planting seeds lesson, the teacher initiated a turn-taking procedure that was the frame of reference in terms of which Jeff's claims were taken up. The turn-taking structure was constructed with the collaboration of the children who knew enough about sitting at the table for the teacher to be able to make very productive use of their actions in terms of her larger goal of getting seeds to grow in the planter, even though at times their patience may have been tested. This structure made it possible for the group of 7 children to participate in a lesson together learning from each other as well as from the teacher. The teacher, for example, used early turns to help teach the planting procedure: "Are you watching so that you'll know what to do when it's your turn?" Each of the turns became part of the larger lesson. The lesson as an externalization of the teacher's plan becomes the interpretive frame for the children's external actions that both help to construct the lesson and are transformed by it.

In what follows, the process of appropriation is seen to provide a key mechanism for cognitive change. It is also seen that group lessons can be structured as interpretive frames to which students contributions can be appropriated. These two observations suggest ways that social context can be used for science teaching and ways that technology can be used as a tool for supporting these interpretive frames.

APPROPRIATING STUDENTS' ACTIONS

The first component in my argument is the idea that a student's actions and utterances can take on meaning depending on what actions follow them. This unique feature of socially constructed actions makes possible a process we have found pervasively in instructional interactions. This process we call "appropriation" may be a key in accounting for the construction of new more complex understandings. Vygotsky's theory is important for focusing on the constructive processes in interaction located in what he calls the "zone of proximal development," that is, the supportive interaction that allows students to tackle problems beyond their individual capacity. Within this zone, the teacher may appropriate a student action in order to have it play a role in the task as the teacher understands it.

A Tutorial on Combinations

I can illustrate appropriation with a task that asks students to find all the pairs in a set of objects. Piaget & Inhelder (1975) used this task in their studies of children's development of formal understandings. We worked with 4th graders who were generally unsystematic in their unaided approach to the problem. As part of a larger study of the relation between classrooms and laboratory settings (Newman, Griffin, & Cole, 1989) we had students go one at a time to the library corner of the classroom where our research assistant presented them with a task involving stacks of little cards, each stack with a picture of a different move star on it. Their job was to make all the pairs of movie stars that he could with no pairs the same. This is a relatively difficult task for 4th-graders to solve systematically. Typically, children of this age go about the problem "empirically" by which Piaget meant that they make whatever pairs they can think of in no particular order until they cannot think of any more. But they lack any sense of certainty that they have all that are possible or even the sense that there is some definite set that constitutes "all." A strategy that is more typical of adolescents is what Piaget calls "intersection." In this method, the child systematically pairs 1 & 2, 1 & 3, 1 & 4, 2 & 3, 2 & 4, 3 & 4, as illustrated in Figure 9.1. When they get to the end of the series they have a clear sense that they have all the pairs. Our sample was consistent with Piaget's age norms in that only 3 out of the 27 students in the class used the intersection strategy on the first trial.

Our interest, however, was in teaching the intersection strategy, not in testing the students' operational level. There were three trials in which the tutor asked the student to make all the pairs. Each trial added another item so the initial trial used four stacks of cards, the second used five stacks, and the last trial used six stacks. Between each trial the researcher conducted a little tutorial in the form of a checking session. When the student made all the pairs he could, the tutor asks, "How do you know you have all the pairs?" The student usually answers with a hint of frustration, "I can't think of any more." The tutor then asks, "Could you check to see if you have all the pairs?" The student usually says little and the tutor says, "Well, I have a way to check. Do you have all the pairs with Mork (or the first star on the left)?" From there she proceeds through a checking procedure asking about each star in relation to every other star, going systematically from left to right across the row of stacks of movie-star cards. As soon as the student got the idea of the checking procedure, the tutor let him complete it on his own.

The tutorial was quite successful considering that 17 students (out of 27) used the intersection strategy in the second or third trials. This is in spite of the fact

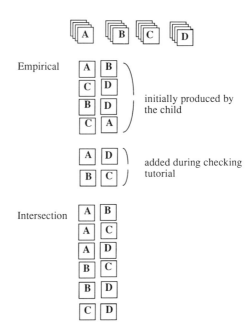

FIG. 9.1. Two solutions to the
combinations task.

that the strategy for producing pairs was never directly taught. What the tutor did was appropriate the column of pairs the student had produced and talk about it in terms of the strategy she had in mind. The tutor's question, "How do you know you have all the pairs?" presupposes that the student was trying to get *all* the pairs. This may have been a false presupposition but it was strategically useful. The question treated the student's column of pairs *as if* it had been produced in an attempt to get all the pairs. The tutor then invoked the intersection procedure as a means to fix up the student's "failed attempt to produce all the pairs." That is, she appropriated the student's pair-making, making it a part of an example of how to achieve the stated goal. It appears that when their own empirical production of pairs was retrospectively interpreted in terms of the intersection schema, students began to learn the researcher's meaning of "all the pairs" and thereby discovered the goal considered as a definite set that can be specified even before pair-making is initiated.

The process of appropriation can be specified in terms of four of steps:

1. An "expert" and a "novice" are engaged in joint activity each with their own understanding of the task in which they are engaged.
2. The novice makes some move that arises out of his own understanding.
3. The expert appropriates that move or attempts to appropriate that move into the system as she understands it and provides feedback about the move in terms of the expert system.

4. The novice comes to understand, in retrospect, what his action meant in the expert system and learns something about that system.

This process depends on the possibility of new interpretations arising in retrospect. The student's action does not have a single meaning for all time. Rather the meaning emerges in the interaction. The process also depends on the student creating some external product that can be reshaped in the interaction. The teacher does not just state his own understanding. Rather, the understanding is shown in relation to the student's action. Furthermore, the teacher may not understand the student's view of the action just as the student, initially, may have little idea of the task that is the teacher's framework. The teacher may not even be explicitly aware of how the process is working or that the student is not initially understanding the task in the same way.

The student is not alone while in the process of discovering the new interpretation in most instructional interactions. Although step 4 above is phrased as an individual cognitive process, the process is often happening in the context of an interaction that the teacher constructs as part of step 3 (Newman, in press). The checking procedure, during which the student's pair making was appropriated, was itself a complex interaction in which the student answered a series of questions structured by the tutor. The student engages in an interaction that presupposes a particular interpretation of his previous actions. Thus the process of coming to understand what the original action meant is, itself, supported in an instructional interaction.

A Lesson on Division with Remainders

Another example of appropriation is found in a set of 4th-grade math lessons analyzed by Andrea Petitto (1985) and, with her collaboration, reported in Newman, Griffin and Cole (1989). The topic was division with remainders that involved problems like that shown in Figure 9.2. The lesson on division with remainders, which was taught to each of the five small "ability groups," was interesting because the explicit lesson did not match the intended outcome. The "expert" performance of long division usually begins with an estimate of the quotient (answer) which is then multiplied by the divisor. Then the product is subtracted from the dividend to check the correctness of the estimate. Successive approximations may be used before a quotient is decided on. In the more complex versions of the algorithm with multi-digit divisors, an estimation process

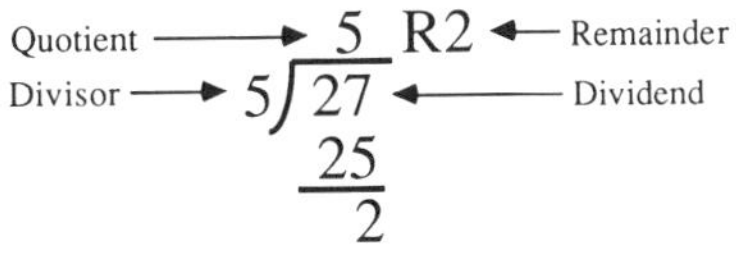

FIG. 9.2. Simple division with remainders.

becomes essential. Estimation is difficult to teach, however, before the goal that the estimate is attempting to approximate is fully understood. In approaching the problem, the teacher began with a mechanical version of the algorithm and allowed the successive approximation procedure to emerge in the interactions around practice problems. Not all the groups successfully learned the target procedure, however.

Instruction for all the groups began with a procedure to find "a number times 5 that has an answer that is close to 27 but doesn't go over it." With an example problem on the board, the teacher began to write the "times table" in a column from 5×1 up to 5×6. At that point it was observed that the answer had gone over, so the number must be 5. This procedure was a means to get to the concept of the constraints on the number and was not intended as a procedure to be learned. Once the group of students had solved several problems with the teacher on the board, they began work on individual worksheets. At this point, at least for the higher-ability groups, the discourse changed. The teacher watched the students as they solved the problems and helped out whenever they had difficulty. Often, the teacher commented on a solution where the quotient chosen was too high or too low. The language used now was that of adjustment rather than of definition and procedure, for example, "Ok, so I think maybe you went too high." Once the students got working on their own problems, the teacher's successive approximation framework began to guide her comments. For most of the groups, without any explicit instruction, the initial mechanical procedure was dropped and a procedure involving jumping in with some number emerged.

Both these cases, combinations and division, illustrate a basic feature of appropriation that ties it to Vygotsky's (1978, 1987) concept of the zone of proximal development. The zone can be understood as an interaction that the teacher constructs in order to engage students at a level above that which they would be able to perform without assistance. Not everything students do can be appropriated into the zone. That is, the actions have to interpretable (appropriate) within the teacher's framework. A response that cannot be interpreted indicates that the teacher has not created a zone in which the student can work.

In one of the combinations tutorials, for example, the student began the task by making a row of cards instead of a column of pairs as the tutor expected. The tutor quickly stopped him and explicitly instructed him to make a column of pairs. Neither we, as observers, nor the tutor understood what he was trying to do. But it was clear that he was not engaged in the task in a way that would make subsequent discussion useful. The tutor had to backup and try to reconstruct the task so that the student played a minimally acceptable role. Similarly in the division lesson of the bottom math group, students occasionally gave answers that were so far off the mark that the teacher could not talk in terms of adjusting them up or down. In these cases, she backed up to the explicit procedure essentially starting the lesson from the beginning. At any point in cognitive change there are a set of activities that students can engage in when the context is structured by a teacher. Being engaged in the task means being able to produce some action that can be appropriated by the task. It is not always possible for the teacher to structure the task so the student's action can be appropriated. Often, adjustments must be made so that students can participate at an appropriate level. Once that is achieved, the students are in a position to see how their actions are reconstructed by the teacher's appropriation.

Discussions of Vygotsky's theory often focus on internalization because this is the process apparently by which cognitive change occurs. But internalization can only occur in the context of externalization as can be observed in many aspects of appropriation. First, the teacher externalizes a frame of reference as in the checking tutorial and the lesson discussions. Second, the student's "mind" is engaged through engaging its externalized products. Third, the construction of new understandings occurs "inter-psychologically" removing some of the burden of cognitive construction from the internal processes. Internalization is not an isolated event but occurs in the context of socially constructed interactions in which both sides have put something on the table to negotiate with. And internalization is not the end point either, because once new understandings have been achieved, new tools, products, and frameworks can be put on the table.

The teacher can externalize a piece of social organization that is more complex than a brief interaction. The social context serves as a tool for the teacher and the basis for reconstruction based on retrospective interpretation. Classroom level organization can also be used as such a tool taking advantage of collaboration among groups of students to set up a framework for reinterpreting the group products.

SOCIAL CONTEXT AS A FRAME FOR APPROPRIATION

Given the notion of socially supported cognitive reconstruction, we can move to the second part of the argument and begin to examine how the process can be applied in teaching larger groups of students. A group of students can tackle a more complex task than an individual student. If the larger task can be used as an interpretive frame, then the same process of appropriation can be applied simultaneously to several students or groups of students. The point is not just greater efficiency. The desired lesson may be contained in the larger task which can only be undertaken when distributed among a group of students. This larger task, for example an experiment that involves several variables distributed among the students or small groups of students, later can become the interpretive frame for the students' separate actions.

In a 4th-grade lesson about chemical indicators, which we designed and conducted in collaboration with the classroom teacher, we distributed the various tests to be done among the students so that each pair of students had to keep in mind only two of a larger set of chemicals that were to be tested using two different kinds of indicators (Newman, et al., 1989). Our interest was both in teaching about chemical properties (i.e., properties that can be discovered by their interactions with other chemicals, in this case red cabbage solution and iodine) and in finding out whether students could use a record sheet to systematically record the results of their tests. (The combinatorial aspect of the task was to be related to the earlier task in which we taught a systematic strategy for getting all the pairwise combinations.) Our initial plan was to conduct the lesson with each of the five math/science groups in the course of one morning. The lesson did not work quite as planned for a number of reasons.

First, although we paired up students in each group, each pair actually divided the work up further so that each student worked with only one chemical, testing it with the two indicators. This was a very obvious solution to the problem of managing the 2 by 2 pairing of chemicals that they were assigned. It simplified the memory and recording task for the students. But as a consequence the "cognitive" task we thought we had given them disappeared into the social context.

Second, the lesson was too long to fit into the 25-minute time slots for the small group lessons. Consequently, the critical summary phase of the lesson covered. We regrouped over lunch and arranged a whole-class lesson that would allow for a tallying and discussion of the findings. The importance of this final phase cannot be overemphasized. It is the "colloquium" of the investigation/colloquium model (Lansdown, Blackwood, & Brandwein, 1971); the phase in which the investigation activity can be summarized and discussed. It is the activity that provides for the appropriation of the individual data into a larger frame (see also Newman, 1989, for discussion of teacher organized reflective activities). The fact that this phase was done as a whole group lesson may have added some interest, because in the larger group some conflicting results emerged and could be discussed. In any case, the data from the whole class was displayed on a large chart and students were able to see the place of their own data in the large scheme incorporated in the whole set of chemicals being tested.

While the students were mixing their chemicals with the indicators, they were quite unaware of the phase of the lesson in which all the results were to be discussed. For the most part, however, the mixing activity made sense as an activity in its own right. The students were interested in what happened to the chemicals and were learning a number of things about handling the apparatus. We did notice, however, that the teacher made an extra effort to remind students to write down their results. Because the need to report the results later was not entirely understood by the students, there was no intrinsic reason for careful recording. It is important for students not to have to know the whole story when they start an activity because it would be overwhelming and likely uninterpretable at the outset. The whole story, however, is likely to put constraints on the first activity, such as consistent record keeping, that have to be "scaffolded" by the teacher. As much as possible, the first activities should be organized to be meaningful on their own in order to minimize the intrusion of requirements external to the first task.

The chemical properties lesson illustrates a feature that allows us to think about group instruction in positive terms rather than as a poor approximation of an individual tutorial. The group situation makes it possible for an individual to be engaged in a task that is larger than would be possible if he were working on his own. This observation goes beyond the clearly related issue of scaffolding which is often thought of in terms of one-on-one teacher (or tester)-student interaction (Wood, Bruner, & Ross, 1976). It is not just that the student is being helped to carry out a task that he is in the process of learning how to do. Rather the student is being helped to carry out a part of a task that is socially distributed among his peers. The investigation of the data then provides the interpretive frame. The individual contribution is appropriated into the larger picture and becomes abstracted in terms of its relations to the other data points.

The point being made is also very different from the arguments for cooperative learning (Johnson, Johnson, Holubec, & Roy, 1984). The very beneficial effects of cooperative learning are usually demonstrated without teacher-group interactions involving appropriation. Usually, the teacher's role is to set up the goal structure and then let the peer interactions take their "natural" course. As shown in our chemical properties lesson, there is much a teacher needs to be aware of in order to design the group task so that it does not become just a set of parallel individual tasks in which the students are simply competing for resources. The appropriation process, however, involves structuring that comes into play *after* as well as during the main action. In this way, it is somewhat independent of cooperative learning groups. That is, the results of either individual or group actions can be appropriated into the larger structure. The additional advantage of cooperative over individual work that has been demonstrated in other studies suggests that a 2-tiered arrangement may be optimal. The "individual" contributions may actually be done by small groups working cooperatively and even the final colloquium discussion might be conducted "intergroup" rather than among individuals.

When we begin considering complex task organizations such as a distributed experiment, it is difficult to maintain a concept of social context as background for individual learning. The individual is part of the context and the cognitive work is distributed. I have argued for the advantages of group organization over individual tutorial on the basis that distributing a larger task makes it possible to consider a wider range of data and a richer set of comparisons. The individual contribution, which the student or small group might view initially in terms of a very elementary comparison, can later be reinterpreted in terms of the larger set of comparisons. The larger group discussion or analytic activity is critical in this process because without this discussion the investigation of the parts into the larger lesson is not possible.

DESIGNING TECHNOLOGIES
FOR GROUP INSTRUCTION

The goal of creating socially organized interpretive frames suggests a new role for computer technology in the classroom. Instead of just designing systems for individual learners, systems can be designed to coordinate the work of individuals or groups and to assist the teacher in synthesizing their contributions. Supporting such coordinated investigations with computer technology requires analysis of the social organization of the activities being supported as well as an analysis of what aspects of the overall process of appropriation the technology is in a position to support.

In a recent project, a group of us at Bank Street College developed and studied a local area network (LAN) system designed for elementary school earth science (Newman & Goldman, 1988; Newman, Goldman, Brienne, Jackson & Magzamen, 1989). The purpose of the technology was to provide communication links among the students and teachers in different classrooms, to help integrate the science instruction with the rest of the curriculum, and to provide means for data to be shared among groups of students in the school. The technology was used in a special earth science curriculum as well as in related curriculum areas such as writing and social studies.

The Earth Lab system consists of a network interface and modifications of existing tool software including the Bank Street Writer, our word processor that was enhanced with an electronic mail system and the Bank Street Filer which makes it possible for students to create databases that could be accessed from any computer in the school. One of our sites is a public elementary school in New York City serving about 600 students in 4th through 6th grades. The current network configuration consists of 45 Apple IIe's and GSs connected via a Corvus network. Computers are spread throughout classrooms as well as in a computer lab concentration.

The Earth Lab system is unique among school networking systems in that it is designed for communication and sharing of data. In most other systems, the central resource of the hard drive is used primarily to store programs; the only data that is stored is information on student progress and other management information. Our decision to use the hard drive to store *students'* data (data files, stories, email, etc.) combined with our interest in supporting the distributed group structure described in the previous section called for a different kind of design. For example, instead of being presented with a menu of programs, students see a menu of the various project groups they are involved with. In this way, the system focuses on shared data that the group is working with. Our initial hypothesis for such a design was the basis for a year long field test. Research is continuing this year and has provided several cases that illustrate how the social distribution of the group work can be supported by technology (Newman, Goldman, et. al, 1989; Newman, 1988).

The networked database made it quite natural to create a single set of data over an extended period of time. For example, one of the earth science activities was about weather. Each day, one of the science groups took a turn going to the roof to collect data on temperature, barometric pressure, wind speed and direction, and cloud type from a weather station. They returned to the computer lab and added the data to a common database stored on the network. After several months, over 60 records had been collected. In the process, the students had learned about measurements, compass points, and other aspects of data collection. Concurrently, in the science lab, they were learning about the properties of air, the causes of wind, and the water cycle. Toward the end of their unit on weather, they were able to analyze the weather data that had been collectively amassed. Each group worked with the database, examining relationships, for example, between pressure and cloud type or wind direction and temperature. This second activity built on the product of the first activity and extended its meaning. That is, while collecting data, students are engaged in an activity which is interesting in its own right and from which they can learn a substantial amount. The analysis activity, now shows their data in a new light since the correlations would not have been apparent on any one day. The way data was organized in the Earth Lab system can thus be seen to support the 2-tiered design in which small groups contribute to a class-wide project.

In some respects, however, the initial Earth Lab design fell short of anticipating ways that LAN technology can support this organization of instruction (Newman & Goldman, 1988). For example, an electronic mail system, which was used quite extensively for both within-school and long distance communication by teachers and students, was designed so that one could send a message to an individual or to a group that had an "alias" defined for it. A message sent to an alias caused copies to be send to everybody on the group list. This is a common feature of email systems in business and research settings. In the Earth Lab school, however, cooperative groups generally worked together at a computer. When a message was sent to the group, each individual received a copy which, of course, meant that there was unnecessary duplication and confusion about which messages to the group were read and where they were saved. A more appropriate scheme, though highly unusual for email systems, would have the co-present cooperative group send and receive mail as though it were an individual. This example illustrates that developing technology to support the organization of classroom science investigations is not a trivial matter. The technology used in research settings cannot directly be used as a model.

The discussion of electronic mail should not imply that classroom discussion in which the teachers appropriates the various contributions into a larger structure can be replaced with written communication. We suspect that this synthesis will require the immediate give-and-take of face-to-face interaction. It will also be facilitated by common reference to blackboards or charts which appropriate the individual contributions. The role of the technology is found during the investigations themselves where teacher-group, group-group communication is useful in supporting the social organization of the distributed task (Goldman & Newman, in press). Technology considered as an extension of the teacher's organizational work can help coordinate the various parts and bring a common product onto the table for discussion.

CONCLUSIONS

Socially distributed cognitive activity is an extension of the idea that we create tools as external devices to support and actually to be part of our cognitive activities. However, the social context introduces some unique features that are not as apparent when considering other kinds of tools. When the external device is a piece of social organization, a reflexivity is introduced into the system that provides for retrospective interpretation. Actions can come to mean new things when introduced into a new social system. The appropriation of actions into new systems of meaning provides an important mechanism for cognitive change, and it is the key to how we can use the social context for instruction.

Within the process of appropriation, the students are not passive recipients of knowledge. The students must be active producers of the externalized actions that can enter into the larger system. Their actions are the material from which the lesson is constructed. At the same time the students are appropriating the conceptual tools provided by the teacher and using them in their own system of actions. This reciprocal appropriation in the instructional interaction means that both sides are externalizing products that can play a role in each others system of understanding, and both can learn from how their products are interpreted and used in the other system.

The social context is not a passive background but is put out there as part of the distributed cognitive activity. As research on educational processes becomes increasingly interested in situated learning, that is, in the social organizational context within which learning functions, it will not be sufficient to view the context as the background for thinking. What is called for is an expansion of the unit of analysis to include the socially organized activity. The process includes the externalization of thought into the actions, tools, and social organization that support and reorder thinking as much as it includes the internalization of the new understandings that result.

Technology can be designed to take advantage of this process. But it requires a new category of software that goes beyond the ordinary assumption of the student-computer interaction. Recently progress has been made in designing learning environments that provide the student with reflective power (Collins & Brown, 1987). That is, programs are able to display for students the paths they took in arriving at a solution. In this way students are able to view the externalized products of their own cognitive activity. The next phase of work anticipated in this chapter is to develop systems that move reflectivity up another notch and reinterpret these products in terms of frameworks that embody different interpretations or that show the product as an important part of a larger whole. This process is a very ordinary result of social interaction where the same action can be viewed from different points of view. Computer systems can be designed that assist the teacher in displaying for students the alternative points of view and meanings of their actions.

REFERENCES

Collins, A., & Brown, J. S. (1987). The computer as a tool for learning through reflection. In H. Mandl & A. Lesgold (Eds.), *Learning issues for intelligent tutoring systems*. New York: Springer.

Dore, J., Gearhart, M., & Newman, D. (1978). The structure of nursery school conversation. In K. Nelson (Ed.), *Children's language* (pp. 337–395). New York: Gardner Press.

Fox, B. (1987). Interactional reconstruction in real-time language processing. *Cognitive Science, 11*(3), 365–387.

Goldman, S. V., & Newman, D. (in press). Electronic interactions: How students and teachers organize schooling over the wires. *Discourse Processes*.

Johnson, D. W., Johnson, R. T., Holubec, E., & Roy, P. (1984). *Circles of learning: Cooperation in the classroom*. Association for Supervision and Curriculum Development.

Lansdown, B., Blackwood, P. E., & Brandwein, P. F. (1971). *Teaching elementary science through investigation and colloquium*. New York: Harcourt, Brace, Jovanovich.

Newman, D. (1988, September). *Sixth graders and shared data: Designing a LAN environment to support cooperative work*. Paper presented for the second conference on Computer Supported Cooperative Work, Portland, OR.

Newman, D. (in press). Cognitive change by appropriation. In S. Robertson & W. Zachary (Eds.), *Cognition, computation, and cooperation*. Norwood, NJ: Ablex Publishing.

Newman, D., & Gearhart, M. (1978, March). The nursery school child's contributions to the social context of teacher-child interaction. In W. Russell (Chair), *Discourse Processes*. Symposium conducted at the annual meeting of the American Educational Research Association, Toronto, Canada.

Newman, D., & Goldman, S. V. (1988). Supporting school work groups with communications technology: The Earth Lab experiment. *Children's Environments Quarterly, 5*(4), 24–31.

Newman, D., & Goldman, S. V. (1987). Earth Lab: A local network for collaborative classroom science. *Journal of Educational Technology Systems, 15*(3), 237–247.

Newman, D., Goldman, S. V., Brienne, D., Jackson, I., & Magzamen, S. (1989). Peer collaboration in computer-mediated science investigations. *Journal of Educational Computing Research, 5*(2), 151–166.

Newman, D., Griffin, P., & Cole, M. (1989). *The construction zone: Working for cognitive change in school*. Cambridge: Cambridge University Press.

Petitto, A. L. (1985). Division of labor: Procedural learning in teacher-led small groups. *Cognition and Instruction, 2*(3&4), 233–270.

Piaget, J., & Inhelder, B. (1975). *The origin of the idea of chance in children* (C. Leake, Jr., P. Burrell, & H. D. Fishbein, trans.). New York: Norton.

Vygotsky, L. S. (1978). *Mind in society: The development of higher psychological processes* (M. Cole, V. John-Steiner, S. Scribner, & E. Souberman, Eds.). Cambridge: Harvard University Press.

Vygotsky, L. S. (1986). *Thought and language.* (A. Kozulin, Ed.). Cambridge, MA: MIT Press.

Wood, P., Bruner, J., & Ross, G. (1976). The role of tutoring in problem solving. *Journal of Child Psychology and Psychiatry, 17,* 89–100.

10 Adolescent Social Categories—Information and Science Learning

Penelope Eckert
University of Illinois at Chicago
and Institute for Research on Learning

Boy 1:   But I just hate to see people, you know, you get a easy class where you just got to write a few things, and they won't do it. Unless, you know, they feel that they're not going to college, so what's the use. Because I, uh, I do have that thought in my mind. If you're going to go to college then open your book and, you know—and if I wasn't thinking that maybe I'd do the same thing. I'd, I'd drop my math. And sciences.

Interviewer: But is there, is there pressure to do that? To drop a math class and stuff like that?

Boy 1:   Yeah. I'm not taking it next year. Whether that's pressure or smart, but—especially senior year, and, uh, I don't really need it.

Interviewer: So you—you'd rather have a good time? Is that, is that what you mean by pressure? I mean it's not like pressure not to take these things?

Boy 1:   Yeah. Just, you know, what's—what's the purpose?

Boy 2:   I just try to keep away from school as much as possible, you know.

Interviewer: Really. Why?

Boy 2:   I hate school. I really do. I really don't—school really is not what it's supposed to be. They don't teach you what you need to know. They just teach you a bunch of bullshit.[1]

The above quotes are from two boys in the same high school graduating class. The first boy is a typical good student. He participates in school activities, does well academically, and is cooperative and friendly with the teachers. The second is the academic teacher's nightmare. He dislikes school, avoids contact with teachers, is uncooperative and truant, and smokes and uses drugs with his friends on school property. Yet while the two quotes show clear differences in attitude toward school, they show the same lack of interest in what actually goes on inside the classroom.

These two individuals can be taken as prototypes of two social categories that appear in public high schools across the country, and that are associated with opposing orientations to school. The speaker in the first quote is a member of the category called "Jocks"—a category of students who are largely college-bound and who center their lives around the school. The speaker in the second quote belongs to the category called "Burnouts," comprised of students who are largely blue- and pink- collar bound, reject the school as the locus of their social lives, and orient themselves to the neighborhood and the local area. Understanding the social dynamics that underlie the formation and maintenance of these categories is essential to any effort to reclaim the intellectual potential of American youth. And the responsibility for these categories lies directly with educators themselves, for category affiliation is not simply a choice that the individual makes from day to day, but the result of something resembling a tracking system fostered by the organization of the school itself (Eckert, 1989).

The following discussion, based on three years of participation in schools in the Detroit suburban area,[2] describes the social dynamics that give rise to categories like the Jocks and the Burnouts, and that lead the Jock and not the Burnout into the science classroom. Much of the discussion will apply to most of the academic curriculum. However, science and math are popularly perceived as the

---

[1]Quotes are from tape recorded conversations with adolescents.

[2]Two of these years were spent in one school following one graduating class through their junior and senior years. The third year, which was divided among other schools representing a range of socioeconomic populations, confirmed the findings of the two-year deep ethnographic study. The student bodies of these schools were predominantly white, following the highly segregated residence patterns of the Detroit area, and while a variety of European and Middle Eastern ethnic groups were represented in these schools, the predominant division among students was along simple class lines. This setting, therefore, provides a set of social dynamics that may be intensified, but which are similar in nature to ethnically more heterogeneous schools. This research was supported by grants from the National Science Foundation (BNS 8023291), the Spencer Foundation, and the Horace Rackham School of Graduate Studies at the University of Michigan.

ultimate in academic pursuit and, as a result, are the most clearly affected by these dynamics.

While what is most striking in the discussion is the way in which social categorization alienates Burnouts from the academic classroom, it is important to point out that they are not the only ones who lose intellectual opportunities in the high school. Intense involvement in the peculiar social dynamics of the high school prevents both sets of students from any more than peripheral involvement in intellectual activity. While the Jocks are far more likely than the Burnouts to appear in the science classroom and to achieve respectable grades once there, this choice is a result of social dynamics unrelated to the individual's interest in what can be learned in that classroom. And although the Jocks' social orientation leads them to participate in the academic activities of the school, they perceive these activities as annoying and secondary to the social business of school. This attitude toward the curriculum is fostered by the school's own emphasis on extracurricular activities. It is not surprising, then, that many Burnouts see their own open hostility to the academic curriculum as simple honesty and see the Jocks' academic cooperation as sycophantic.

SOCIAL POLARIZATION IN THE HIGH SCHOOL

In public secondary schools across the country, school orientation corresponds to a clear socioeconomic difference. Those students who center their lives around the school, participating in extracurricular activities and performing successfully in academic programs, are by and large from the upper end of the local socioeconomic hierarchy; those who are alienated from curricular and extracurricular activities are by and large from the lower end.[3] On closer examination, the statistics of school participation and success translate (in virtually all public high schools) into two opposed and polarized social categories that mediate adult socioeconomic class for the adolescent age group. In the Detroit area, as in many other areas across the country, people who center their social lives in the school's extracurricular sphere are called (and call themselves) Jocks, while those who reject the school as a basis for social identity are called (and call themselves) Burnouts. It is important to note that while the term Jock normally means athlete, in these schools a Jock is anyone who is involved in school activities. In the same way, while the name Burnout refers to those students who use drugs, there are Burnouts who do not use drugs. While the names of these categories, and the specific styles that signal their opposition (clothing, musical tastes, territorial specialization, etc.), change through time and between regions and localities, the fundamental status of this opposition is close to universal in our culture.

---

[3]This division shows up in the early high school studies. See Hollingshead (1949) and Coleman (1961). The pattern of difference is intensified in the minority population (Coleman, 1966).

One can characterize the opposition in any number of ways. While the exact differences may vary from school to school, the opposition in the school where this work is based is typical. The Jocks are overwhelmingly in the college preparatory curriculum, and the Burnouts are in the vocational curriculum. The Burnouts smoke and Jocks do not. The Burnouts are fairly open users of hard and soft drugs, while the Jocks are only open about alcohol use. The Jocks dominate extracurricular activities, while the Burnouts avoid such involvement. The Jocks favor preppy dress styles, while the Burnouts war jeans, rock concert tee shirts, and a variety of urban dress symbols. The Jocks listen to mellow rock, while the Burnouts listen to heavy metal. The Burnouts hang out in the school smoking area and the Jocks do not.

The differences between Jocks and Burnouts are obviously too great and too mutually opposed to be accepted with equanimity. The Jocks' and Burnouts' social norms are embedded in diametrically opposed attitudes toward school, and the behavior of the students in each category threatens the balance that the other is trying to maintain. The Burnouts' rebellion threatens the trust that Jocks work to establish with school staff, while the Jocks' acceptance of the school's authority threatens the Burnouts' fight for personal autonomy. As a result, there is a considerable hostility and, in some schools, violence between Jocks and Burnouts. This hostility feeds the oppositional nature of the two categories, so that almost any aspect of behavior becomes symbolic of category affiliation: dress, territory, musical taste, language, activities, substance use, and—of particular salience to the current discussion—curricular choice.

Headed for the blue collar workforce, Burnouts constitute a large proportion of enrollment in vocational curriculum, while the Jocks populate the various college preparatory and advanced placement (AP) classrooms. This division is shown in Fig. 10.1 which shows, for each category, the degree of involvement in different kinds of courses. These figures are based on enrollments of 102 students in one graduating class during the winter semester of their junior year. For each social category, the number of enrollments in a given kind of course is divided by the total number of enrollments in courses overall. The division into upper, middle, and lower level courses (A, B, & C) within each discipline is based on values assigned in the school catalog. These figures show a clear division between Jocks and Burnouts, with Burnouts specializing in vocational courses (these include clerical courses) and Jocks in college preparatory courses. In addition, within each academic discipline, the Jocks populate the middle- and upper-level courses, while the Burnouts populate the middle- and lower-level courses. This specialization is based not simply on different interests and aspirations, but on a variety of social dynamics related to social polarization. Jocks and Burnouts strive to avoid contact with each other and to avoid activities of any sort that are associated with the other category. This polarization, along with the common preference for high school students to take classes with their friends, encourages the curricular specialization that accompanies category affiliation.

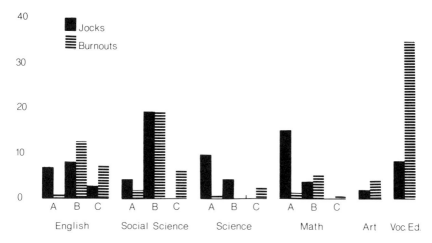

FIG. 10.1. Degree of involvement in various courses—Jocks vs. Burnouts.

This specialization is, in turn, related to a differential contact with faculty. It is well known that the vocational curriculum is generally marginalized in the public schools, and that those participating in it are themselves marginalized (Stinchcombe, 1964). The most obvious aspect of this marginalization is the overwhelmingly academic mission of the school, which denigrates the value of vocational subjects; and the middle class orientation of the school, which denigrates the working class norms associated with these courses, their students, and frequently their teachers. Indeed, the structure of teacher networks frequently shows a separation between vocational and academic teachers, and generally it is the latter who supervise extracurricular activities (at least the nonathletic activities). Academic teachers' attitude towards Burnouts is affected not only by Burnouts' attitudes towards their subjects, but by the loss of status that academic teachers assigned to "general" courses suffer within the teacher hierarchy. The vocational teachers by and large regard Burnouts as simply an important part of their student constituency and, for a variety of reasons, their experiences with Burnouts are more rewarding than are those of the academic teachers. This has obvious consequences for the Burnouts' attitudes towards the various subjects. But, in addition, it also has important consequences in the further development of teacher attitudes.

The gossiping that takes place within teacher networks becomes socially specialized. Insofar as teachers tend to gossip about their own students, vocational teachers will gossip about Burnouts. This gossip is a means of transmitting important information about students' personal or family problems, their interests, sensitivities, or their aspirations. Academic teachers, on the other hand, are privy to less constructive gossip about Burnouts, both because they have less contact with them and because, for the most part, Burnouts will confide only in their vocational teachers. And since Burnouts are their "unwanted" students, academic teachers are more likely to exchange negative gossip about them and positive gossip about the Jocks. Finally, insofar as the science and math teachers tend to interact among themselves rather than with humanities teachers, positive gossip about Burnouts is most rare among them.

Territoriality is another factor that contributes to curricular specialization. Much of the school is divided up into Jock and Burnout territories. The most important territories are the leisure time spaces, where members of each category go before school, during lunch, and between classes. But, in addition, much of the teaching space is loosely recognized as territorialized. Jocks command most of the space associated with athletics and with extracurricular activities, while Burnouts command space associated with vocational training. While much of the classroom space is shared by students in both categories, Burnouts generally avoid areas around academic classrooms.

These invisibility factors are relatively minor in the greater range of factors keeping Burnouts out of the science classroom. However, they loom large in the individual's conscious rationale. If a student is to step into foreign academic territory, there must be some support. It requires great daring for a Burnout to go into a science or math classroom other than the few designated for vocational students. They do not know anything about the teachers, the teachers do not know them, and they do not know any of the other students. And clearly the latter is the most important. A Burnout going into a Jock class will be isolated from the other students, will sit on the edge of the room, and will have no one to talk to or study with. One Burnout told me that he had signed up for advanced algebra because he had always liked math, but after just a few days in the classroom he dropped it. He said that although he liked the teacher and the material, he felt out of place in the class, and while all the other students seemed to know each other, he had no one to talk to and feared he would have no one to work with when the going got tough.

It is not only the social isolation between Jocks and Burnouts that contributes to curricular specialization, but a far more subtle phenomenon. There are differences in social network structure and orientation to information between the categories that play a key role in students' attitudes to classroom learning and particularly to those classrooms seen as embodying certain social norms. These more subtle dynamics involved in the curricular specialization and in the differences in orientation to school lie in the subtle histories of the social networks and cultures that constitute the real differences between Jocks and Burnouts.

These social categories are not superficial; they are not simple aggregates of teenagers who like or don't like school. Indeed, school personnel tend to see them as "good" kids and "bad" kids or as those who "care" and those who "don't care." But underlying each category is a culture with its own history and its own orientation not only to school, but to the acquisition and use of information. Although the Jock and Burnout categories arise overnight upon entrance into junior high school, their polarization is the continuation of a process of social differentiation that begins in childhood, and an understanding of the cultural differences that underlie the Jock-Burnout split requires an understanding of this process.

NEIGHBORHOOD NETWORKS

The class differences between the Jocks and the Burnouts are well known in the school, and the Burnouts sometimes refer to themselves as poorer or not having as much. These class differences, however, are statistical rather than absolute, and a variety of other factors indirectly related to class enter into individuals' category affiliation. Nonetheless, the categories themselves are built upon social networks that extend back into childhood and that differ from neighborhood to neighborhood. The Jock and Burnout categories are class-based social prototypes, and these prototypes dominate adolescents' perceptions of social structure and identity. A social tracking system that leads children from family to neighborhood social network to school peer group to high school social category, and finally to an adult class membership matching that of their parents represents, perhaps, a path of least resistance. I will not discuss here the many forms that resistance takes; rather, I will focus on the aspects of this path that affect responses to information in general and, by extension, to the science curriculum.

It is well established that social life in the working class is neighborhood oriented. This is particularly true of old established neighborhoods where kin, work, and friendship ties frequently come together in residence patterns forming dense, multiplex social networks. In such neighborhoods, where Burnout networks have their origins, the intertwining of associations creates a sharing of interests that makes it both possible and necessary for residents to trust and rely on each other, and relative lack of material means necessitates a sharing of resources—temporal or material. Most crucial to this discussion is childcare. Neighborhood parents tend to cooperate in supervising children, in homes and outside in the neighborhood. The outdoor territory of the neighborhood—yards, streets, and parks—becomes the locus of neighborhood life where cooperation among parents provides regular but loose supervision. Another important source of supervision is among siblings where older siblings are responsible for younger ones and are expected to include them in many of their activities. The networks that form in the neighborhood, therefore, are built around friends, friends' siblings, siblings, and siblings' friends and include kids from various age groups. The children's networks that emerge from these neighborhoods are typically ascriptive, that is, as in a family, coresidence (in this case in the neighborhood) qualifies one for membership. With the intensity of peer activity outside the home in the neighborhood, a separate peer culture develops early in which children learn to share information, guidance, emotional support, and material resources. This egalitarian and open flow of resources frequently compensates for a lack of parental resources, and leads to strong age-group alliances, wherein children view themselves as an age-class in opposition to a class of parents rather than simply as members of individual child-parent sets.

The middle class neighborhood, home of the typical Jock, contrasts sharply with this picture. The more sparse child population that characterizes most affluent neighborhoods precludes the development of extensive neighborhood-based child networks. Furthermore, the orientation of the typical middle class adult away from the neighborhood taken together with the more insular nature of the middle class family, precludes the kinds of ties among parents that would provide regular supervision. Children are encouraged to spend more time in the home under parental supervision, or under paid supervision, and are expected to make a sharp distinction between family and neighborhood possessions, ties, and space. By virtue of their greater material resources, middle class parents can provide for their children's needs and desires in such a way that prevents the kind of exchange of goods, services, and support that occurs among working class children. Furthermore, middle class parents typically organize social and educational activities for their children and may even drive them around to play with children of acquaintances in other neighborhoods. As a result, these children do not develop the kinds of extensive local peer networks and strong ties that are characteristic of working class neighborhoods. Their parents are, to some extent, gate-keepers to the world outside the home.

Children from the two kinds of neighborhoods experience a very different transition when they enter elementary school. School is the first opportunity for many middle class children to develop independent extended networks—a place where they can freely meet and choose large numbers of friends on their own. Because the friends they make in school may not live near by, school is the primary locus of their independent interaction. School, therefore, represents independence and social opportunity for these children. To those from working class neighborhoods, on the other hand, school represents a disruption of social networks.

The age grading of the school determines that the friendship networks developed among the middle class children will be age-homogeneous. In addition, social age grading in schools is not simply a byproduct of educational practice, but is part of a school (and middle class parent) ideology that children should stick with others their own age. The networks developed in school by the middle class children, therefore, have the sponsorship of the school while children from working class neighborhoods are expected to redefine their networks in school and to gradually give up their age-heterogeneous associations. Thus not only does the school separate these children from their networks, the ideology stigmatizes working class networks. Parents' and teachers' arguments against age-heterogeneous networks in childhood are that older friends are a "bad influence" and that younger friends will retard one's development. Certainly there is some truth to the former and within a large age-heterogeneous network it probably cancels out the latter. By virtue of constant interaction with older children, many of the children in the typical working class networks are exposed to a variety of activities and tend to want to do what their older peers are doing. As a result, their own timetables are generally ahead of adult norms for childrens' development. Meanwhile, those from middle class neighborhoods remain, for the most part, within the general range of adult expectations for their developmental timetable.

The most obvious difficulty arising from age-heterogeneous networks and the consequent speed in development is that these children are more likely to get into trouble or at least alienate adults in positions of authority. As these children mature, their earlier contact with adult experiences and prerogatives (e.g., cigarettes, alcohol, drugs, sex, the law, and the emotional difficulties that may accompany these) creates a need for certain kinds of information earlier than adults are willing to give it to them. Information about such things as drugs, sex, birth control, legal rights, etc. is generally not openly available from adults until they feel that children "need" it. This means that from a fairly early age, adults cease to meet the most pressing informational needs of these children, and these needs are then being met, however inadequately, by older members of the peer group. The middle class child's homogeneous network, on the other hand, leaves its members dependent on adults for information of all kinds. But insofar as these childrens' developmental timetables correspond more closely (though rarely entirely) to adult norms, they will find adults to be more adequate purveyors of relevant information and will not resent their ultimate control of it. In fact, the delivery of information may function in a system of personal rewards. In school, the teacher comes simultaneously to share the middle class parent's role of appropriate and major source of adult-controlled information and the working class parent's role of arbitrary safekeeper and denier of adult information.

While class differences in social networks are well known (Bott, 1957) these differences have not been connected with class differences in school performance. But it is clear that the network differences described above would give rise to important class differences in approaches to information. Because of the greater autonomy of working class youth networks, a whole range of information, not simply stigmatized information, are acquired from friends rather than from adults. And no doubt the peer social context in which this information is imparted and shared has an important influence on attitudes toward that knowledge. With the complexity of peer relations within a heterogeneous extended network, information does not come from a single source, and it does not come from an authority figure who can or would apply norms or strictures on the acquisition and use of that information. Indeed, the free-flowing nature of this information no doubt predisposes members of such peer networks to prefer this kind of learning because it is available and geared to their needs, and to oppose the clearly vertical and one-way acquisition of knowledge from adults.

SOCIAL NETWORKS IN THE HIGH SCHOOL

The difference in network structure has clear consequences for individuals' receptivity to school learning. These consequences are intensified by the process of social polarization that takes place in the secondary school. As an age cohort approaches secondary school, the discrepancy widens between working class childrens' and adults' perceptions of informational needs, particularly since a variety of adult prerogatives come into play. At the same time, the difference in childhood social networks changes in significance upon entrance to junior high school, when a clear and intense social structure develops within the age cohort. As this structure develops, the relation of each individual to the school is interwoven with relations within the cohort, and what may in elementary school have been a matter between the individual and relevant adults becomes the material of social ideology within the peer society. The social networks arising in different neighborhoods emerge in junior high school as the basis of the Jock and Burnout social categories, and the fundamental differences in norms that characterize these networks become associated with radically different orientations to school.

The beginning of secondary school represents an official entrance into adolescence, accompanied by an emphasis on personal autonomy and separation from parental authority. This autonomy can only be found in a social structure other than the family unit, and the eagerness for autonomy coupled with the accompanying anxiety over separation from the family imbues the resulting "adolescent society" with a special intensity and rigidity. While the extended peer networks in working-class neighborhoods already provide an extensive social structure outside the home, the prototypical middle class adolescent is dependent upon the school to provide the locus for such a structure. Building on this need for autonomy through the performance of meaningful social roles, the school offers an unwritten three-way "contract" with students and parents. In this contract, parents relinquish to the school a certain amount of authority over their children, and the children agree to respect the quasi-parental role and ultimate authority of the school in exchange for the opportunity to participate in adult-like activities outside their parents' purview. These activities involve limited self–governance and a moderate amount of control over the conception and execution of social, business, and athletic activities within the school. Participation in these extracurricular activities brings a variety of rewards. It provides qualification for college entrance, it gives some control over the environment, and it is rewarded with a certain amount of freedom and privilege in the school; the value here derives from the restrictions inherent in any large institution and from all adolescents' search for autonomy.

Public high school activities comprise a corporate setting very much like that of the business world or, for that matter, the academic world. The school is a closed, age-graded community. Like a corporation, the school has clear external boundaries, expects loyalty and full participation within its walls, and engages in competition with other comparable institutions. The sharply defined population within these boundaries is hierarchically organized, with school personnel at the top controlling resources for education and for the development of extracurricular activities. Students are expected to compete for management of the extracurricular resources that teachers control and to build "careers" through the strategic distribution of these resources among the student body. Careers and personal identities are defined in corporate terms as a function of roles, offices, and influence within the extracurricular organization. The distribution of resources is facilitated by a student hierarchy defined by relative access to teacher networks and resources. The individual's place in the hierarchy is a function of concrete roles, of relations with teachers, and of the size and breadth of the individual student's constituency. Each role (cheerleader, student council member or officer, team captain, honor society president, etc.) has a fairly specific social value, and these values are cumulative. The well-managed individual career, then, involves strategic choices aimed at the greatest accumulation of value. In other words, activities within this hierarchical system constitute, in Bourdieu's (1977) terms, "symbolic capital." Within such a hierarchy, friends and associates also are an important component of one's symbolic capital, and they are chosen and dropped to some extent according to their hierarchical value.

Career building within the school's social system defines personal worth and accomplishment in terms of the individual's place in the hierarchy. It is important to note that extracurricular roles serve somewhat the same function as college board examinations, insofar as they provide a standardized means of evaluating the individual's "leadership qualities." Thus Jocks are told, and have reason to believe, that their adult career interests are identical to their high school career interests. Many Jocks participate in activities not because they are inherently interesting or entertaining, but because they are the means for career building. To a great extent, the available roles are set from the beginning, and Jocks are satisfied to compete for and fulfill these roles. In fact, it is the conventional nature of the roles that makes them valuable in the career system. As a result, Jocks tend to be extremely conservative about activities and practice little creativity in developing activities. Rather, they generally react to the discontinuation of a traditional activity as a threat to their high school careers. Innovations normally come from student activity directors who themselves have a career stake in their students pursuing certain activities learned about and valued in national educational networks. While Jocks obviously prefer engaging in activities that are personally satisfying and valuable, they are frequently more likely to engage in an activity for its value to their careers. The very existence of a hierarchy implies upward mobility, which sets the value of what is at the top.

One might want what is accessible at a particular point in a hierarchy not because of its inherent value but for its value in the hierarchy.

SOCIAL NETWORKS AND THE
VALUE OF INFORMATION

The Jocks' basic hierarchical view of social life, and the intimate place of this hierarchy in the school institution, leads naturally to a hierarchical approach to knowledge and information. The childhood acceptance of adults as sources of information, as previously discussed, feeds into the structure of the secondary school, where adults also become the sources of social resources and rewards. For most Jocks, a group frequently including the best students in the school, intellectual activity is inseparable from the social activities, although it plays a somewhat minor role in the school hierarchy. One could say that they typically acquire knowledge for its value in its hierarchical system, and the hierarchical usefulness of a particular bit of information determines its desirability. Within this framework, one might consider that scientific information might not capture the Jock's attention. An individual near the top of the Jock hierarchy would have use for gossip about the prom queen, but not about an insignificant ninth-grade "wallflower." In the same way, that individual would have greater use for information about chemistry originating with the chemistry teacher than from a nerd with a chemistry set. Information about the prom queen is useful because its display shows that one has access to such information. If it is negative, it can also be used to lower the prom queen's status, thus disturbing and freeing up the upper end of the hierarchy. Information about chemistry from the chemistry teacher can be advertised in situations (normally devised by that very teacher) in order to accumulate points in the academic area of career-making. Ultimately, the reason that information about chemistry in general is less in demand than information about the prom queen is its relatively low value in trade in the high school social hierarchy.

For the Burnout, functioning in an egalitarian network with mutual support as the trademark, information has no trading value. Rather, information—like cigarettes and personal possessions—is freely shared. Just as information must have career (i.e., hierarchical) value for the Jock, however, it must have some kind of value for the Burnout. For the Burnout, an important aspect of information-value is the degree to which it can provide solidarity and support within the peer network, autonomy for the network as a whole, and autonomy for its members. Information about such things as coping in the urban environment, resources for juveniles, drugs and birth control, and the legal system can all provide such autonomy. The Burnouts regard the Jock courses as centers of received knowledge of no great relevance. They perceive students in these courses as giving themselves over to adult authority and competing with each other as they perform for adults in order to win favor. Burnouts view vocational courses, on the other hand, as providing useful skills.

The characteristics of Jock and Burnout social networks and the norms governing information embodied in these networks dispose the members of the two categories to different responses toward teaching as it is commonly practiced in our schools, and particularly in the academic curriculum. While the Jocks are predisposed to accept information from a single hierarchical source and to compete in its display, the Burnouts are predisposed to prefer cooperative learning and to shun both the competitive and the authoritarian aspects of classroom teaching. In addition, the clash of these norms in the school leads to a social polarization that intensifies the two categories' differing orientations to the curriculum. The two sets of dynamics together conspire to keep Burnouts out of the academic classroom and, particularly, out of the science and math classroom except when these courses are compulsory.

It is notable that music and art courses, for the most part, escape the Jock and Burnout label, and it is also worthy of note that Burnouts in significant numbers choose art courses as electives. It is not simply relevance that makes these courses desirable, but the nature of learning in the vocational classroom, which is similar to that in the art classroom. The social dynamics in these classrooms are clearly more congenial to the Burnouts, as they do not fit the authoritarian stereotype of the Jock classroom. The reasons that Burnouts give for liking art courses indicate that the assumptions about learning in these courses do not conflict with the norms embodied in Burnout social networks:

1. Students can talk among themselves about what they choose while they are working.
2. The teacher's help is always constructive and provided primarily at the student's initiative. In other words, the teacher is a guide at the service of the student.
3. What students produce is a reflection of their active creativity and skill rather than docile receptivity.
4. There is no right way or ideal product; the goal is for the student to develop and use creativity, rather than to receive authoritative knowledge.

It is interesting to note that the well-run science laboratory does not conflict with Burnout social norms and in many ways resembles the vocational or art classroom more closely than the humanities classroom. In fact, Burnouts gave reasons similar to those above for having chosen a course in horticulture to fulfill their science requirement. The fact that Burnouts do not generally perceive the science classroom as congenial is no doubt because laboratory work is reserved for the later years in secondary school when social polarization has already taken its toll. Burnouts' experience in earlier years tells them that science and math are not only the most difficult, but the most authoritarian of the school subjects. Not only is there always a right and a wrong answer, but the right answer can only be learned directly from the teacher or the teacher's resources. This is the ultimate in received knowledge.

CONCLUSION

The previous discussion concentrated on differences in relation to knowledge that arise from the kinds of social networks and network histories that characterize the two major social categories. The Jocks, who generally profit in childhood from adult domination of resources and information, find positions of power in the asymmetrical flow of goods and services through participation in the school-based peer social hierarchy. Because they view the school simultaneously as the main locus of their social lives and as the main locus of learning, learning in school becomes part of a narrowly defined social system, and it has value primarily in the context of that system. They accept information as tokens in a hierarchy to be acquired and used within that hierarchy. This kind of global learning context could be seen as an obvious basis for the development of Perry's "received" knowers (Perry, 1970).

The Burnouts, by virtue of their social network histories, reject adults as sources of information. They see the school curriculum, therefore, as an ulti-mately authoritarian exercise. The school decides what they should know and forces it upon them without providing any motivation for its acquisition. In addition, the school's status as the legitimized provider of information for chil-dren and adolescents challenges not only the sources of information upon which Burnouts traditionally rely, but also better sources that they might otherwise consult. Therefore, the acceptance of the school's asymmetrical learning situa-tion would amount to a denial of their autonomy and a rejection of their own information sources.

What is particularly unfortunate is that by the time students reach high school, the Jocks and the Burnouts are all too generally perceived as representing good and bad, cooperation and rebellion, success and failure, intelligence and stu-pidity, respectively. Most of the Burnouts question all or some of these evalua-tions, but the power of the opposition and the school's acceptance of it ultimately leads to resignation. Rather than asking themselves how they can succeed in spite of the school, Burnouts discard goals along with the available means to achieve them.

REFERENCES

Bott, E. (1957). *Family and social network*. New York: Free Press.
Bourdieu, P. (1977). *Outline of a theory of practice*. (R. Nice, Trans.). Cambridge: Cambridge University Press.

Coleman, J. S. (1961). *The adolescent society*. New York: The Free Press of Glencoe.

Coleman, J. S. (1966). *Equality of educational opportunity*. Washington, DC: U.S. Government Printing Office.

Eckert, P. (1989). *Jocks and burnouts*. New York: Teachers College Press.

Hollingshead, A. B. (1949). *Elmtown's youth*. New York: John Wiley & Sons.

Perry, W. G. (1970). *Forms of intellectual and ethical development in the college years*. New York: Holt, Rinehart, & Winston.

Stinchcombe, A. (1964). *Rebellion in a high school*. Chicago: Quadrangle.

11

The Interplay Between Children's Learning in School and Out-of-School Contexts

Geoffrey B. Saxe
University of California, Los Angeles

As part of efforts to investigate social processes in cognitive development, researchers have found it useful to distinguish between two general types of social contexts for learning. Out-of-school contexts are those that are generated in everyday interactions, interactions not purposely contrived to foster learning. For a child, such contexts may include playing with games or building a toy model, and the learning that may occur in such contexts includes learning the consequences of violating a game's rules or learning about mechanical relations among a model's parts. School contexts are those specifically designed to further learning. In school contexts, a more knowledgeable individual arranges materials and communications in a way specifically designed to facilitate the learning of particular concepts that are potentially applicable to a wide variety of activities, concepts like government or the physics of gear ratios.

Research across a range of cultural settings has revealed that children develop fundamental intellectual skills as a result of participation in both school contexts (Rogoff, 1981; Stevenson, 1982) as well as out-of-school contexts, contexts that emerge in such cultural practices as weaving (Greenfield & Childs, 1977), pottery-making (Price-Williams, Gordon, & Ramirez, 1969), and tailoring (Lave, 1977). To date, we know little about the interplay between learning that occurs in school and out. In the context of the topic of this paper—mathematics—we can ask, does formal schooling in mathematics impact on the way children conceptualize and solve mathematical problems that emerge in informal practices? Conversely, does participation in a cultural practice in which mathematical problems emerge impact on the way children conceptualize and solve formal school mathematics problems? Clearly, answers to such questions would be of importance for an understanding of social processes in cognitive development as well as for the improvement of the practice of teaching.

A NATURAL EXPERIMENT ON THE INTERPLAY BETWEEN LEARNING IN OUT-OF-SCHOOL AND SCHOOL CONTEXTS

In the studies I describe below, I summarize the results of an investigation in which it was possible to take advantage of a "natural experiment" that bears on the questions outlined above. The site of the investigation is Recife, an urban center in Brazil's northeast. Public schooling is available to all children in Recife, though many poor children either attend school only for one or two years or do not attend school at all. In their everyday activities, many poor children attempt to generate an income through selling (in the streets) such commodities as candy, fruit, popcorn, vegetables, and shopping bags. The children that I targeted for study were candy sellers.

The conditions for a natural study of the interplay between learning in school and out-of-school contexts is remarkable in Recife because (a) same-aged candy sellers vary in their extent of schooling, and (b) schooled children vary in whether they have had street vending experience. Because schooling experience varies among same-aged candy sellers, it is possible to study the influence of schooling on the mathematical understandings children construct in the street. Conversely, because school children vary in their participation in out-of-school practices involving mathematics—some are vendors and some are not—the conditions in Recife also lend themselves to studying the influence of participating in an out-of-school context like candy selling on the mathematical understandings children construct at school.

In the following section, my concern will be to present first an analysis of the mathematical problems that emerge in the candy selling practice. Guided by this analysis, I then address the issue of whether or not and in what way schooling affects the way in which sellers conceptualize and solve these problems. Finally, I turn to the issue of whether candy selling experience affects the way children approach and conceptualize problems at school.

Mathematical Problems that Emerge in the Candy Selling Practice

In selling, venders address mathematical problems of some complexity. To gain insight into the character of these problems requires not only an analysis of the general structure of the candy selling practice, but also an analysis of the way a variety of social processes are interwoven with and support this structure.

The Structure of the Candy-Selling Practice

Candy sellers, virtually all of whom are boys, do not work for a boss, but regulate their business themselves. The inner rectangle in Fig. 11.1 depicts the practice's cyclical organization. In a *purchase phase* (right upper corner of inner rectangle), sellers buy boxes containing 30 to 100 units of candy from 1 of about 30 wholesale stores. These boxes are of many different types and include various types of chocolate bars, lifesavers, and lollipops. In a *prepare-to-sell phase,* sellers price their units of candy for retail sale. In the *sell phase,* the units are sold to people while they wait in line for buses, shop, or sit in outdoor cafes. In a *prepare-to-purchase phase,* sellers prepare for the purchase of a new box of candy, estimating what candy type is most in demand and coordinating those considerations with possible comparative pricing at different wholesale stores. During each phase of the practice, sellers must perform mathematical problems, the structures of which are interwoven with a web of social and economic processes.

Social Processes Interwoven with the Structure of the Practice

Various social processes are interwoven with the form mathematical problems take in each of the phases of the candy-selling practice. Consider three such processes depicted in Fig. 11.1: (a) macro socioeconomic processes like infla-

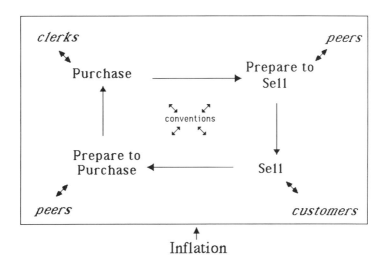

FIG. 11.1. The candy selling practice.

tion, (b) social conventions that have emerged over the history of the practice, and (c) social interactions that occur during the practice.

Brazil's runaway inflation is a social process that has influenced the structure of seller's mathematical problems both in terms of the magnitude of the values that sellers address in their everyday calculations as well as the denominational units of exchange. When data collection for this project began in the summer of 1985, the wholesale price for the least expensive box of candy was very large in magnitude, about 3600 cruzeiros (Cr$3600), and when data collection was terminated five months later, its value was even larger, Cr$5600. The government's effort to deal with this inflation rate has been to issue new currencies on a frequent but irregular schedule. For instance, just before the study began, the government issued a Cr$50000 bill; during the study, the government issued new coins in values of Cr$100, Cr$200, and Cr$500 and issued a new bill of Cr$100000; and, just after the study was completed, the government altered the currency system by eliminating three zeros from the cruzeiro and calling the new unit the cruzado. Thus, inflation itself affects the magnitudes of values that sellers address in their everyday arithmetic as well as the numerical units that are the objects of their computation in their exchanges.

A retail pricing convention that has emerged over the history of the candy-selling practice also is interwoven with the mathematical problems sellers address in their practice. The convention is to offer a variable number of candy units for a single bill denomination—most typically a variable number of bars for Cr$1000 (e.g., three chocolate bars for Cr$1000). The use of the convention minimizes the need to perform on-the-spot arithmetic calculations by eliminating the manipulation of odd values. The pricing convention does lead, however, to other types of mathematical problems in the practice. For instance, sometimes sellers will offer units for more than one pricing ratio such as five for Cr$1000 and two for Cr$500. Such pricing gives rise to ratio comparison problems either when one compares one's own pricing ratios or when one compares ratios with one's competitors.

Finally, social interactions during the practice often alter the structure of the problems that emerge. For instance, store clerks sometimes help sellers with the mathematical problems in their purchase; other sellers may collaborate and/or compete in setting prices that require the incorporation of additional constraints in pricing calculations; and customers may bargain with sellers requiring re-determinations, on the part of the seller, of what constitutes a reasonable retail price.

Thus, the candy-selling practice is a social context in which problems are configured as a result of the structure of the practice itself, a structure that is supported by a variety of social processes. If a child is to participate in the practice, the child must form solution strategies that address problems of the representation of large numerical magnitudes, the arithmetical manipulation of large numerical magnitudes, and ratio comparisons.

The Influence of Schooling
on Child Sellers' Mathematics

I will now briefly summarize two types of findings that provide some insight into the way schooling in mathematics may influence the kinds of mathematics sellers generate to address problems in their practice. The first was generated in an observational study of sellers as they practiced their trade. The second, and the type on which I will focus more attention here, was produced through individual interviews with candy sellers as they addressed standard tasks designed to reveal core properties of their mathematics.

Observational Study

In this study sellers, who ranged between 6 and 15 years of age and who had varying levels of schooling, were observed as they practiced their trade. We recorded all transactions sellers had with customers as well as with peers. In addition, during lulls in their selling activities, sellers were queried about how they had priced their boxes for retail sale. The data from this observational study are reported elsewhere (Saxe, in press). However, to provide some insight into the way school experience may influence the mathematics children construct in the street, it is useful here to consider two sellers' responses to the queries about price translation.

Two Approaches for Translating Wholesale
Box Price to Retail Unit Price

Consider first a seller with little schooling and how he translates his wholesale purchase price for a box of candy to a retail price for units. Note particularly how he arranges his calculation by use of the units/Cr$1000 selling convention, a calculation strategy quite commonly used by sellers.

MARCOS#11 (12 years, completed first grade). Marcos begins his day with a full box of candy bars. The box contains 30 units, and he paid Cr$8000 for the box. He's selling the bars at three for Cr$1000.

Observer: How much will you sell the full box for?
Marcos: I'll sell the full box for Cr$10000.

Marcos was questioned about how he determined his prices, and he explained the following, referring to his box illustrated in Fig. 11.2.

Marcos: I count like this (illustrating a count of the bars in groups of three by a value of Cr$1000). These two (two groups of three) bring Cr$2000, these two (two groups of three) Cr$4000, these two Cr$6000, . . . these two Cr$10000. I count

223

30 unit box, Wholesale price: Cr$8000

Retail price: 3 Bars for Cr$1000

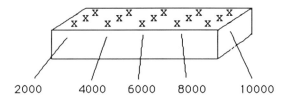

FIG. 11.2. MARCOS' pricing strategy.

like this 'cause I'm going to sell the chocolate at three for Cr$1000, and this way the full box will bring Cr$10000.

Observer: How much will you profit after selling the full box?

Marcos: Since the box cost me Cr$8000 and I'll sell the full box for Cr$10000, my profit will be Cr$2000.

Observer: Do you think your profit will be good?

Marcos: It's not going to be very good. But, if I sell two for Cr$1000, it's going to be hard to sell, and if I sold four for Cr$1000, I'd lose too much.

Observer: (Who taught you to sell chocolate like this?)

Marcos: Nobody. I taught myself.

Now consider the seller who had been to school for some time and the alternative problem structure that the boy constructs as he approaches the problem.

LUCIANO#07 (13 years, completed fifth grade). Luciano begins his day with a box containing 50 candy bars for which he paid Cr$7000. He's selling to customers at four bars for Cr$1000.

Observer: How much are you going to sell the full box for?

Luciano: I'm going to sell the box for Cr$12500.

Observer: How much will you profit after selling the full box?

Luciano: I'll make Cr$5500.

Luciano was asked to explain his calculations. He used a paper and pencil and performed a variety of calculations by standard algorithmic methods (see Fig. 11.3a–f): First, to determine (or confirm) the wholesale worth of each unit of his candy bars, he multiplied 50 bars (the number in the full box) by Cr$120 (a value he believes is the wholesale price per unit he paid; Fig. 11.3a). His computation was correct (Cr$6000), but he believes that he's wrong because he wants to obtain a product of Cr$7000—his wholesale purchase price. He then proceeds to multiply 50 by Cr$140 to obtain Cr$7000—a revised value he believes is the wholesale price per unit he paid for the box (Fig. 11.3b). He appears pleased by his result.

50 unit box, Wholesale price : Cr$7000
Retail price: 4 bars for Cr$1000

<u>a</u> 120 (Cr$)	<u>b</u> 140 (Cr$)	<u>c</u> Concludes candy cost
x 50 (bars)	x 50 (bars)	Cr$140 per unit.
6000	7000	<u>d</u> 250 - 140 = 110 prof/unit
(S.A.)	(S.A.)	<u>e</u> 4 x 110 = 440 prof/sale

<u>f</u>   440 (Cr$ profit per Cr$1000)
x 12 (number of sales in box)
────────
5280 (net profit for box)   (S.A.)

FIG. 11.3.   LUCIANO'S pricing strategy. "S.A." indicates that a school algorithm was used to effect the multiplication.

Luciano: Each candy costs me Cr$140 (Fig. 11.3c). Since I sell one candy for Cr$250, I profit Cr$110 per unit: Cr$250 − Cr$140 = Cr$110 (Fig. 11.3d). If I sell four for Cr$1000, I profit Cr$440 (Fig. 11.3e).

In order to calculate his net profit, the boy then multiplied Cr$440 (his profit after each time he sold 4 bars) by 12 (the number of potential sales in his 50 bar box using the selling convention 4 bars for Cr$1000) using the standard algorithmic approach and achieved the appropriate product, Cr$5280 (Fig. 11.3f).

Observer: What about the remaining two bars in the box (those that would be left after 12 sales)?
Luciano: The last two I'll sell for Cr$500 or less.

Both the first and fifth graders offer an adequate solution, and both accomplish this by constructing and coordinating mathematical relations between the wholesale purchase price and the retail unit price. Clearly, however, the nature of their constructions differed markedly. The first grader conceptualizes the mathematical relation with respect to the selling convention—3 bars for Cr$1000—a convention linked to the retail transaction itself. The first grader's solution has a direct mapping on the actual operations of exchange in that each count of three by Cr$1000 represents a transaction. In contrast, the schooled child conceptualizes the mathematical relation with reference to an intermediate value that is not a part of any aspect of the actual transaction itself—the wholesale price per unit, a value that is accessible by the use of the standard division or repeated application of a multiplication algorithm. Using this approach, the schooled seller marks up the derived wholesale unit price and performs another computation to translate the mark-up into the street convention. Such a solution strategy is both distant from the seller's anticipated transactions and it is also powerful;

the solution strategy is one that can be used across many types of problem regardless of the particular selling convention.

Interview Studies with Sellers of Different Levels of Schooling

To investigate more systematically the character of sellers' mathematical understandings and whether they are influenced by school experience, child sellers were interviewed individually during off-hours in a rented room near where they practiced their trade. Unlike the observational study, all sellers were presented with identical problems under standard conditions. In this study, sellers differed systematically in their amount of school experience, ranging from no schooling to completion of the seventh grade. All sellers were about the same age (between 12 and 15 years) and approximately matched for age across different levels of schooling. In all, 52 sellers were interviewed with about equivalent numbers of children who had no schooling through second grade, third through fourth grade, and fifth through seventh grade.

In the interviews, sellers were asked to solve mathematical problems that were designed to reveal characteristics of their mathematics. The problems selected for discussion here involved number representation, arithmetical problem solving, and reasoning about ratios.[1]

Number Representation

The first approach to studying children's representations of number was to determine sellers' abilities to read numerical values (in our standard number orthography). Such an ability would have clear uses in the practice itself. The prices of candy are often posted in the wholesale stores, and in arithmetical computations the ability to read and write numbers has some clear advantages as illustrated in Luciano's (the fifth grader) approach to retail pricing cited above.

To determine their skills, I asked sellers to read and compare 20 numerical values that ranged from 146 to 5000, values that were within the range that they addressed in their practice. Sellers' abilities and error patterns varied dramatically. The children with little or no schooling read, on average, less than 50% of the values correctly. These errors often involved *partitioning errors* in which sellers read multidigit numbers as if they were single- or double-digit numbers (e.g., 146 read as 1 and 46, or 935 read as 9 and 3 and 5) and *place value errors*

---

[1]A more complete report of sellers' solutions to these and other problems, contrasts between sellers and nonsellers, and sellers at different age levels is contained in other sources (Saxe, 1988a; Saxe, 1988b; Saxe, in press).

in which sellers mistakenly added or subtracted zeros in reading the values (e.g., 146 read as 1,046 or 5000 read as 500). In contrast, the third and fourth graders read, on average, about 70% of the values correctly, and the fifth to seventh graders read virtually all of the values correctly.

Schooled sellers' skills provide them with the option of using our standard number orthography in computations. On the basis of these data, however, it was not at all clear how nonsellers accomplish representational aspects of their everyday computations. Further analysis of observations of the sellers at their practice suggested that sellers might be representing values quite effectively by an alternative to the standard orthographic system for number representation. Sellers' manipulations of bills as they added values suggested that they used bills themselves as tokens for number. Such a representational system, coupled with a reliance on more capable others (when necessary) to read prices (e.g., store clerks or peers), would appear to provide a powerful system to accomplish practice-linked problems.

To determine whether children had the capability to use the currency system as tokens for number, I administered to sellers additional tasks specifically designed to determine whether sellers could identify denominations of currency without reference to the standard orthography. I also administered tasks to determine whether they had knowledge of numerical relations between currency units. In the currency identification tasks, sellers were presented with bills of currency with their printed numbers occluded by tape (12 bills) as well as photocopies of cutouts of the numbers. (In an additional condition sellers were presented with the standard bills without numbers occluded). In the currency comparison tasks, sellers were presented with problems in which they had to tell the larger value of two currency units (ordinal relations) and tell how many of the smaller units were equivalent to the larger unit (multiplicative relations). If they had acquired both the ability to identify and compare currency units numerically, sellers would have the basis for a representational system to manipulate large numerical values, a system that could serve them well in their everyday computations.

Sellers' performances on these currency tasks were striking. Virtually all sellers, regardless of schooling experience, solved all the bill identification tasks with numbers occluded and most of the currency comparison tasks involving both ordinal and multiplicative relations. In contrast, sellers' performances on the photocopied number identification tasks were again related to level of schooling.

Thus, these findings provide strong evidence that with little or no school experience, sellers construct a representational vehicle linked directly to their medium of exchange—knowledge of the currency system. Such a representational system is indeed functional in the contexts of the practice. The findings also show that schooling provides new options for representing number, options that provide a more specialized system for representing numerical values.

Arithmetic

To analyze whether schooling influenced sellers' practice-linked arithmetical problem solving, sellers were presented with various problems that required them to add bills and subtract currency denominations. In one addition problem, for example, sellers were handed a stack of 17 bills (in a standard but haphazard order) totaling Cr$17300 and told, "Suppose you started the day with this amount. Would you add the money for me?" The child was provided with a paper and pencil and told that he could use them or anything else he liked to help him solve the problem.

Childrens' solution strategies to the addition problems revealed no marked differences as a function of schooling, and most children achieved accurate summations. A typical strategy involved rearranging a stack into an order that could be partitioned into convenient values—such as assembling like-value bills together. For instance, consider the following typical strategy to the addition portion of the problem.

> *IDMAR#65 (15 years, completed fourth grade).* Idmar first attempts to add the values of bills in the standard order, but interrupts his addition after the seventh bill at which point he appears confused (Cr$1000, ◯$500, Cr$1000, Cr$200, Cr$500, Cr$500, Cr$200, stops). He then tries again in the same manner, stops after the seventh bill, and begins a third time. With this attempt he completes the addition of all 17 bills with the standard order and produces the incorrect result of Cr$10,300. He then requests one final attempt where he reorders the bills as his calculation proceeds. He selects the following: (a) two Cr$1000 bills, (b) four Cr$500 bills, (c) one Cr$1000, (d) two Cr$500 bills, (e) one Cr$10000, (f) Cr$100, and finally (g) the six Cr$200 bills to produce the correct answer, Cr$17,300.
>
> *Observer:* Why did you solve the problem that way?
> *IDMAR:* I separated them to be more sure and not to get confused; the money was all mixed up.

While there were no differences documented in children's addition strategies, there were schooling differences in children's subtraction strategies. Sellers who had greater levels of schooling were more likely to use paper and pencil solution strategies involving standard algorithms than sellers with less schooling experience. Nonetheless, schooled children did not produce more accurate answers than non-schooled children. The absence of paper and pencil solution strategies by both schooled and unschooled sellers for addition problems points again to sellers' use of currency itself as a representational vehicle to mediate problem solving.

Ratios

To analyze whether schooling experience affected sellers' ability to compare and reason about ratio comparisons, sellers were presented with ratio problems in which they had to determine which pricing ratio would yield the larger profit. In a typical problem, a seller was told the following: "Suppose that you bought this bag of *Pirulitos* and you must decide the price you will sell the units for in the street. Let's say that you have to choose between two ways of selling: selling one Pirulito for Cr$200 or selling three for Cr$500 (one Pirulito was placed next to a Cr$200 bill and three were placed next to a Cr$500 bill). Which way do you think you would make the most profit?"

Sellers, in general, regardless of schooling level demonstrated an understanding of how to compare ratios. Consider the following excerpt that was typical of sellers' performances.

PAULO#53 (12 years, dropped out of first grade).

Observer: Which way do you think that you would make the most profit—one for Cr$200 or three for Cr$500?
Paulo: Selling one for Cr$200.
Observer: Why?
Paulo: Because by selling one for Cr$200 I sell three for Cr$600 and that way (referring to the three for Cr$500) it brings only Cr$500. I would lose Cr$100.

Not only does this child demonstrate an understanding of the need to generate a common term to compare ratios, but he spontaneously offers the relative gain at preferred selling price.

SUMMARY

In summary, some sellers used mathematics learned in school to address problems that emerge in the selling practice. The influence was most important when sellers addressed problems involving the standard orthography. Schooling did not appear to impact upon the conceptual problems of ratio comparison and addition of bills, problems for which most sellers used adequate solution strategies.

THE INFLUENCE OF PRACTICE PARTICIPATION ON THE MATHEMATICS LEARNED AT SCHOOL

In school, mathematical problems take on specific forms. They are removed from the pragmatic contexts in which they often emerge in everyday life and are often presented as computations represented in the standard orthography or in stories (word problems). In the study summarized here, my concern was to understand whether sellers made use of knowledge generated in their out-of-school practice to help conceptualize and solve problems of the classroom. To this end, I interviewed second and third grade sellers and nonsellers. In the interview, children were presented with 12 arithmetic problems involving addition, subtraction, multiplication, and division (Saxe, in press, for a description of problem types). For the purposes of this presentation, I will merely focus on some general features of the results.

Children's solutions to the arithmetical problems differed as a function of their selling experience. At Grade 2, the mean number that sellers achieved correct was about 3.5 times greater than the mean of the nonsellers; at grade 3, this difference was considerably attenuated. The solution strategies that children used are revealing about what led the sellers to higher levels of competence than nonsellers. Consider, for instance, three types of solution strategies that could be used to solve the problem $26 + 28 = ?$.

The *standard strategy* is the algorithm taught in school. To solve $28 + 26 = ?$, a child uses our Base-10 numeration system, one that requires knowledge of place value structure and uses a column format to help effect a solution. Using this procedure, an individual proceeds from right to left, first adding the digits in the units column, carrying 10s values, and then adding the values in the tens column.

The *regrouping strategy* is one that is most closely linked to strategies sellers generate as they attempt to compute transactions involving currency. For instance, recall the sellers' computations involving the addition of multiple bills, computations in which bills were grouped into convenient values as sellers produced a progressive summation of values. When this strategy is applied to school problems, numbers are similarly grouped into convenient values, this time, however, by conceptually decomposing and recomposing them. Consider, for instance, the two variants of childrens' solutions to the problem $28 + 26 = ?$, each of which involves decomposing and recomposing numbers into convenient values that are then composed to yield a final sum. In Variant #1, a child operates first by decomposing the two terms into $20 + 6$ and $20 + 8$, and then composes the two 20s to yield 40. Then, the 8 and 6 are composed to yield 14, which is then recomposed with the 40 to yield 54. In Variant #2, a child takes from one term and adds to another to produce a convenient value, for instance $26 -2 = 24$ and $28 +2 = 30$. The two resulting terms $24 + 30$ are recomposed to yield the sum 54.

An analysis of childrens' strategies revealed that sellers used the regrouping strategies more frequently than nonsellers. This was especially apparent when the Grade 2 sellers' and nonsellers' successful efforts were contrasted. This difference narrows at Grade 3, a phenomenon that suggests that as the nonsellers attempt to solve the problems presented in school, they are beginning to structure more adequate solution strategies which, like those of the sellers, favor the regrouping solution procedures.

Some sellers illustrate well the way standard and regrouping strategies can

<u>c</u> <u>b</u> <u>a</u>

/

7 9 0

+ 4 7 0

──────────

1 2 6 0

a. 0+0=0.

b. 9-3=6,   3+7=10; 6+10=16, writes 6, carries 1.

c. 4-3=1,   3+7=10; 10+1+1=12, writes 12.

d. Reads 10,260.

FIG. 11.4. ENRICO's solution to a school arithmetical problem. The letters *a–d* indicate the order of Enrico's operations.

support one another in problem solutions. Consider the following sellers' efforts to solve a multidigit addition problem.

ENRICO#3109 (Third grade candy seller). Enrico is presented with the problem 790 + 470 = ? in computational form and solves it correctly (see Fig. 11.4).

Observer: How did you solve this problem?
Enrico: 0 + 0 = 0 (referring first to the units column); 9 (of 790) − 3 = 6, 3 + 7 = 10, 10 + 6 = 16 (referring to the tens column and using a regrouping strategy).

He writes down the six and carries one to the hundreds column. Again he uses a regrouping strategy to add the numbers in the hundreds column.

Enrico: 4 − 3 = 1; 3 + 7 = 10; 10 + 1 + 1 = 12

He writes down the number 12. While his final answer is correct, he reads the value as ten thousand two hundred and sixty, making a place-value error described earlier.

In sum, childrens' solutions to the school-linked problems provide clear support for the thesis that children are creating novel strategies linked to their out-of-school practices as they approach problems that emerge in the school context.

CONCLUDING REMARKS

Children are engaged in a multiplicity of different contexts in which occasions for problem solving emerge. Whether and how they may be able to use or "transfer" knowledge constructed in one context to address problems in another context is a critical question for both educators as well as students of cognitive development. In the present investigation, it was clear that sellers did make use of knowledge generated in one social context to solve problems in another: Sellers used school-linked knowledge to generate new pricing strategies in the selling practice, and they used solution strategies from their practice to address school-linked problems. In these concluding remarks, let me sketch three current models of transfer and the bearing that the findings summarized here have for these models.

One approach to transfer is what Michael Cole and colleagues at the Laboratory of Comparative Human Cognition (LCHC) term *general processor* models (Laboratory of Comparative Human Cognition, 1986, p. 331). Under this type of model, learning in a social context affects some executive process, whether it is a cognitive structure, a developmental level, or some general information processing mechanism. Under this model, in its extreme form, learning that affects the central process should affect subjects' problem solving across all contexts. LCHC cites various studies that provide, at best, only limited support for this type of model—studies ranging from experimental studies across problem types of the same logical structure (Gick & Holyoak, 1980) to studies in natural contexts as individuals address similar problem types in different settings (Lave, 1977). In these studies, there was only limited transfer of knowledge gleaned in one context to problem-solving efforts in another, and thus little support for a general processor model.

LCHC points out, however, that the extreme form of a context-specific view of learning, the second model type, is also untenable. Individuals do make connections across contexts, though perhaps not with the immediacy expected by general processor models. LCHC has argued for a third model in which learning is local to situations, and transfer is mediated by social processes. In their view, cultural life provides a wide range of supports for transfer including assistance from others as well as socially-elaborated representational vehicles (e.g., language, numeration), which lead individuals to use knowledge acquired in one situation to solve problems in another.

The candy-selling practice provides some indication that social processes may facilitate transfer. For example, (a) the appearance of number orthography (a socially-based mediational system) on bills, coins, and candy boxes in the practice may have helped sellers to bring school-linked knowledge to bear on problems of the practice in their arithmetical computations; (b) assistance provided to sellers from wholesale store clerks may have helped some link school-based knowledge with problems of the practice; and (c) peer collaborations in price setting may have provided support for sellers with greater levels of schooling to help less-schooled peers use their school-based knowledge in addressing problems of the practice. However, even if these social supports did facilitate transfer, individual candy sellers ultimately must make these links, and we need an analysis of their constructive efforts. The LCHC model neglects an analysis of the interplay between the cognitive activities of the child and social contexts in the processes of transfer.

I view transfer as a developmental process, one that involves a shifting relation between cognitive forms and cognitive functions (Saxe, Guberman, & Gearhart, 1987; Werner & Kaplan, 1962). Recall the prepare-to-purchase phase of the selling practice—the phase in which sellers address problems of translating wholesale to retail prices—and the unschooled seller's (MARCOS, Fig. 11.2) and schooled seller's (LUCIANO, Fig. 11.3) efforts to solve this problem. To accomplish the arithmetical functions inherent in this translation problem, these sellers brought to bear or transferred cognitive forms initially constructed in other functional contexts. Thus, MARCOS, the unschooled child, adapted the pricing convention used in the sell phase of the practice (3/Cr$1000) in a repeated addition operation mimicking sales transactions to determine an appropriate retail price, whereas LUCIANO adapted the multiplication algorithm acquired in school in this strategy to achieve the intermediate term of wholesale price per unit value.

The sellers' uses of prior cognitive forms to serve the function of price translation can be understood as an extended process of repeated construtions rather than merely an alignment of prior knowledge to a new functional context. In the case of the unschooled seller, MARCOS is beginning to make an abbreviated use of sales transactions in his additions by adding sales two by two—a process that could evolve into even more economical cognitive forms (e.g., additions by larger values). In the case of the schooled seller, LUCIANO is using a trial and error multiplication approach to determine an intermediate wholesale unit price value, an approach that may eventually lead to the construction of a more systematic use of a division operation. In both cases, transfer may be best understood as an extended process in which the relations between cognitive forms and functions are shifting. Each seller is using prior forms—forms initially not well-suited to accomplish the function of pricing. In the process of their application, children are reworking these forms into function-specific pricing strategies, a transfer which itself is a dynamic process of construction.

The view that emerges from the present discussion is that learning in both school and out-of-school contexts can present important opportunities for cognitive growth. Further, an inherent part of this growth may entail using knowledge generated in one context type to serve functions linked to others. For the student of social processes in cognitive development, this means that an investigation of children's learning in any one social context should in part be guided by an analysis of related contexts in which children participate. For the educator, this means that helping children to use strategies they have constructed in out-of-school activities to address novel problems that emerge in school is an ingredient of good pedagogy.

ACKNOWLEDGMENTS

This paper is based on a presentation at a conference sponsored by the Lawrence Hall of Science and Graduate School of Education, University of California,

Berkeley, and the Institute for Research on Learning, Xerox Palo Alto Research Center, January, 1988.

Various individuals aided in the completion of the research described here. I am particularly grateful to Luciano Meira and Regina Lima who aided in preliminary ethnographic work and in the administration of interviews; to David Carraher, Terezinha N. Carraher, and Analucia D. Schliemann who provided a supportive research environment at the Universidade Federal de Pernambuco during the conduct of the project; and to Joseph Becker, Maryl Gearhart, and Jim Greeno who provided comments on a prior draft of the chapter.

This chapter is based on work supported by the Spencer Foundation and the National Science Foundation under Grant No. BNS 85-09101; any opinions, findings, and conclusions or recommendations expressed in this publication are those of the author and do not necessarily reflect the views of the Spencer Foundation or the National Science Foundation.

REFERENCES

Gick, M. L., & Holyoak, K. J. (1980). Analogical problem solving. *Cognitive Psychology, 12,* 306–355.

Greenfield, P. M., & Childs, C. P. (1977). Weaving, color terms, and pattern representation: Cultural influences and cognitive development among the Zinacantecos of Southern Mexico. *Interamerican Journal of Psychology, 11,* 23–48.

Lave, J. (1977). Cognitive consequences of traditional apprenticeship training in West Africa. *Anthropology and Education Quarterly, 8,* 177–180.

Laboratory of Comparative Human Cognition. (1986). Culture and cognitive development. In W. Kessen (Ed.), *Manual of child psychology: History, theory, and methods* (pp. 295–356). New York: Wiley.

Price-Williams, D., Gordon, W., & Ramirez, M. (1969). Skill and conservation: A study of pottery-making children. *Developmental Psychology, 1,* 769.

Rogoff, B. (1981). Schooling and the development of cognitive skills. In H. Triandis & A. Heron (Eds.), *Handbook of cross-cultural psychology* (Vol. 4, pp. 233–294). Boston: Allyn & Bacon.

Saxe, G. B. (1988a). Candy selling and math learning. *Educational Researcher, 17*(6), 14–21.

Saxe, G. B. (1988b). The mathematics of child street venders. *Child Development, 59,* 1415–1425.

Saxe, G. B. (in press). *Culture and cognitive development: Studies in mathematical understanding.* Hillsdale, NJ: Lawrence Erlbaum Associates.

Saxe, G. B., Guberman, S. R., & Gearhart, M. (1987). Social processes in early number development. *Monographs of the Society for Research in Child Development, 52*(2, Serial No. 216).

Stevenson, H. W. (1982). Influence of schooling on cognitive development. In D. A. Wagner & H. W. Stevenson (Eds.), *Cultural perspectives on child development* (pp. 208–223). San Francisco: W. H. Freeman.

Werner, H., & Kaplan, B. (1962). *Symbol formation.* New York: Wiley.

12 Cooperative Learning in Science, Mathematics, and Computer Problem Solving

Vincent N. Lunetta
The Pennsylvania State University

The development of modern science has been set in social and cultural contexts that have influenced greatly the development of scientific conceptual schemes and problem-solving processes. Similarly, the development of scientific concepts and problem-solving skills in school science is embedded in social and cultural contexts that are very powerful in shaping the beliefs, attitudes, and behaviors that influence what is learned (Cole & Griffin, 1987). The peer group has an especially great influence upon attitudes and behavior among adolescents and some ethnic minorities (Ogbu, 1986) in mid-American culture. What is the nature of that influence? How can that influence be utilized productively to promote more effective science education? How can that influence be engaged in promoting access to science and mathematics for women and ethnic minorities? Education in science will itself become more scientific as research reveals more of the complex intersections between cognitive development and the social contexts that are so significant in school learning.

SOCIAL GOALS

Cognitive research in science education in the recent past has focused on the concepts and structures the individual learner brings to the study of science and mathematics. A growing body of evidence indicates that the learner's prior concepts and structures have great influence on what is learned (Linn, 1986). There is also a growing body of evidence that effective student-student transactions can enhance achievement and the development of positive attitudes. In the late 1960s and 1970s concerns about the implications of ethnic, racial, and language diversity stimulated research on the effects of group interactions and learning environments. The resulting studies have produced a substantial body of research about student transactions in school learning in several academic disciplines. The research has implications for the development of concepts, problem-solving skills, and attitudes consistent with broad goals of school science education. Certainly it is very important for researchers in science education to focus on the optimal development of scientific concepts and structures. It is also important in responding to complex needs in multivariate classrooms to examine ways to promote effective group interaction, an important element of the scientific problem-solving skills implicit in statements of science education goals.

Activities in school science and mathematics provide unique and special opportunities for the development of social interaction skills. Laboratory practical activities have frequently been described in the literature as synonymous with good science teaching (Hofstein & Lunetta, 1982), and small-group activities are an important part of many school science labs. Small groups can plan, share laboratory equipment, gather data, process data, and discuss results as a team and with the larger class. Some labs are especially amenable to the sharing of tasks so that different team members investigate different aspects of a lab problem. Sometimes it is facilitative for a team to assign the investigation of different variables to specific team members. Yet there is evidence that school science and mathematics curricula and teaching practice tend *not* to address the development of group and social skills (Stake & Easley, 1978).

DISCREPANCIES BETWEEN GOALS AND PRACTICE

In stereotypic school laboratory activities, one student in a lab group may make most of the decisions while another records data. This second member of the group, and sometimes third and fourth members as well, often has little involvement in the decisions that are made by the group leader. Often there is a ritualistic following of directions to get correct answers. Laboratory activities provide important messages to learners about the nature of science, yet the ritualistic labs that are so commonplace may have more in common with some religious practices than with science. They often provide misleading ideas about science. The important point, however, is that small-group lab activities and the interactions of individuals within those groups provide very important opportunities to develop social skills concurrently with appropriate scientific concepts and problem-solving skills. Appropriate group activities can also promote understanding about the nature of the scientific disciplines and positive attitudes toward science. An array of applied research is needed to provide information and direction to teachers and curriculum developers regarding the appropriate use of laboratory practical activities.

In spite of the importance of good laboratory activities and the opportunities they provide for working in small groups, small-group activity occurs infrequently in science classes (Gallagher, 1985; Goodlad, 1984). This reality may be due in part to the mixed messages and confusion that surround the use of laboratory activities.

> Students commonly work as technicians following explicit instructions and concentrating on the development of lower level skills. Relatively few questions or instructions are presented to stimulate higher order skills in analysis, conceptualization, application, and experimental design. [There is an absence of engagement in post-laboratory discussions and in sharing the results of class members.] In general, the laboratory handbooks of contemporary curricula do not provide students with extended opportunities to investigate and to inquire consistent with goals of communicating the method and spirit of scientific inquiry that have been espoused by science educators and curriculum developers. (Tamir & Lunetta, 1981, p. 483)

Similarly, in mathematics classrooms problem solving in small groups can be a normal and important part of class activity. Just as in science classrooms, however, opportunities to develop and promote appropriate student-student interaction and to facilitate small-group learning activities are generally not addressed. Ritualistic and individualistic activity is commonplace (Siegel, 1983).

The advent of the microcomputer in education provides special challenges and opportunities for small-group activity. Yet, there has been a generally unchallenged assumption that microcomputers in schools are to be used only by individuals. Social isolation of that kind can create loneliness, boredom, and frustration that have negative influences on a student's motivation and ability to pursue learning activities (Showers & Cantor, 1985). The isolation also inhibits opportunities to discuss, explain, and summarize activities that could result in higher-level reasoning and conceptual understanding. Social isolation is also common in many science and mathematic classroom activities; student interaction is more frequent in classrooms in certain other disciplines. Social isolation may be a factor that influences the negative attitudes of many students toward mathematics and science.

To promote improved science education, researchers must clarify the discrepancies between goals and teaching practice. Teachers, curriculum developers, and educational leaders must have research-based information that will enable them to address these significant problems. Researchers must also examine how group processes influence the development of problem-solving skills, concepts, and attitudes toward science. Research on learning must result in the development of models that will clarify cause-effect relationships in the teaching and learning of science if the practice of science education is to become more effective and more scientific in its own right. The research can shed light on important instructional questions. What kinds of roles should be assigned to students and groups to facilitate the development of positive attitudes and group and individual achievement? What student evaluation procedures should be employed to promote maximum learning for the group and for individuals within the group? What kinds of social contexts can increase motivation and the perception of relevance? What teaching strategies and behaviors are effective and appropriate?

In constructing concepts and structures, students are influenced by their observations and explanations and by their ability to perceive and cope with the alternative explanations of others. Thus, important science learning outcomes are influenced by skills, work habits, and group dynamics. Cognitive learning is a function of attitudes, motivation, perceptions of relevance, locus of control, and other variables that are shaped by peer and environmental factors. Yet low motivation and perceptions of limited relevance have been special problems in school mathematics and science. There is much evidence in the literature that school mathematics and science often result in student alienation especially in communities outside suburban areas. It is reasonable that research studies should examine specific cognitive elements of concept development. At the same time, it is also important to examine the social contexts in which school learning occurs. The control and examination of selected variables should occur within a sense of the whole. There is a need to define more explicit goals for selected instructional activities within the context of larger more global goals. The development of concepts and problem-solving skills in science and mathematics is embedded in a context of attitudes and social interactions that occur in school classrooms. The sections that follow focus on selected elements of social interaction. They review a subset of studies that has special implications for change and for clarifying research needed to promote more effective science and mathematics learning.

STUDENT-STUDENT COOPERATION

Several hundred studies have compared effects of competitive, individualistic, and cooperative instructional strategies on cognitive and affective learning outcomes. These studies have covered a range of academic disciplines, age and readiness levels, and tasks. Deutsch (1962), Johnson and Johnson (1983), Sharan (1980), and Slavin (1977) have provided comprehensive reviews of research on *cooperative learning*. In cooperative groups students have interest in each other's learning as well as in their own learning. They support one another and work cooperatively toward shared goals. Meta-analyses were conducted by Johnson, Johnson, and Maruyama (1983) and by Johnson, Maruyama, Johnson, Nelson, and Skon (1981). Several subsequent studies examined the effects of cooperative learning and interaction on specific variables that influence science, mathematics, and computer education. Critiques of the cooperative-learning research have been an important part of the scholarly dialogue and have contributed to more precise and valid research studies. Reviews of the research studies on cooperative learning have reported the following general results.

1. Cooperative learning promoted greater achievement than competitive or individualistic experiences. Achievement appeared to be greater when the learning tasks were more difficult and involved problem solving, divergent thinking, or conceptual learning, for example. The lower one third of students appeared to make the greatest gains in achievement from cooperative experiences, but the middle and upper thirds also maintained consistent learning gains in a cooperative environment. Information retention and development of specific skills and strategies were enhanced for all students in cooperative settings. With more difficult subject matter, the cooperative learning environment resulted in greater achievement differences from one student to another than did group environments.

2. Cooperative learning promoted higher motivation to learn. In cooperative groups, students tended to be urged on and encouraged by their peers. A consistent finding in the research has been that cooperative learning produced more positive attitudes toward the instructional experiences and toward the instructors. Cooperative learning also resulted in higher levels of self-esteem than did alternative competitive or individualistic learning experiences. In addition, cooperative learning promoted greater acceptance of interpersonal differences among students with different ethnic backgrounds and handicaps; it provided more interaction among students with handicaps and nonhandicaps and among different ability groups.

Cooperative environments influenced gender differences in learning and may, in fact, reduce them, conclusive studies on these effect were not identified. More thorough study on the effects of cooperative learning on gender differences appears to be needed. Appropriate interactions within a heterogeneous cooperative group can promote learning for both the skilled and less-skilled student. The less-skilled student profits from the insights and explanations of the more talented student. The more talented student can enhance learning by talking through the material with others and responding to questions. According to Johnson and Johnson (in press), the failure to provide opportunities to promote student cooperation has especially hurt gifted students. When students talk through what they are learning, they not only learn more but they are also more likely to develop a strategy for learning the material. The Johnsons also point out that effective teachers intervene to assist groups in developing appropriate student-student behavior. The learning of science is enhanced by the development of collaborative skills including communication, leadership, trust, and conflict resolution. In addition to conducting and sponsoring a large number of research studies on the subject of cooperative learning, the Johnsons have also been active in publishing materials outlining implications for effective teaching (e.g., Johnson & Johnson, 1987).

Johnson and Johnson have promoted research on the appropriate use of groups and cooperative learning in the context of computer-assisted learning. Johnson, Johnson, and Stanne (1986), for example, reported that computer-assisted cooperative learning promoted: (a) higher quantity and quality of daily achievement, (b) accuracy of recognition of factual information and ability to apply facts in test questions requiring higher-level reasoning and problem solving, (c) more success in a complex problem-solving task involving mapping and navigation, and (d) success in operating a computer program when contrasted with students operating in competitive and individualistic environments. In that research report, the authors noted:

> students in the cooperative condition addressed far fewer remarks to the teacher and more remarks to each other than did students in the competitive and individualistic conditions. The student-student interaction within the cooperative condition was almost all task-oriented, consisting of statements concerning the completion of assigned work whereas in the competitive and individualistic conditions there were significantly more social and off-task statements. Within the individualistic learning condition relatively few comments were made and many were directed at the teacher. (Johnson, et al., 1986, p. 390)

The Johnsons have also elaborated implications of their extensive research for computer-assisted learning in certain papers written for teachers (Johnson & Johnson, 1986).

Thus far this brief review has focused primarily on research and publications surrounding the work of Roger T. Johnson and David W. Johnson at the University of Minnesota. While their publications are particularly visible in the cooperative learning literature, many other individuals have also been involved. Robert Slavin (1983) has been a significant contributor and has raised important criticisms about the research in cooperative learning. He has criticized and questioned generalizations that cooperation is most effective for achievement and productivity when compared with competitive and individualistic structures. Slavin has suggested that it is unwise to generalize across widely divergent tasks, outcome measures, and settings. He has encouraged instead a focus on research in small well-defined areas of some special theoretical or practical importance. He has noted that groups are superior to individuals for solving problems because of the reality that they share answers in those circumstances and not because of any unique group interaction property that results in increased productivity. (Hill, 1982). Slavin pointed out that two or more individuals who take a test together will get a better average score than an individual who takes the test by himself. He also suggested that the important question is how much each person will learn from the experience. "If a group produces an excellent lab report, but only a few students really contributed to it, it is unlikely that the group as a whole learned more than they might have learned had they each had to write their own (perhaps less excellent) lab reports" (Slavin, 1983, p. 430). Slavin also pointed out that the achievement outcomes of group study depend entirely on the incentive structure used. Cooperative learning methods have had positive effects on "a wide range of social and emotional outcomes, such as self-esteem, race relations, and acceptance of mainstreamed academically handicapped students . . . These non-cognitive outcomes do not appear to depend to the same extent on particular incentive or task structures, and for many practical applications, these outcomes might justify the use of cooperative learning methods as long as they do not reduce student achievement" (Slavin, 1983, p. 443).

Similarly, researchers should examine more carefully the interaction effects of cooperative learning strategies with the development of attitudes toward science and toward the school science classroom. Criticism like that found in the Slavin (1983) paper has focused research activities and enhanced precision; it has stimulated a variety of subsequent studies examining specific factors that promote group learning. Some studies have examined, for example, incentive structures and evaluation that enhance group learning and achievement. Other studies have examined the effects of specific meta-cognitive student behaviors on learning.

Webb (1983, 1985) examined student interaction in small groups learning mathematics at several levels. The material ranged from topics in general mathematics to algebra, geometry, and probability, and she wrote that "giving explanations was consistently and positively related to achievement. Giving terminal responses on the other hand, was not related to achievement" (p. 36). She has related her results to Wittrock's (1974) model on generative learning. Giving explanations not only involved verbalizing associations between new and learned information but may also involve forming new elaborations.

Other related studies have examined the "cognitive benefits of teaching" (Baragh & Schul, 1980, p. 593), by analyzing specific effects of peer tutoring and of giving and receiving help. Webb (1982) noted that giving and receiving help in peer tutoring are beneficial but that the helping behavior must be sensitive. Subsequently, she examined distinctions between solicited and non-solicited help and whether calls for help were answered. She suggested that explaining to others may be more beneficial to the explainer when the material is complex requiring integration or reorganization than when the material is simple or straightforward. Some studies in the literature report contrary findings. Other studies have examined more explicit activities leading to metacognition, suggesting that metacognition occurs principally through dialogue and interaction with other people. Johnson, Johnson, and Stanne (1985) suggested that cooperative learning promoted more of such interaction than did competitive and individualistic learning. Several studies suggest that there are correlations between achievement and the student's gender and between attitudes and gender.

The extensive body of research on cooperative learning in the literature has implications for science and mathematics teaching. At the same time, explicit connections between the general research and the development of scientific concepts and problem-solving skills still need to be researched more thoroughly. There are different points of view, for example, on the effects of grouping and reinforcement with increasingly complex tasks. Derry and Murphy (1986) have suggested that strategies for promoting the learning of complex problem-solving skills with groups are not well understood. However, the organized body of information on cooperative learning now has some underlying theoretical schemes that can provide direction for further research. It is important also to pursue understanding of the intersections between the research on cognitive development, cooperative learning, and learning environments in contemporary schools and communities.

Ultimately, research in science and mathematics education should lead to specific recommendations for curricula and teaching practice. Research that may have classroom utility generally will be focused on relatively precise goals, but also must be based upon broader theoretical paradigms. Specific studies are needed to elaborate on details about the kinds of group interactions needed to enhance achievement and to promote the development of positive attitudes toward science. Johnson and Johnson (1985), for example, reported that controversy can result in uncertainty or cognitive conflict. That particular study provided some support for the hypothesis that controversy (as defined in that study) results in an active search for more information and cognitive restructuring. When should small group activities be promoted and included? What specific kinds of student behaviors should be encouraged?

1. What can be done to make science problem solving more interesting and engaging for students?
2. What kinds of teacher intervention strategies are most appropriate in fostering the development of cooperative learning skills?
3. What kinds of interaction patterns are most appropriate for specific types of learning tasks?
4. How are learning groups best structured? For example, to what extent should they be homogeneous or heterogeneous?
5. Should groups be encouraged to reach consensus or merely to identify areas of disagreement?
6. What kinds of student evaluation schemes are most appropriate in promoting cooperative learning and individual achievement?
7. For what kinds of science and mathematics learning activities and goals are small groups most appropriate?

LABS, PROBLEM SOLVING, AND COMPUTERS

There are special opportunities in science and mathematics education to promote social goals consistent with general education in schooling, especially through the small groups that are a normal part of school science labs and science, math, and computer problem-solving activities. Laboratory and computer-based activities provide special opportunities for the control of variables and for study of science learning (Krajcik, Simmons, & Lunetta, 1988). Software and laboratory manuals contain specific instructions that can shape the behaviors of groups and individuals. The instructions can be carefully controlled, and the effects of the instructions on behavior and learning can be examined.

Although it is relatively easy to examine and control variables in software and in laboratory manuals, there is evidence that writers and publishers of lab handbooks and software have not been particularly responsive to opportunities to do so. For example, in only 3 out of 47 laboratory activities did the precedent-setting *PSSC Laboratory Guide* (Haber-Schaim, Cross, Dodge, & Walter, 1970) suggest a division of tasks among groups of students facilitating pooling data for group analysis and interpretation. Furthermore, little or no specific guidance was offered regarding the nature of prelaboratory or postlaboratory discussions. Explicit ". . . post-laboratory discussions were suggested in 3 out of the 40 *Project Physics* (Rutherford, Holton, & Watson, 1970) experiments. Although some teachers use the results of laboratory work for discussion, others assume that the students' written responses to the questions in the laboratory exercises are sufficient. Teachers often feel too pressed for time to conduct post laboratory discussions" (Lunetta & Tamir, 1981, p. 638). There is much evidence that when students follow instructions in laboratory guides, for example, their behaviors are inconsistent with many of the goals espoused by the curriculum developers. Discrepancies between goals and behavior were especially visible in content analyses of laboratory handbooks of large curriculum development projects of the 1960s and 1970s (Tamir & Lunetta, 1981, p. 482). Yet the practices of teachers and curriculum decision makers are not likely to change until the complex issues that underly their decisions are clarified through extended research and through interpreting and disseminating research results.

Small-group activities provide important opportunities to develop problem-solving skills and concepts in ways that are not possible in most large group or individual instruction. A group in a science class can wrestle with ways to solve a problem, and competing alternative ideas can be brainstormed, shared, and evaluated in small groups. An investigation itself can be broken into component tasks to be shared by different members of a group or team; in other activities it may be appropriate for team members to share a common task and compare results. The data gathered in an investigation or even simulated data presented to a class can be shared and discussed in small groups. The group can share in interpreting the data and in raising questions about inferences, error in measurement, meaning, and application. The resulting group discussions (postlab) can assist in developing understanding and in developing problem-solving skills. The sharing of tasks within a class or group has other benefits. Task sharing can enable a science or mathematics class on occasion to address more complex and realistic problems that may enhance the sense of relevance about school science.

Research is needed to provide guidance to teachers and curriculum developers about specific strategies that will promote meaningful learning and positive attitudes. Research and development are needed to develop software and systems that can facilitate the sharing of data for group interpretation. What data base software can be used in school science and mathematics to facilitate the sharing and interpretation of data? How are these systems properly incorporated in instruction that leads to meaningful learning? Similarly, what existing data bases can be modified for appropriate group interpretation?

Science and mathematics teachers can use explicit activities and intervention techniques to enhance the development of problem-solving skills while facilitating the development of group interaction skills consistent with goals for education in a democratic society. Enhancing concept learning, promoting positive attitudes toward science, and developing a more internal locus of control are other anticipated outcomes of applying relevant research to classroom teaching.

CULTURAL AND SOCIAL CONTEXTS

An entire formal education will be a healthy mix of large-group and small-group activity. To that end, it is important for researchers to help clarify the kinds of tasks and goals that are most appropriate for large-group and for small-group activity. While this paper has reviewed selected elements of research related to small-group learning, certainly some learning objectives will be efficiently and appropriately accomplished in large-group settings. It is important for the growing body of research to help teachers and curriculum developers determine when large-group and small-group activities will be optimally effective. Research in science and mathematics classrooms can provide information about perplexing questions facing teachers and curriculum developers: What kinds of activities and teacher and learner behaviors can promote the development of specific concepts and problem-solving skills? What can be done to enhance the development of skills in observing, questioning, interpreting data, explaining, and applying? What are appropriate models for teacher-student interaction? When is instructor intervention appropriate? What kinds of instructor intervention are most facilitative? These questions warrant careful research, but they are embedded in social and cultural contexts that also warrant careful study and analysis.

Fig. 12.1 represents the limited intersection between the contemporary concepts and models of academic science disciplines and the concepts and models of learners. There is much evidence in the alternative concepts research literature (Novak, 1987) that students come to the study of science with a rich array of concepts; those concepts are represented by people's science in Fig. 12.1. The learner comes to school science with concepts that generally do not coincide with the concepts of the formal disciplines. These people's science concepts are

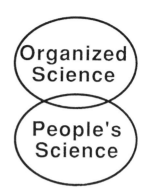

FIG. 12.1. Limited overlap between organized science concepts and learners' concepts

embedded in cultural and social contexts that are very powerful in learning. People's science can retard school learning if school science is perceived to be out of phase with local values, causal explanation, and reality; people's science can enhance understanding and development if teaching and the learning environment are sensitive and responsive to the learner's entry concepts, world views, and social contexts. People's science incorporates values of the subculture regarding the place of people in nature, explanation, the role of authority figures and of questioning, and perceptions of what is relevant.

Effective schooling brings the disparate conceptual worlds of organized science and people's science into more complete intersection. Effective teaching and curricula intersect both worlds, enabling individual students to see discrepancies between their concepts and those of the academic discipline; subsequently, the learner builds more appropriate concepts and models in response. Schools and teachers connected principally to the world of students and people's science are not optimally prepared to promote greater intersection; thus, teachers must be competent in the science discipline they hope to teach. Similarly, schools and teachers attending principally to the concepts of the discipline are not in a position to assist the student optimally to see the discrepancies between his or her concepts and those of the academic discipline. Failure to help students see discrepancies between their own conceptual world views and those of the organized science they are studying results in research findings that have reported, for example, that some physics students can respond successfully to science test questions on Newtonian physics while they respond in interviews with unscientific Aristotelian world views. To promote meaningful learning, effective teachers must understand concepts of their discipline, and they must be sensitive and responsive to the social and cultural contexts in which their students are immersed. Thus, to promote better science and mathematics education, teaching strategies and curricula must incorporate research findings not only from the cognitive sciences, but from sociological and anthropological studies of schools and students. Sociological and anthropological research is needed to enable teachers, teacher educators, and curriculum developers to respond effectively to very important and changing social and cultural contexts of science learning.

Since the great curriculum revolution stimulated by the launch of Sputnik, great investments have been made in new curricula and to a lesser extent in science teacher education. These investments resulted in the development of excellent new resources and in an array of experiences that provide important information and data regarding effective schooling in science and mathematics. It is generally recognized that the influence of the curriculum projects was less powerful than their developers had hoped, and the limited influence can be attributed, in large measure, to the failure of these curricula to address social and cultural contexts of learning and schools. Nevertheless, there is much new information as a result of these experiences, and it is extremely important for people involved in educational decision making, in curriculum development, and in teacher education to understand that information. Much research needs to be conducted to clarify the results of these rich experiences. Unfortunately, in the United States, at least, the efforts to develop and change curricula have not been studied with care, and the effects of these expensive investments in curriculum reform have not been well understood. Some of the data are still available and can yet be gathered and analyzed.

In the past few years in the United States there has been increasing recognition that teachers play a critical role in promoting excellence in science and mathematics education. To that end, the Presidential Awards for Excellence program supported by the National Science Foundation, by national scientific organizations, and by states and local organizations has sponsored national searches that have identified exemplary teachers in mathematics and science in each of the states. These teachers can serve as an excellent data base for important research, but unfortunately the characteristics of these exemplary teachers have not yet been examined carefully and objectively. What are the special characteristics of teachers who have been identified as exemplary? What kinds of support systems have enabled them to grow and to survive? How are they sustained? These are very important questions that should be examined; we know far too little about them at this time. One recent study of exemplary practice in science and mathematics education conducted in western Australia and reported by Tobin and Fraser (1987) provides some examples of what can be done.

Similarly, in a small number of schools and school systems science programs and teaching practices approach the exemplary. Where are those schools? What occurs there? What social and organizational factors have enabled the exemplary programs to develop? What factors sustain exemplary programs? What behaviors and conditions terminate them?

That science and mathematics educators must promote more positive attitudes toward school science is a logical inference from the contemporary research literature on attitudes toward school science. They must do so to promote the development of scientific conceptual understanding and problem-solving skills and to promote the level of scientific literacy in the society. In a free society, students tend to enroll and invest effort in courses they value because of perceived reward, relevance, interest, or enjoyment. Science and mathematics educators should be able to promote a higher level of interest in science and in science classes while maintaining high standards and quality in learning. Research reviewed in this paper suggests that promoting cooperative learning is one medium through which positive attitudes toward science and science classes can be enhanced for girls as well as boys and for students from a variety of ethnic subcultures. If cooperative learning were only to promote more positive students' attitudes as the research suggests, it would be valuable to incorporate cooperative learning strategies in school science and mathematics. However, the research also suggests that properly applied cooperative learning promotes the development of concepts and problem-solving skills. The body of research on cooperative learning is already large, and there are some obvious intersections between the general research on cooperative learning and special needs in science and mathematics education. The desirability of appropriate small-group activities in the school laboratory, in problem solving, and in computer activities was examined briefly in this paper. Many questions about these topics, however, need additional scholarly analysis in the pursuit of better education in science.

REFERENCES

Bargh, J. A., & Schul, Y. (1980). On the cognitive benefits of teaching. *Journal of Educational Psychology, 72,* 593–604.

Cole, M., & Griffin, P. (1987). *Contextual factors in education.* Madison, WI: Center for Education Research.

Derry, S. J., & Murphy, D. A. (1986). Designing systems that train learning ability: From theory to practice. *Review of Educational Research, 56,* 1–39.

Deutsch, M. (1962). Cooperation and trust: Some theoretical notes. In M. R. Jones (Ed.), *Nebraska Symposium on Motivation* (pp. 275–319). Lincoln: University of Nebraska Press.

Gallagher, J. J. (1985). *Secondary school science* (Interim Report). East Lansing: Michigan State University, Institute for Research on Teaching.

Goodlad, J. L. (1984). *A place called school: Prospects for the future.* New York: Macmillan.

Haber-Schaim, U., Cross, J. B., Dodge, J. H., & Walter, J. A. (1971). *Laboratory guide PSSC physics.* (3rd ed.). Lexington, Massachusetts: Heath.

Hill, G. (1982). Group versus individual performance: Are N + 1 heads better than one? *Psychological Bulletin, 91,* 517–539.

Hofstein, A., & Lunetta, V. N. (1982). The role of the laboratory in science teaching: Neglected aspects of research. *Review of Educational Research, 52*(2), 201–217.

Johnson, D. W., & Johnson, R. T. (1983). The socialization and achievement crisis: Are cooperative learning experiences the solution? In L. Bickman (Ed.), *Applied Social Psychology, 4,* pp. 119–164.

Johnson, D. W., & Johnson, R. T. (1986). Computer-assisted cooperative learning. *Educational Technology,* January 1986, 12–18.

Johnson, D. W., & Johnson, R. (1985). Classroom conflict: Controversy versus debate in learning groups. *American Educational Research Journal, 22*(2), 237–256.

Johnson, D. W., & Johnson, R. T. (1987). *Learning together and alone: Cooperative, competitive and individualistic learning* (2nd ed.). Englewood Cliffs, NJ: Prentice-Hall.

Johnson, R. T., & Johnson, D. W. (1988). *Cooperative learning and the gifted science student.* In Brandwein, P. F., & Passo, A. H. (Eds.), *Gifted Young in Science,* Washington, D.C.: National Science Teachers Association.

Johnson, D. W., Johnson, R. T., & Maruyama, G. (1983). Interdependence and interpersonal attraction among heterogeneous and homogeneous individuals: A theoretical formulation and a meta-analysis of the research. *Review of Educational Research, 53,* 5–54.

Johnson, R. T., Johnson, D. W., & Stanne, M. B. (1985). Effects of cooperative, competitive, and individualistic goal structures on computer-assisted instruction. *Journal of Educational Psychology, 77*(6), 668–677.

Johnson, R. T., Johnson, D. W., & Stanne, M. B. (1986). Comparison of computer-assisted cooperative, competitive, and individualistic learning. *American Educational Research Journal, 23*(3), 382–392.

Johnson, D. W., Maruyama, G., Johnson, R. T., Nelson, D., & Skon, L. (1981). Effects of cooperative, competitive, and individualistic goal structures on achievement: A meta-analysis. *Psychological Bulletin, 89,* 47–62.

Krajcik, J. S., Simmons, P. E., & Lunetta, V. N. (1988). A research strategy for the dynamic study of students' concepts and problem solving strategies. *Journal of Research in Science Teaching, 25*(2), 147–155.

Linn, M. C. (1986). Establishing a research base for science education: Challenges, trends, and recommendations. *Journal of Research in Science Teaching, 24*(3), 191–216.

Lunetta, V. N., & Tamir, P. (1981). An analysis of laboratory activities, project physics and PSSC. *School Science and Mathematics, 81*(8), 635–642.

Novak, J. (Ed.), (1987). *Proceedings of the Second International Seminar, Misconceptions and Educational Strategies in Science and Mathematics.* Ithaca, NY: Cornell University.

Ogbu, J. U. (1986). The consequences of the American caste system. In U. Nesser (Ed.), *The school achievement of minority children* (pp. 19–56). Hillsdale, NJ: Lawrence Erlbaum, Associates.

Rutherford, F. J., Holton, G., & Watson, F. G. (1970). *The project physics course handbook.* New York: Holt, Rinehart & Winston.

Sharan, S. (1980). Cooperative learning in small groups: Recent methods and effects on achievement, attitudes and ethnic relations. *Review of Educational Research, 50,* 241–271.

Showers, C., & Cantor, N. (1985). Social cognition: A look at motivated strategies. In M. Rosenzweig & L. Porter (Eds.), *Annual review of psychology, 36,* 275–306.

Siegel, M. H. (1983). The statistical survey: A felt need. In G. Shufelt & J. R. Smart (Eds.), *The agenda in action.* Reston, VA: National Council of Teachers of Mathematics.

Slavin, R. (1977). Classroom reward structure: An analytic and practical review. *Review of Educational Research, 47,* 633–650.

Slavin, R. E. (1983). When does cooperative learning increase student achievement? *Psychological Bulletin, 94*(3), 429–445.

Stake, R. E., & Easley, J. A. (1978). *Case Studies in Science Education* (Vols. 1 & 2). Urbana: University of Illinois at Urbana-Champagne, Center for Instructional Research and Curriculum Evaluation and Committee on Culture and Cognition.

Tamir, P., & Lunetta, V. N. (1981). Inquiry related tasks in high school science laboratory handbooks. *Science Education, 65*(5), 477–484.

Tobin, K., & Fraser, B. J. (1987). *Exemplary practice in science and mathematics education.* Perth, Australia: Curtin University of Technology.

Webb, N. M. (1982). Student interaction and learning in small groups. *Review of Educational Research, 52,* 421–445.

Webb, N. M. (1983). Predicting learning from student interaction: Defining the interaction variables. *Educational Psychologist, 18*(1), 33–41.

Webb, N. M. (1985). Student interaction and learning in small groups. In R. Slavin, S. Sharan, S. Kagan, R. H. Lazarowitz, C. Wells, & R. Schmuck (Eds.), *Learning to cooperate, cooperating to learn* (pp. 147–172). New York: Plenum Press.

Wittrock, M. C. (1974). Learning as a generative process. *Educational Psychologist, 11,* 87–95.

13 Views of the Classroom: Implications for Math and Science Learning Research

Jean Lave
University of California, Berkeley

Usually a comment on a set of papers is a critical appraisal of those papers, but in this case I would like to propose a synthesis rather than a critique. There are two senses in which the attempt at synthesis seems worthwhile. First, the papers themselves converge on some new directions in research on math and science education. Each one emphasizes the socially constructed character of learning activity. Second, it seems important, from an anthropological perspective, to lay out the socio-logic with which social studies of learning and sociological studies of schooling can be brought together to illuminate research on math and science learning. The main point to be emphasized in these remarks is that a synthesis of primarily psychological with sociocultural analyses of learning is not just feasible, but is also crucial. I will try to explain why I think so and what sort of synthesis I have in mind.

The place to begin is with a review of each contributor's way of conceiving of learning in social terms. Saxe's research is focused on arithmetic in socially-organized everyday activity and its settings. He finds evidence of the transfer of knowledge between these settings—both from school lessons to candy sellers' arithmetic in the streets and from vending activity to school math classes. But his conception of transfer is quite different from conventional "cerebral transportation" views, in which internalized knowledge is supposed to be carried from one setting to another, taken out and applied, and put back into memory unaffected by its "use." Instead he argues for a socially-constructed form of transfer: Both objects and people, in this case the currency system and the candy sellers' peers, act as social mediators between activities in different contexts; "transfer may be best understood as an extended process in which the relations between cognitive forms and functions are shifting . . . . Transfer . . . is a dynamic process of construction" (Saxe, in press).

Newman argues, similarly, that we should change the unit of analysis from cognitive processes to the "socially situated cognitive system" (Newman, this volume). He begins his analysis with an interesting proposition: "What is outside the head is just as much a part of the cognitive system as what is inside the head" (Newman, this volume). This leads him to reconsider the meaning of internalization, customarily based on the assumption that learners take in what is being transmitted to them, so that it is inside or outside the mind, but not both at the same time. Newman proposes that "internalization is not an isolated event but occurs in the context of socially constructed interactions in which both sides have put something on the table to negotiate with. And internalization is not the end point either, since once new understandings have been achieved, new tools, products, and frameworks can be put on the table" (Newman, this volume). "The process includes the externalization of thought into the actions, tools, and social organization that support and reorder thinking as much as it includes the internalization of the new understandings that result" (Newman, this volume).

Lunetta argues that we must investigate the powerful social forces and social contexts that influence concept development as well as their cognitive elements (Lunetta, this volume). He documents the missed opportunities for science laboratory learning activities to utilize cooperative learning techniques (Lunetta & Tamir, 1981). He also raises a number of important questions for which the socially situated contextual analysis of classroom practice, and of the school more broadly, may provide some answers: What can be done to make science problem solving more interesting and engaging for students? What kinds of interaction patterns are most appropriate for specific types of learning tasks? How are learning groups best structured? (Lunetta, this volume).

If we ask ourselves where we might go for answers, I would suggest it is not to psychological experimentation on motivation or achievement, but to the social structuring of the school—its classes, students, knowledge, and everyday practice. Thus, Eckert's study of a Detroit suburban high school offers a potential source of insight into possible answers. She clearly insists that learners are their own social context, and she gives us a rich and detailed ethnographic description of the sociology of American high schools.

To begin with, I am struck by how different the view of high school life and learning Eckert describes is from the one that is assumed by most researchers for purposes of discussing math and science learning. The latter view of schools and school students' lives is not based on research intended to produce an analysis of the school. Nor is an analysis of the social organization of schools central to discussions of learning in school, judging by the absence of discussion about it in papers in other sessions of this conference. So I shall speculate on, rather than describe, the "world view" of science and math researchers concerning the institution of school.

Figure 13.1 shows the school as I think learning researchers conceive of it. It is a set of identical, socially undifferentiated classrooms filling all of the "social space" of the school—the sum total of the social institution called "school." Within a classroom the teacher is very important. At the same time, teachers are viewed by learning researchers, I think, as an aspect of the process of transmitting knowledge. There is a pedagogical unit consisting of the teacher, the subject matter, and the pedagogy employed. Teachers do not seem to be thought of as actors with complex and conflicting goals, activities, and relations within and beyond the school. We assume that these activities and relations have no impact on teaching, and that the teachers' practice is mainly shaped by subject matter and pedagogical considerations. (Such an observation raises the question of how much the teacher, subject, and pedagogy shape each other, and what else might have significant effects on all three.) In the researchers' world view students appear to occupy a peripheral role as objects or clients on whom services are to be performed. In short, the researchers' view conveniently makes the school into a single-purpose institutional client for researchers' products in a hierarchical social system of which they too are a part.

Students are not viewed as powerfully influential on teacher, subject, pedagogy, or the learning that transpires in the classroom. Math and science researchers at this conference haven't mentioned students' extracurricular activities, sports, dating, and only rarely ethnic group differentiation or gender

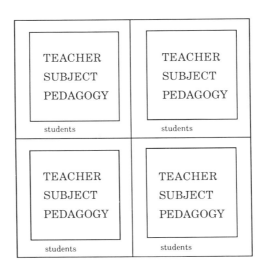

FIG. 13.1.   The school: A learning researcher's view.

differences. There are different types of students in their world view, including good students, slow learners, and/or A,B,C,D, and F students. Dropouts exist outside the school, but only in the background. Only recently has this collective implicit "map" been changed because of work by Saxe (e.g., this volume), Carraher, Carraher, and Schliemann (1982, 1982), Scribner and Fahrmeier (1982), Hutchins (in press), Murtaugh (1985), de la Rocha (1986) and others. From the vantage point of math and science learning research, the most frequent social identity given for the targets of the research is "novices," and sometimes "learners" or "students." These are not, however, the social categories Eckert talks about in discussing the social organization of schooling.

Figure 13.2 is my schematic rendering of the social landscape described in Eckert's work. The school is divided into two opposed social categories—Jocks and Burnouts. The same division creates separate geographical areas (including classrooms), and different subject matters, groups of teachers, and folk epistemologies about the nature and value of knowledge and its forms of circulation in and across social groups. The division is one of social class and reflects class divisions in the society as a whole. Burnouts, who generally come from working class families, configure their minimal (educational) activities in school in vocational classrooms around vocational courses and vocational education teachers. They resist participation in school extracurricular activities including sports. Middle class jocks configure their lives around extracurricular activities within

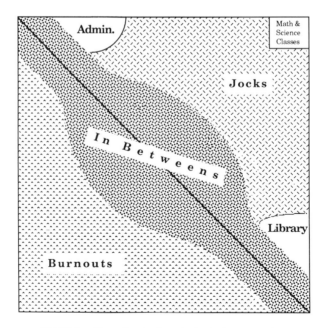

FIG.13.2. The school: A socio-logical perspective.

the school and take academic classes from jock teachers. In neither case is learning in school the central activity students are engaged in, nor is learner or novice a recognizable social category.

Indeed, the term novice which is so common in math and science learning research deserves a bit of discussion. A novice is someone who is deeply committed to a vocation. The term has connotations of a person who is so committed as to strip away all other worldly concerns to concentrate on this one. Paired with the word expert instead of the word "nun," the vocation seems a little less world-rejecting and drastic, but at the very least it implies that someone wants (and intends) to learn what the expert will teach. This characterization, applied to junior high and high school students, is a simplification for learning researchers that has three seriously misleading effects on their research. One is to separate the category of learner from the real social world and the social categories within which people act in the world. Second, it implies that there are no crucial prior conditions (e.g., whether students have agreed to learn) that teachers must negotiate with students before they can teach the subject matter. Third, by assuming that there is nothing about learners of a sociocultural, economic, or political nature that would lead them to treat learning-in-school as a low priority within the broader contexts of their lives, when researchers find comparatively poor understanding or performance, it is easy to attribute it to a permanent and intrinsic psychological inability of the learner. It seems worth reiterating that there are no novices, students, or learners in the social world described in Eckert's research. If, therefore, current theories of the learner and the conventional interpretations of variations in students' performance in research settings are irrelevant and erroneous, we might worry about the power of research on learning to broaden our understanding of effective methods of teaching and learning in schools today.

This worry grows in part from a theoretical perspective which asserts that learning and knowing are socially situated; especially that learning is generated in relations among people, their activities, and the social world in and with which they act. That activity is generated in context implies that it is historically and culturally specific. It is important to understand the actual sociocultural categories in use among the actors in the activity system called school if we wish to infer anything about the intentions of learners or the meaning of their activities.

Referring again to Fig. 13.2, other aspects of the social organization of the school deserve mention. Classrooms are peripheral to most of life in and out of school for Jocks and Burnouts, respectively. In general, they view what goes on in classrooms as less important than other aspects of school or neighborhood life. But math and science classrooms and subject matter are more distant than all the others, for they are seen as the apex of a hierarchy of increasingly authoritative subject matters, and hence ones to be more strongly avoided (by the Burnouts, at least, and by many Jocks, though the two groups have different reasons). Math and science educators should see themselves as especially strongly marginalized in the school social world for a particular set of reasons which it might be useful to understand. And finally, as shown in Fig. 13.2, the figure of teacher, *as teacher* looks very small and parenthetical to the main arenas of action and interest in the school. They loom large as resource controllers and gatekeepers rather than as sources of valued opportunities to learn.

Jocks and Burnouts treat information circulating among them—including academic, classroom, teacher-dispensed information—quite differently (Eckert, this volume). The Jocks experience information as coming downward through a social hierarchy and hence control of information is a resource in climbing up that hierarchy. Information has more instrumental than intrinsic value. Academic knowledge is of only mild and indirect value in the social world in which these children trade—primarily in their knowledge of social relations. Treating knowledge as intrinsically interesting and of value is a violation of the terms of trade. Burnouts share rather than trade information, and resent the hierarchical form which would have to be submitted to in order to gain value from either classroom or Jock social arrangements.

The bad news, then, is that with the exception of small numbers of "In Betweens," pupils in high schools in the United States cannot afford to take school learning as central to their lives or of unconditional interest. That is, there is no room in the social organization of the school for direct and passionate commitment to learning, for it violates the social organization of both Jock and Burnout worlds. That means teachers are faced with students who are not engaged in learning. Or if the teacher is effective it may be that students will have to defend themselves by actively engaging in "not learning". The good news is that the stereotype that many students do not learn because they cannot or because they have no interest in learning, is wrong. High school students, perhaps especially the Burnouts, have lively and pressing interests in knowing about lots of things—though not in the context of the implicit bargain between the school and students, with its attendant hierarchy and privileges through the dispensation of adults.

Since it would be useful to demonstrate some of the implications of this sort of understanding for research on math and science learning, let me reiterate as a set of propositions some of the more important points made by Eckert. Then I'll suggest some examples of questions math and science learning researchers might wish to consider for which the propositions would help to illuminate possible answers. In the United States, at least, schools are total social institutions in which students are expected to carry out the primary activities of their lives and give meaning to them. Jocks conform to these expectations to a greater extent than Burnouts. The differences in social class basic to the powerful and pervasive division that permeates all aspects of school life for all participants in it (including teachers and administrators), are ubiquitous in this country and have been for many years. Variations exist among schools, but the basic system is present virtually universally. Both Jock and Burnout cultures are conservative and opposed to change partly because the opposition between them is a major factor in maintaining the categories in the first place. Jocks are conservative about changes that disturb the distribution of rewarded positions in the student/teacher hierarchy. Changes in classroom modes of organization and reward may be met with resistance from them for this reason. The salient social categories by means of which people organize and give meaning to their lives and activities, including their academic work in school, are not academic ones. If a child does not wish to learn in school, it is easy not to do so. If a child wants to learn something, but not a lot, this is also easy to do. If a student is very interested in learning, it may be easy to learn a whole lot, but it is also socially costly. The choice here is far more the student's than that of the teacher, the school, or the curriculum. Only a few In–Betweens can afford (or are willing) to pay the costs of participation in what might be called "the life of the mind." Neither Jocks nor Burnouts can do so without violating the fundamental criteria for being a Jock or a Burnout. Our public school system is organized as a social institution in which sustained learning stands in conflict with and opposition to the major categories, intentions, and activities with which all members of the institution operate. Schools are basically antilearning institutions. The authoritarian character of the school personnel's management of access to information is a major source of learning rejection by Burnouts. And the math and science curricula play a special role intensifying these problems.

It seems useful to make plain what I am assuming, for purposes of this discussion, as goals for secondary education. I would hope that children learn through experience that working with the mind is not mere toil. The stronger goal is to hook children on the life of the mind or some piece of it. A bookworm is a good example. It would be nice if children left school fascinated and unable to resist learning and doing more about *something* for the rest of their lives. From this point of view "success" for the school would not to be measured in numbers of college-bound graduates or average SAT scores, but by the prevalence of magnificently obsessed kids. According to this view, Jocks, who resist learning less overtly than Burnouts and conventionally appear more promising allies to teachers in the enterprise of knowledge transmission, may in the end be tougher "nuts to crack" than the Burnouts because of the commodity-like instrumental character of information in the Jock social milieu.

There are some general questions concerned with math and science education that appear to have plausible explanations within the framework of socially-situated analysis and in terms of Eckert's analysis of United States schools in particular. I will discuss five of them. First, there has been a great deal of discussion in this conference about cooperative learning (Lunetta, this volume), reciprocal teaching (Brown & Palincsar, in press), distributed cognition and the external reconstruction of understanding (Newman, in press). Cooperative learning methods seem to be effective because it becomes possible for the learner to "own the problem," "understand what the learning is about," "comprehend the story," or "do what the experts do" (Brown, Collins, & Newman, in press). Various of these researchers have emphasized that the finegrained decomposition of tasks, the remedial reaction of teachers when the child's problem is general in scope, leads away from such general forms of comprehension. (Newman gave an example from Petitto's work on long division that beautifully illustrates the point.) Cooperative learning methods seem to work, as Brown and Palincsar (in press), Schoenfeld (1985), and others have demonstrated in robust ways. Why they work is a question that may be addressed in sociological as well as psychological terms. This in turn raises questions about what kinds of differences cooperative learning processes make as the social class division solidifies at higher levels in the public school system. I would predict that it should work as well in high school as in junior high and elementary school, but that students would increasingly resist the changes required to *implement* it as they move higher in the Jock hierarchy. Burnouts' very general rejection of the knowledge-commodity hierarchy might lead them simply to resist, but if resistance could be overcome, they might learn distinctly more and better. The general sociological mechanism by which cooperative learning works lies in the fact that this method lowers the social costs of learning in the school setting. The process is notably less hierarchically organized; teachers give up some control when using the cooperative learning methods and knowledge becomes more available for appropriation in egalitarian terms. For Jocks who are ashamed to be seen liking to learn, groups offer a certain cover and anonymity perhaps. Some combination of these effects (and others) would lead to the prediction that the findings of its effectiveness should be robust, but given that the method requires a shift away from teacher-controlled didactics, high schools would be hostile habitats for its sustained use because of the cost to teachers (McNeil, 1986). (Herein lies a clue to the process by which LOGO, the movement to teach problem solving, and other attempts at less teacher-controlled, more improvisational approaches to instruction have consistently been transformed into more rigidly didactic forms as they have become more widely used.)

It may seem that I have overdone claims for a sociological explanation and moreover have not integrated it with a psychological explanation. I will argue that processes of knowing and learning and the structure of knowledge itself are *made* in socially organized activity: If a teacher teaches didactically, prescribes the way to do something, owns the problem, and goes back to more elementary details as the method of choice in dealing with a befuddled student, then the transmission process, the relations of teachers and students, the relative power of each, and the relative value of the portions of knowledge each controls, will be asymmetrically distributed between teacher and students, and among students. The organization of schools as hierarchical structures, especially the centrally exemplary Jock culture, Jock teachers, and Jock academic subjects, is a single structure of space, people, knowledge, and learning experience. I believe that there is no division to be made between psychological explanations and sociological ones. And there is no value-free knowledge or pedagogical form. The circulation of power is central to knowledge, knowing, and learning. It follows that, if teachers rely on processes of externalization, distributed learning, collaboration, and sharing of information (Saxe, this volume; Newman, this volume; Lunetta, this volume), student ownership and reflection about what is being known, knowledge, students, and classrooms will be organized in less hierarchical ways. And except for some Jocks who are likely to be disaffected because they are more skilled at eliciting resources from teachers through academic negotiations than their peers or who insist on treating knowledge as a commodity to compete against one another for, the result may be a more equitable distribution of knowledge.

Suppose cooperative learning and coaching in the classroom were introduced into a typical high school. What proportion of the variance in what, how much, and who learns would be affected by this (or any other curricular/pedagogical invention of educational researchers)? And what proportion would be accounted for by the social organization of the school as it configures the experience of students actively engaged in negotiating their lives in relation to the institution? I asked Denis Newman this question, and he guessed 2% of the variance in learning might be accounted for by what and how teachers teach and how curricula are organized. Suppose that cooperative learning approaches are less at cross-purposes with the social organization of Jock and Burnout culture(s) than more hierarchical, didactic processes. And imagine the possibility of enhancing the effectiveness of curricula and pedagogy by trying to work through and around the social organization of schooling. It seems much easier to lower the social costs of learning to Burnouts than to convince Jocks that information is more than instrumentally useful. Because so much more of the variance in learning is lodged in the organization of students' lives within the school than in teachers' organization of lessons in the classroom, I should think educational researchers would want to make possible, through sociological analysis, more powerful ways of engaging students in learning (even at the expense of some of the hierarchical asymmetry from which teachers, too, benefit.)

The other four questions for which sociological analysis amplifies or offers novel explanations may be mentioned briefly, though each is a major problem in itself.

1. Eckert's analysis suggests an explanation for why more than once in this century an ethnic minority has found a niche rapidly and strongly as the intellectual elite of the high school. The intellectual apex and the social apex of the school social organization do not coincide. The former is clearly not highly valued within the organization of the school and is essentially "up for grabs," and hence is available to students other than popular Jocks. At the same time, no

ethnic group lodged clearly within either Jock or Burnout class structure can or would move into such a niche because they are locked in a mutually conservative opposition.

2. It puzzles me that there appears to be long-term, perennial dissatisfaction with schools in educational policy making and research circles. I see this dissatisfaction as a structural feature of the social order. It seems inevitable, given that schooling rests on ubiquitous contradictions between socialization and ranking/selection functions and between the ostensible mission of the school to educate children and the concrete reality that the school is organized around the reproduction of the domestic group cycle—a process in which it plays quite a successful, if unacknowledged, part. That is, the school has a key role in the project of controlling the ability of children to gain (Jocks) or guard (Burnouts) independence from the family at a phase in the life cycle (in our nuclear-family oriented society) where to make equivalently strong ties with one's own spouse and children requires a radical break in adolescence and young adulthood with parents. This independence is jeopardized by engaging seriously in academic pursuits, which are organized so as to require docile compliance within the school setting. It seems to me that the constant official dismay with our educational system fuels the cyclical structure of searches for new approaches to curriculum and pedagogy within the educational research community. And the more dire the problem at any given point in history, the more likely are increasing (and arguably self-defeating) efforts at didactic control.

3. So long as 2% of the variance in who-learns-what is all that can be affected by the research and teaching community, while its own view of its mission is much more ambitious and comprehensive, the discrepancy and its attendant crisis mentality and dissatisfaction should certainly drive research in a big and long-term way. But the sociological analysis discussed here maintains that taking a more encompassing view of the problem might reduce the dissatisfaction or increase the possibility of greater than 2% changes, or at least begin to make the dissatisfaction (and its dangers of increasing authoritarian moves) more comprehensible to us all.

4. Under what circumstances could we get more children hooked on knowing something or everything? General pedagogical strategies might aim at lowering the social costs to students (as in various forms of cooperative learning), trying to modify social barriers within the school, trying to mute the antagonism that supports the social division between groups of students and groups of faculty, and helping students to see the value of doing so. This would require teachers who are socially analytic and who are aware of, and try to gain respite from, ways in which institutional structures push them to fulfill strongly hierarchical functions in the system. Administrative structures need to be persuaded to desist or even offer support so that teachers are not forced to deal in destructive didactic control. At the same time, teachers need a clear and steady vision of goals for

what would be of value for students to know. This is congruent with the cooperative, improvisational character of approaches to learning discussed in this conference.

Certainly prospective teachers should understand that if they conform to existing teaching practice in United States schools, they are signing an implicit contract with students to offer authoritative knowledge and distribute resources of freedom in return for differential compliance from students, a contract which is embraced by some and rejected by others. Teachers need to know that they will hear informal reputational information selectively about different categories of students (Eckert, in press). They should know that the agreements to learn made by good students are likely to be highly conditional and partial commitments to learning. Teachers also should know that school work is of only peripheral relevance in schools; that Jocks and Burnouts are hostile to one another and to any creative activity introduced into the school by In-Betweens. They should know how the school systematically destroys Burnout age-heterogeneous networks and allocates social issues between but not across the Jock/Burnout division. Then they might, at the margins, be slightly more knowledgeable participants with a bit more choice and certainly more understanding of the implications of their actions within the institution.

I earlier proposed to call attention to the ramifications of the social organization of the school, beyond the school, and into, specifically, the world of educational research. I think we must analyze how our socially-organized enterprise of research on learning is an extension of the hierarchical control and downward distribution of access to knowledge from the university community to teachers. Much of the discomfort and sense of alienation between researchers on learning and teachers may revolve around these issues. Apple (1979) has documented the increasingly tight scripting and standardization of curricula and the lessening autonomy in teachers' work—the equivalent of going back to drill on old skills for hopeless learners of long division (see Herndon's (1971) brilliant essay on teachers as the "dumb class" of American society). In any event, and without attempting an analysis here, it is clear that we are part of the problem even though we hope to contribute to its solution.

There are several direct connections between the general issue of comprehensive sociocultural forms of explanation of learning and the distribution of knowledge and existing research paradigms in math and science learning. When learning researchers describe their results, I find the possibilities for explaining them in terms of the social organization of knowledge and learning in the school well nigh irresistible. Much more could be done with such an approach, I believe. Second, I would use analyses of learning in and out of school like those of Saxe, Lunetta, Newman, and Eckert as a basis for generating research designs, rather than continuing to rely on the misanalyzed social categories and classroom scenarios that are currently customary in the field. Third, it might be useful to locate the learning research community itself in relation to the social organization of the school. In the process of designing curricula, it would be useful to employ an understanding of the social organization of the school, the meaning of math and science for those who inhabit it, the social significance of learning in a classroom, the meaning of learning in one way rather than another, and the nature of knowledge as a commodity or alternatively as a resource so that new curricula have the meaning and accessibility that we as researchers desire. Finally, it seems cruel to send teachers into high schools unarmed—without a socio-analytic understanding of the school and its impact on their mission there. If, in fact, a substantial part of the variance in which children learn what lies in the social organization of schools, teachers should not have to do without the resources of research and analysis that might allow them, marginally at least, to shift their territory and allies and to have an effect on their own accessibility to different kinds of students and on the cost to different kinds of students of learning what teachers want to teach. These goals can be met only through a deep sociological commitment to the project of improving math and science learning.

There is an old saying that to talk of love is to make love. It may be generalized to the proposition that to focus attention on differences in social class in research contexts is to participate in the reproduction of those categories. If there are ways to resolve this paradox, they are to be found at a different level of organization of our activities as researchers and teachers than the level at which we actually employ those categories in specific acts of research, writing, and teaching. Thus, to both understand the way in which social class is reproduced in school institutions and to use that knowledge with the intention of trying to mitigate its effects on the distribution of knowledge, requires thoughtfully-designed general goals and intentions to act in ways that blur class divisions. If the analysis here were to lead to a new variable in studies of learning, say a "Jock/Burnout index," where the goal was to show that Burnouts do not learn as well as Jocks or Jocks go to college more often, it would indeed contribute to the preservation of those categories. Such research will not illuminate social processes in a way that makes it possible to make learning more powerfully accessible without regard to class and social class itself a less salient organizing principle in schools. It may be more risky, but of greater educational value to try to subvert the divisions that prevent learning in school in the hope that all children might enjoy the benefits of getting "hooked" on some piece of the life of the mind, which seems a lot more fun and worthwhile than not doing so.

ACKNOWLEDGMENTS

I would like to express my appreciation to James Greeno for this opportunity to participate in ongoing debates about math and science learning and to Brigitte Jordan and Jere Confrey for their very helpful comments and suggestions for improving the paper.

REFERENCES

Apple, M. (1979). *Ideology and curriculum.* London: Routledge and Kegan Paul.

Brown, A., & Palincsar, A. (in press). Reciprocal teaching and comprehension strategies: A natural history of one program for enhancing learning. In J. Borkowski and J. Day (Eds.), *Intelligence and cognition in special children.* Norwood, N.J.: Ablex.

Brown, J., Collins, A., & Newman, S. (in press). Cognitive apprenticeship: Teaching the craft of reading, writing and mathematics. In L. B. Resnick (Ed.), *Knowing, learning, and instruction: Essays in honor of Robert Glazer.* Hillsdale, NJ: Lawrence Erlbaum Associates.

Carraher, T., Carraher, D., & Schliemann, A. (1982). Na vida dez, na escola, zero: Os contextos culturais da aprendizagem da matematica [Life 10, School 0: The cultural contexts of math learning] *Caderna da Pesquisa, 42,* 79–86.

Carraher, T., & Schliemann, A. (1982). *Computation routines prescribed by schools: Help or hindrance?* Paper presented at NATO conference. Keele, England.

de la Rocha, O. (1986). *Problems of sense and problems of scale: An ethnographic study of arithmetic in everyday life.* Unpublished doctoral dissertation, University of California, Irvine.

Herndon, J. (1971). *How to survive in your native land.* New York: Simon & Schuster, Inc.

Hutchins, E. (in press). Learning to navigate. In S. Chaiklin & J. Lave (Eds.), *Situated Learning.* New York: Cambridge University Press

Lunetta, V. & Tamir, P. (1981). An analysis of laboratory activities, project physics, and PSSC. *School Science and Mathematics, 81*(8), 635–642.

McNeil, L. (1986). *Contradictions of Control: School structure and school knowledge.* New York: Routledge and Kegan Paul.

Murtaugh, M. (1985). A hierarchical decision process model of American grocery shopping. Unpublished doctoral dissertation, University of California, Irvine.

Schoenfeld, A. (1985). *Mathematical problem solving.* New York: Academic Press.

Scribner, S., & Fahrmeier, E. (1982). *Practical and theoretical arithmetic: Some preliminary findings* (Industrial Literacy Project, Working Paper No. 3). New York: City University of New York, Graduate Center.

IV THE IMPACT OF TECHNOLOGY

Andrea A. diSessa
Alan H. Schoenfeld
University of California, Berkeley

There are good reasons to be entranced by technology. Some, perhaps even most, of the major social revolutions in history have been technologically propelled. The Iron and Stone Ages bear the names of the technologies that characterized them. The Middle Ages may have been made by armor, the Industrial Revolution by mechanical power technology, and current popular culture by television and radio. And now those of us who are concerned with education have arrived at a time when the cutting edge of information technology (computers, videodiscs, etc.) spices up our intellectual world—which hadn't really changed much technologically since the invention of movable type. We eagerly await and modestly try to help fashion the revolution that we expect will ensue.

Interestingly and importantly, none of the three papers in this section are technophilic. Not one is filled with unbounded optimism about what technology will automatically accord us in educational terms. In surprising ways, none of the three papers are about technology, per se. Instead, they tell us about clashes of cultures, about developing new cultures, and about how critically the impact of technology depends on our ability to make progress in understanding long-standing educational problems.

Goodstein describes the production of a video-based, lavishly and meticulously computer-animated attempt to "propel the teaching of physics, in one breathtaking step, into the twentieth century." But not very far beneath the surface, we find the intellectual struggle for dominance between video production, which has been nurtured almost entirely by interest and the public's willingness to pay, and the academics' insistence on integrity and on their own values and instincts as to what works educationally. We find that the production itself draws heavily on an instructional tradition at the California Institute of Technology that is deeply embedded in older media—lecturing and textbooks. And we find that some of the central expectations for how the produced materials would work with one of its major targets—the woefully under-populated and under-educated ranks of high school physics teachers—needed considerable revamping including support from other sources, tutorials, and supplementary textual material.

Schoenfeld takes as his central theme the exquisite complexity involved in any kind of understanding; his example is understanding functions and graphing. We see in his principles of design old educational goals—making the content visible and manipulable and encouraging students' reflections about their learning and about the nature of the subject they are learning. Of course technology puts a new slant on these issues; some things can be made manipulable and visible in new ways. The fundamental mode of interaction between teachers and students can change: The computer, rather than the teacher, becomes a source of mathematical observations whose meaning must then be negotiated by the student and teacher. In these interactions, we have a new window on mathematical cognition. Yet, at root, we are still dealing with questions of what constitutes knowledge and what is the character of human routes to its acquisition.

diSessa claims that neither the central issues for design of educational technology nor even what and how students will learn with computers can be seen in the machines and software themselves. Instead, he is concerned with the activity structures that surround the use of the technology and with the social niches, both old and new, into which technology will fall. He aims to provide us a view of the system he is designing, Boxer, by imagining those structures rather than by being concerned directly with the technology. We are challenged to consider constructivism as a principle for activities as well as a principle for intellectual structures: We must seek to understand and engineer a continuity of goals and meaningful activities for our students just as much as we seek to build a continuity of ideas and concepts.

So, while technology is the label for this section, we should appreciate that technology is really not the name of the game. Information technology will certainly have a revolutionary impact on our society, but whether or not it will spark a revolution in education depends on another quieter advance. That is the theme of this conference: the advance in our scientific understanding of learning and of the social contexts for it and our concomitant ability to engineer new and better kinds of learning contexts with or without new technological resources.

14

The Mechanical Universe and Beyond: Physics Teaching Enters the 20th Century

David L. Goodstein
California Institute of Technology

The Mechanical Universe and Beyond the Mechanical Universe[1] are the two semesters of an introductory college-level physics course presented by means of television. As I discuss later, it is intended for a wide variety of audiences, but its central target audience is the high school physics teacher. The reason can be seen from the following starkly simple argument: There are about 25,000 high schools in the United States. The number of fully qualified high school physics teachers—those with the equivalent of an undergraduate major in the subject—is not known with precision, but all authorities seem to agree that it is fewer than 2,000, not even 1 for every 10 high schools. It follows that the majority of American students pass through high school without encountering a competently taught physics course and arrive at college having already foreclosed the possibility of majoring in physics in order to become part of the next generation of teachers. Thus the problem is not only critical, it is also self-perpetuating. There are, to be sure, other problems associated with physics education in America, but reversing this situation must be the first order of business in any serious attempt to improve matters. That is the central mission of The Mechanical Universe project. The medium chosen to accomplish this purpose is, perhaps, the only one

---

[1]Produced by Caltech and the Corporation for Community College Television (CCCT). The academic staff included: David Goodstein, project director and host; Richard Olenick, associate project director; James Blinn, computer animation; Tom Apostol, mathematics and academic content editor; Steven Frautschi, academic content; Dave Campbell, Robert Westman, Judith Goodstein, academic consultants; Don Delson, project manager; and Judy Post, Renate Bigalke, Gwen Anastasi, and Debbie Bradbury, project secretaries. The production staff was led by Sally Beaty, executive producer and president of CCCT, and included Peter Buffa, producer; Mark Rothschild, associate producer; Jack Arnold, story editor; and Robert Lattanzio, production manager.

that has any real chance to do so: the use of network quality television with all its built-in appeal and existing technical infrastructure.

The organizers of this conference and this volume have chosen to place the description of this project in a section called Uses of Technology in Education. Presumably that is because they regard the use of television as a technological innovation in education. They are undoubtedly correct in that view. The classroom is one of society's most unchanging creations. The dominant technologies of today's classroom are the chalkboard (traceable directly to prehistoric cave paintings) and the printing press (15th century). Television, however, has certain advantages as a contender for the next technology to sweep into the business of education. For example, according to a Roper poll conducted in January 1987, 93% of all American households own color television sets. The number of households that owns books is unlikely to be any higher. Moreover, as educators, we must be aware that the young consumers of our product have, to a remarkable extent, become acquainted with the world through the medium of television. Undoubtedly, television already plays a vastly larger role in learning (broadly defined) than do the more direct ancestors of cave painting and the printing press. Moreover, it has also begun to find its way into more formal education.

In the Fall of 1986, college television courses broadcast on public television stations had an estimated audience of 2,360,000 households out of a potential 87,000,000 (Dirr, 1987). Furthermore, broadcast television is by no means the only way televised instruction can reach its audience. The Roper poll previously mentioned indicated that 47% of American households own VCRs, and 50% are connected to cable television outlets. In the 1984–85 academic year, some 902 college-level institutions (32%) offered telecourses to students, and 87% of all public four-year institutions used video for in-class instruction, (Riccobono, 1986). It is obvious that television has already become a visible part of American higher education. Television in higher education is part of a quiet revolution that has seen the traditional college student (18–22 years old, enrolled full time) become a small minority (less than 20%) of those enrolled in college credit courses. Nevertheless, before The Mechanical Universe and Beyond, the list of courses available through television did not include physics.

The Mechanical Universe and Beyond is, by any standard, an extraordinary project in education. It is a rigorous mathematically sophisticated college course in physics, prepared with all the technique and dazzle made possible by television, together with textbooks (Frautschi, Olenick, Apostol, & Goodstein, 1985; Olenick, Apostol, & Goodstein, 1985, 1986), teachers' manuals, student study guides (Apostol, Campbell, Dukes, & Sirko, 1988; Campbell, Dukes, & Sirko, 1988), and a separate and complementary version for use in high school classrooms. In short, it is an attempt to propel the teaching of physics, in one breathtaking step, into the twentieth century. A few words about how this all came about seem to be in order (Goodstein, 1986; Goodstein & Olenick, 1988).

Dedication to the teaching of physics began at the California Institute of

Technology (Caltech) more than 50 years ago with a popular introductory text-book written by Robert Millikan, Duane Roller, and Earnest Watson, (1937). Millikan was Caltech's founder, first president, first Nobel prizewinner, and all-around "patron saint." Earnest Watson was dean of the faculty, and both he and Duane Roller were distinguished teachers.

Twenty five years ago, the introductory physics courses at Caltech were taught by Richard Feynman, who was not only a scientist of historic proportions, but also a dramatic and highly entertaining lecturer. Feynman's words were lovingly recorded, transcribed, and published (Feynman, Leighton, & Sands, 1963–1965) in a series of three volumes that have become genuine and indispensable classics of the science literature. Unfortunately, they have also proven virtually unusable as introductory textbooks.

The teaching of physics at Caltech, like the teaching of science courses everywhere, is constantly undergoing transition. Caltech's latest effort to infuse new life into freshman physics was assigned to me around 1980 and eventually led to the creation of The Mechanical Universe and Beyond. Word reached the cloistered Pasadena campus that a fundamental tool of scientific research, the cathode-ray tube, had been adapted to a new purpose and, in fact, could be found in many private homes. Could it be that a large public might be introduced to the joys of physics by the flickering tube that sells us spray deodorants and light beer? About the time when the idea of using television to teach physics started to take on serious proportions in Pasadena, it was announced that Walter Annenberg, publisher and former Ambassador to Great Britain, had established a gift of $10 million per year for 15 years to make use of telecommunications in higher education. The ultimate outcome of this happy concurrence of events is a series of 52 half-hour television programs plus the previously mentioned materials at a cost of nearly $6 million.[2] (The high school version, separately financed by the National Science Foundation, cost an additional $3.5 million.)

The genesis of The Mechanical Universe and Beyond explains why it is not (as many telecourses are) a course prepared by a committee. Although a very large number of people have made indispensable contributions to the project (see Footnote 1), and although it takes every possible advantage of the opportunities offered by the television medium, it is nevertheless firmly rooted in the original Caltech course in style, emphasis, choice of subject matter, and many other ways. The raw material from which the television series and textbooks were produced was a complete set of verbatim transcripts of the lectures in the Caltech course delivered by me during the academic year 1980–1981.

The series covers the standard subjects of introductory physics: mechanics, electricity and magnetism, relativity, thermodynamics, and modern physics.[3]

---

[2]The fund is administered through the Corporation for Public Broadcasting (CPB) by the Annenberg/CPB Project.

[3]For more detailed information of these programs, see Goodstein and Olenick, 1988.

Algebra and trigonometry are prerequisites, and students should be prepared to use them in the study of physics. We do not assume that they are familiar with differential and integral calculus, but we do believe that physics cannot be studied seriously without those techniques. We therefore teach them as a part of the science of mechanics, which is indeed how they were discovered originally. This is not the standard practice in American higher education, but The Mechanical Universe was not designed to reaffirm standard practices.

Under the guidance of Caltech Mathematics Professor Tom Apostol (well known for his highly regarded calculus textbooks), the calculus is handled in an intuitive way, designed to make its use seem a simple, easily learned procedure, but without losing sight of its remarkable power, and without teaching crudities that would have to be unlearned by students who want to learn mathematics more thoroughly. As plans for The Mechanical Universe evolved, the decision to include calculus drew much fire from critics who had suffered the painful experience of teaching the standard course without the added difficulty of calculus. Accordingly, once the first semester of The Mechanical Universe had been offered as a course, a preliminary evaluation was made to see how the idea worked.[4] Results showed that, while some students did have severe difficulties with the mathematics required of them, the calculus was not their downfall. In what some regard as one of the major surprises and triumphs of the project, they were able to master derivatives and integrals much as we had hoped they would. Even algebra offered relatively little difficulty. The principal problem was that they had never mastered the trigonometric functions in high school. Once this lesson was learned, a trigonometry primer written by Professor Apostol (Apostol et al., 1988) was quickly added to the arsenal of printed aids available to students and teachers.

If the choice of mathematical level is the most important curricular innovation in The Mechanical Universe, the use of computer animation is the jewel in the crown of teaching technique offered by the series. Dr. James Blinn, who teaches at Caltech and toils in the Computer Graphics Laboratory at Caltech's Jet Propulsion Laboratory (JPL),[5] is the superstar of the computer graphics field, and The Mechanical Universe is his magnum opus. The series contains an astonishing average of more than eight minutes of new animation per program, and in the words of one knowledgeable observer, "These materials will constitute the principal visual image of physics for decades" (C. H. Holbrow, personal communication, April, 1986).

There are some 550 animated scenes in all, comprising about 7½ hours of screen time. To set the scale, consider that the commercial market rate for high-

---

[4]This telephone survey of users was done by Dr. Peter Combes, an independent consultant hired by the Annenberg/CPB Project.

[5]The JPL Computer Graphics Lab was headed by Robert Holzman who paved the way administratively for Blinn to work on The Mechanical Universe.

quality computer animation (the flying logos on television network broadcasts) is about $4,000 per second. All of this work was done by Blinn, with help from his assistants Sylvie Rueff and Tom Brown. During the last 14 months of the project, Blinn turned out new animation at the astonishing rate of close to 3 minutes per week.

Blinn's undergraduate degree is in physics and his PhD is in Computer Science. In daily (7 days per week) consultation with me, the computer animated scenes were designed and developed to take maximum advantage of Blinn's vivid visual way of understanding physics as well as the techniques at his disposal (many of them invented by him). A discussion of visual, numerical, and other techniques used in animation for The Mechanical Universe is itself a book-length treatise (Blinn, 1987). I mention here only a few details in the hope of conveying some of the flavor of the enterprise.

Most of the scenes that will appear to casual viewers as merely attractive illustrations in fact use scientifically valid, sometimes very elaborate models. For example, the many scenes of atoms forming states of matter—solid, liquid, and interacting gas—are in fact two-dimensional molecular dynamics calculations, in which the atoms, interacting via Lennard-Jones potentials, are given initial conditions and set free to obey Newton's laws. (Of course it always took some experimenting with initial conditions to get the results to look right.)

Blinn used colors in a way that may have a subliminal effect, even if viewers don't notice his schemes explicitly. For example, differentiation makes quantities redder. Thus, position is green, velocity yellow, and acceleration red; angular momentum is pale blue and torque is lavender (red + blue). These are the colors of both the vectors and the algebraic symbols representing each quantity. In the program Low Temperatures there is a phase diagram in the pressure-temperature plane. Here, solid is earthy brown, liquid is watery blue, and gas is transparent white. Above the critical point, where there is no boundary between liquid and gas, the color shades smoothly from blue to white according to a formula based on the Van der Waals equation of state. This sort of meticulous attention to scientific correctness, even where no one but the makers will notice it, is characteristic of all the animation in the series.

The problem of how to present detailed mathematical derivations is confronted in the animated scenes. The dilemma here is that one risks losing the audience either by skipping steps or by presenting each step in an excessively didactic way. Skipping steps would rob the presentation of its rigor and continuity and leave the impression that there are things about physics that we consider too difficult or arcane to show to our audience. Presenting each step would have the same effect it often has in real lectures. In producer Peter Buffa's memorable phrase, it would "put their brains in a 60-cycle hum." It would also be a woeful misuse of the television medium. The compromise solution of this problem, invented at the outset of the project, is called the "algebraic ballet."

In an algebraic ballet, an animated derivation is done in detail, but rapidly and entertainingly. The viewer is not expected to absorb every operation merely by watching the symbols move on the screen, but every step is displayed correctly, and the mathematical argument is presented without losing the viewer's attention. Our own (strictly informal) evaluation of how well this works indicates a strongly age-related effect. Younger viewers—perhaps because they are more attuned to television or possibly because they are not accustomed to understanding everything they see—seem to enjoy them much more than older viewers, who are made uncomfortable by the algebraic manipulations they cannot quite follow. In any case, the algebraic ballets do serve their primary function, viewers are unanimous in agreeing that their attention is never lost during these mathematical passages. And for students who need to study them in detail, the same derivations are to be found in the textbooks (Olenick et al., 1985).

Each program opens and closes with me lecturing to what is supposed to be a Caltech class. As we'll see later, the lecturing professor has long been anathema in educational television, and I may very well succeed in bringing that proscription back forever. The lecture hall sequences were shot in blocks, 2 programs per day, 2 or 3 days per week, during the summers of 1983 (for The Mechanical Universe) and 1985 (for Beyond the Mechanical Universe). In each case, this was before most of the scriptwriting had begun, much less finished, and so the scripts had to conform to the already shot book-end sequences. The material in the scenes was all taken from real Caltech lectures, and the scenes were shot without scripts because producer Buffa wanted a sense of realism and spontaneity. However, because each shot was repeated five or more times, spontaneity sometimes suffered.

The scenes were shot before an audience of about 20 extras, whose entrances and exits were carefully choreographed and rehearsed. The production staff felt they were combining the best parts of the then hit series Paper Chase and Hill Street Blues. In fact, after a futile search for a classroom that resembled the one in Paper Chase, Buffa reluctantly settled for the real one used by Caltech physics classes. Audience reaction shots of real Caltech students were sprinkled in later during post-production editing.

The writing of the scripts posed a problem of formidable proportions. At first, story editor Jack Arnold organized a string of free-lance writers to whom he assigned scripts. Later, writer Don Button was hired full time, and most scripts were written in-house. Early in the project, the writer of a given script would be given edited portions of relevant lecture transcripts to work with supplemented by long earnest discussions in which the academic staff would seek to teach the writer the necessary physics. Later on, a new procedure was adopted; the writers were given drafts of all the physics scenes, which I wrote in advance. Their task was thus to weave a program between the bookends and around the physics scenes.

Each script would start with the choice of a "concept location" (e.g., a magician's performance for Static Electricity, a gas station in the desert for

Temperature and the Gas Laws and so on. Some programs were purely historical in theme and did not have a location in this sense). With this starting point, the writer would successively submit (for Caltech's approval) a one-page outline (the concept), a multipage detailed prose description (the treatment), and finally a draft script. At this point, the script was polished by Arnold, who rewrote the narration into what became a distinctive "Mechanical Universe style," and the result was reviewed once again by the academic staff and sometimes by outside consultants to check for accuracy. If it passed this test, it was declared "in production" and the production team would be free to start planning scenes. Of course, in reality the process seldom worked this smoothly, and, as we'll see later, every script underwent extensive rewriting at later stages.

The scriptwriting process was the principal stage for a classic conflict between the academic and production sides of the project. In succinct but exaggerated terms, the production people were in favor of fewer equations and more "beauty shots," and the academics wanted more physics and less time wasted on filler material. The outcome of the fray may have been influenced by the fact that contrary to the case of most telecourses, the final decision was in the hands of the academics. However, it was much more strongly influenced by the fact that each side had genuine respect for the professional competence of the other, each understood that the ongoing debate itself was healthy so long as it remained in reasonable balance, and perhaps by the fact that, with separate offices in Pasadena and Hollywood, the two sides did not meet often enough to get on each other's nerves.

Lurking behind that debate, however, was a deeper question: For what audience was The Mechanical Universe intended?

According to the charter of the sponsor, the primary audience was to be "nontraditional students," especially "distance learners," earning college credit by watching television. In reality, of course, physics is far too difficult to learn merely by watching television, but it was hoped that with a resourceful, dedicated local teacher, a physics course by television would be possible, and that the teaching of introductory physics at any level could be enriched using these programs in or out of the classroom. It was also hoped that a large, casual, nonstudent audience would watch the programs for pleasure and instruction. In making specific decisions, however, it's often useful to have a clearer target in mind. For The Mechanical Universe, that ideal target was the high school physics teacher. We shall return to this point again later.

The Mechanical Universe, from beginning to end, is intensely historical in its approach to physics. This orientation is a direct reflection of the Caltech course from which it arose. In essence, the earlier Feynman course had sought to make physics exciting by relating each subject, wherever possible, to contemporary scientific problems. The new course took a different position and attempted to recreate the historical excitement of the original discovery. For example, classical mechanics—a notoriously difficult and uninspiring subject for students—is treated as the discovery of "our place in the universe." Accordingly, celestial mechanics is the backbone of the subject, and its climax is Newton's solution of the Kepler problem.

Given this point of view, and the central imperative of television—that there must be something on the screen, preferably moving, at all times—historical recreations inevitably become a staple part of the project.

The historical scenes in The Mechanical Universe are generic in nature. Young Newton strolls through an apple orchard, old Newton testily refuses a cup of tea from a servant, and so on. Attempting to recreate a specific moment or event (e.g., Newton inspired by the apple falling) would have been silly and presumptuous (and expensive). The scenes that were made could be used in many programs, whenever the person in question came up in the narration. Aside from Newton, scenes were made of Galileo, Leibnitz, Benjamin Franklin, Michael Faraday and James Clerk Maxwell. To film these scenes and others, crews were sent to England, Holland, and Italy as well as Dearborn, MI and Philadelphia, PA. However, young Newton's apple orchard was in Yucaipa, California; Galileo confronted Vatican authorities at the Clark Library in south central Los Angeles; and Leibnitz's visit to a German royal court unfolded in the main lounge of The Athenaeum, Caltech's faculty club. In addition, about 11 minutes of Kepler footage was purchased from an earlier Public Broadcasting Services (PBS) series, Cosmos. For more recent personages (Einstein, Michelson, Rutherford), stock footage—some of it rare archival material—was turned up. But Robert A. Millikan posed a special problem.

Early in the series, a program is devoted to Millikan's oil-drop experiment, partly as an application of Newton's second law, and partly to introduce some philosophical ideas about how science is really done. To make this program, some sort of Millikan footage was needed, but an actor playing Millikan did not seem to be a good idea, especially at Caltech where many people still remember him well. The solution was to create a meticulously detailed Millikan Museum, combining elements of his laboratory and his office in a room in the Norman Bridge Laboratory where he had worked. The set involved thousands of artifacts, many of them Millikan's own (the iron canister of the oil-drop experiment, his desk, letters, papers, notebooks, etc.). After 3 days of shooting the scene under the supervision of associate producer Mark Rothschild, the museum was captured on videotape and then disassembled.

In addition, the series includes a large selection of stills (usually shot with a moving camera), many of them portraits or pages from rare first editions of important books. These were generally dug up by researcher Carol Buge. Extensive use was also made of stock footage, much of which was taken from previous telecourses (Oceanus, Project Universe, etc.). There are also a small number of original cartoon (or cell) animations executed by artist Mike Shaw.

A substantial portion of the work done on each program took place very quickly, in the (usually) frantic final stages, after scriptwriting and shooting of all scenes was completed. The key element was the computer animation. When the animation for each program emerged from the JPL graphics lab, the final assembly process could begin.

Assembling the program was the responsibility of the editor. He would read the narration onto the soundtrack of a ¾" tape, imitating the professional narrator's normal speed, and then using in-house editing equipment in the Hollywood offices, he would choose and transfer scenes onto the tape. The resulting reel, called a "rough cut" would be sent to Pasadena to give the academic staff its first clear picture of what had been created. The result was usually an intense period of rewriting under considerable time pressure because the subsequent editing stages had to be scheduled in advance, and the final programs had to be delivered on a remorseless schedule that was tied to contract payments.

The first consideration were the animated scenes. Entirely new narration, timed to the actual scenes, almost always had to be written because, rather than follow the script he was given, Blinn tended more to treat it as a source of inspiration, taking off from there to create something entirely new. The second priority was the overall timing of the program. If it was more than one minute long or short (almost always the case), serious emergency surgery had to be performed. Often, one or the other of these kinds of changes required work that affected every part of the original production script. This rewriting was generally done by Buffa, Arnold, and me, with Apostol and others pitching in on some occasions. The program was then reedited into a second rough cut that would be subject to detailed scrutiny by the entire academic staff. At this stage, however, changes were generally limited to the wording of narration, not to the order or length of scenes.

The final stages of preparation of each program involved: (a) recording narration by actor Aaron Fletcher (in some of the more technical passages, we hear the female voice of executive producer Sally Beaty); (b) transfer by computer of the narration and scenes onto 1" broadcast tape; and (c) the final video and audio editing, including sound effects, music (composed and synthesized for the Mechanical Universe by Sharon Smith and Herb Jimmerson), dissolves, and other special effects. These final stages, which cost close to $20K per program, were done in the same rental facilities used by many commercial television programs. Mistakes were sometimes caught and corrected in this process, but each fix became progressively more difficult and expensive as the process unfolded. After a final check in Pasadena, the program was declared "in the can," and was sent off by overnight courier to the sponsor in Washington.

Although producing the television programs was a dramatic and supremely demanding task, work of comparable magnitude was proceeding simultaneously on other fronts. Textbooks were being written at a rate that would have been impressive if the authors had no other responsibilities. Richard Olenick, working from the Caltech lecture transcripts, wrote the first draft of each chapter, with Tom Apostol and me revising and editing. Steven Frautschi would then rework the manuscript for the advanced edition. In addition, there were the teachers manuals, study guides, and so on. But from the point of view of the purpose for which The Mechanical Universe was intended, no part of the project was more important than the high school adaptation.

As remarked earlier, the ideal audience for The Mechanical Universe was high school physics teachers. All the participants in the project have been impressed with the quality, dedication, and professionalism of the high school physics teachers with whom we have had contact, but those tend to be among the minority who are genuinely well qualified to teach the subject (Yager, 1980). To facilitate use of The Mechanical Universe to attack the teacher shortage problem, a new initiative was undertaken to adapt material from the programs for use in high school classes. The idea was to induce teachers to study the college-level version so that they could use the high school materials with poise and confidence in the classroom.

The high school adaptations of The Mechanical Universe were begun during the summer of 1984. Under a materials-development grant from the newly revived science and engineering education directorate of the National Science Foundation, 12 teachers were chosen from the many responses to an ad placed in *The Physics Teacher*.[6] They convened for the first time in the summer of 1984 on the Caltech campus under the direction of Richard Olenick and Kathleen Martin, and have been at work ever since. The group is called the Materials Development Council (MDC).

Starting from scripts, rough cuts or, where available, final programs from the broadcast version, the MDC chooses topics suitable for presentation in high school curricula and fashions, from the available footage, video modules averaging 15 minutes in length. These are accompanied by extensive written materials for the teacher, including suggestions, background material, demonstrations and sample questions to be used in conjunction with the video tapes. Each package is then field tested under professional direction[7] before being revised and issued for distribution. By the 1986–1987 school year, 70 teachers and thousands of students in 30 cities and 10 states were participating in the evaluations alone. Twenty four modules have been completed and four more are planned.[8]

As originally conceived, The Mechanical Universe was to be "physics in a plain brown wrapper" for high school teachers. Crossover teachers from other disciplines could brush up their physics in the privacy of their own living rooms. In practice this view proved to be naive. Teachers want and need special courses with considerable expert help, well organized workshops and, where possible,

---

[6]For a complete listing of the teachers selected, see Goodstein and Olenick, 1988.

[7]Evaluation was conducted by Geraldine Grant and J. Richard Harsh, Educational Evaluation Consultants, Long Beach, CA.

[8]The high school materials are distributed in "Quads." Each Quad consists of four 15-minute video adaptations and related print materials for the student and the teacher.

tuition and stipend support. Efforts to provide these services have been organized in a number of states (Staff, 1987). Reports from the field indicate that teachers are willing to work very hard to improve their skills when given reasonable support, and that even those teachers who are well qualified in physics find new inspiration in the unusual approach and materials offered by The Mechanical Universe.

The diffusion and impact of The Mechanical Universe beyond the realm of high school physics teachers has been difficult to assess. One reason is that although all PBS stations receive the programs via satellite, they do not have to pay for their use and therefore do not have to report whether they have broadcast them. Nevertheless, it is known that in the Fall of 1986 alone, The Mechanical Universe was carried by approximately 100 PBS stations and had an estimated casual audience in excess of 400,000 households. It has been purchased or licensed for use as a telecourse in roughly 300 institutions of higher education, and the high school adaptations have been distributed to more than 2,000 high schools in all 50 states. In addition, there have been numerous private reports of teachers at all levels taping broadcast programs at home for use in class, a procedure of noble purpose but questionable legality. Incidentally, neither Caltech nor the Corporation for Community College Television (CCCT) receive any part of the revenues generated by distribution of the series.

At the insistence of its sponsors, The Mechanical Universe and Beyond is designed so that it can be used as a complete stand-alone telecourse. A college telecourse administrator choosing to offer the course for credit identifies a teacher of record, uses the prepackaged catalog description, press releases and so on, and arranges for airing on the local PBS station. For the teacher there is a manual containing advice on how to run a telecourse, plus solutions to all the problems in the textbooks, banks of test problems, and so on. If all this sounds a little strange to those of us accustomed to more traditional ways of going about the business of higher education, it has to be put in the context of educational television.

In the early days of television, an early rising professor could often be found in front of the video camera starting the new day with "sunrise semester." As time went on, that invaluable treasure of the academic profession, the professorial lecture, came to be known among the television professionals as the "Talking Head." Although they neglected to say from what other part of the body they preferred speech to originate, the talking head went out of fashion as a means of television presentation. However, the television pros may have had a point. For all its virtues, the live lecture was hardly novel by the time of Aristotle, just as the tutorial predates Socrates. It may be that neither is the final word in education technique. But no one was quite sure what to do instead.

As a result, television education evolved in two separate directions. One is the prime time (8 p.m.–10 p.m.) television series on PBS stations, which are designed primarily to attract a broad audience (by PBS standards), but later equipped with enough apparatus (textbook, study guide) so that some colleges are willing to offer credit in connection with them. Recent examples include a number of science-related series, such as Cosmos, The Brain and Planet Earth (two Annenberg products), and earlier, the memorable Ascent of Man. The other direction is the set telecourse. Designed specifically to be offered for college credit, today's telecourses are much more highly produced than the old sunrise semester classes, although they still typically cost less than what even PBS prime time material costs to produce. These telecourses are much more nearly legitimate college-level education than are many of the warmed-over prime time series, and in fact some are quite good. But even at best they have tended to be designed by an assembled committee of academics headed by a television producer. If the essence of college education is to learn about a subject from someone who has spent a lifetime mastering it and, if possible, adding new knowledge to it, the traditional telecourse lacked that essence.

To be sure, there is a legitimate—perhaps even a very important—role for telecourses to play in American higher education. The population contains a vast reservoir of people who, for one reason or another, are or were unable to attend college in the conventional way and who are genuinely eager for the chance to learn. That rich pool of potential talent should not be ignored by a nation whose greatest strength has always been the ingenuity of its people. It also should not be ignored by colleges and physics departments faced with the problem of anemic enrollments. The intent and hope of The Mechanical Universe is to serve that need, not merely with adequate education, but with the very best that can be created.

Two principal factors seem to govern whether The Mechanical Universe succeeds as a telecourse at any given institution. One is how it is advertised— whether students are clearly told what they are getting into. Truth in labeling does tend to lose some students, but it also picks up others who would not otherwise be interested. The other factor is the infrastructure that is set up to aid the student. When possible, on-campus, in-person recitation (and laboratory) classes are probably the most helpful, but that often defeats the very purpose of a telecourse. Depending on local circumstances, limited on-campus meetings, mailings, telephone, and even radio and cable television are being used. The bottom line is that The Mechanical Universe can succeed as a telecourse, but only where a resourceful telecourse administrator works in full cooperation with a committed physics professor.

An alternative use of the material that has proved more consistently successful is as an audio-visual aid to conventional classroom physics courses. For example, one professor who uses the videotapes in class for a course taken by pre-med and other pre-professional students says the students are delighted to find that the most dreaded course of their college careers has turned out to be unexpectedly pleasant. The mode of use varies widely with the tastes of the instructor. At Caltech, where all students have a higher level of mathematical preparation than is assumed for the telecourse, the advanced edition of the textbook has been adopted, and the television material has been used as support for the standard lecture format. Snippets, usually computer animation, are shown in class, and full programs are available on reserve at the library, where a playback system is kept in the next room and in a comfortable lounge near the dormitories much frequented by the students. There are various other means by which students can get to see programs (including, of course, broadcast by local stations). However, professors at other institutions report success with showing entire programs in class, using the remainder of the period to discuss, expand, illuminate, or refute what has been seen. Remarkably, wherever the series has been seen (on broadcast television), there has been much acclaim including numerous awards at film festivals,[9] and ecstatic reviews, letters, and phone calls. The series has even registered Nielsen ratings in some cities. Some of this may be due to sheer surprise that anyone would attempt something so difficult. An editorial about The Mechanical Universe in the Los Angeles Times said " . . . if differential calculus is not television's Supreme Test, it would certainly make the semifinals in any competition" (Los Angeles Times, 1986 editorial page, Thursday March 27). Nevertheless, there is no mistaking the genuine pleasure and enthusiasm in the responses of viewers of all ages and backgrounds. I take the meaning of this reaction to be that our programs succeed, at the very least, in being physics lessons that do not produce hostility in students toward the subject. The LA Times editorial took a more sanguine view. The editorial, titled "More Than a Promise," maintained that The Mechanical Universe has finally fulfilled the educational promise that has always been inherent in television.

I hope they were right.

REFERENCES

Apostol, T. M., Campbell, D. A., Dukes, T. S., & Sirko, R. J. (1988). *The mechanical universe: trigonometry primer and student study notes.* New York: Cambridge University Press.

Blinn, J. F. (1987). *The mechanical universe: An integrated view of a large scale animation project.* Unpublished manuscript. California Institute of Technology, Pasadena.

Campbell, D. A., Dukes, T. S., & Sirko, R. J. (1988). *Beyond The mechanical universe: Student study notes.* New York: Cambridge University Press.

Dirr, P. (1987, July). *Report on Usage* (Unpublished Annenberg/CPB Report).

Feynman, R. P., Leighton, R. B., & Sands, M. (1963–1965). *The Feynman lectures on physics, Vol. I, Vol. II, Vol. III.* Reading, Massachusetts: Addison-Wesley.

Frautschi, S. C., Olenick, R. P., Apostol, T. M., & Goodstein, D. L. (1985). *The mechanical universe: Mechanics and heat, advanced edition.* New York: Cambridge University Press.

Goodstein, D. (1986). *Physics teaching today* (Annenberg CPB Project Report to Higher Education). Washington D.C. p. 8.

---

[9]The series has been presented with a variety of awards from—to name a few—the American Film Festival, International Film & TV Festival, the National Educational Film Festival, and The Japan Prize.

Goodstein, D., & Olenick, R. (1988). The making of the mechanical universe. *American Journal of Physics,* V. 56 p. 779.

Unsigned editorial (1986, March 27). More than a promise *The Los Angeles Times,* editorial page.

Millikan, R., Roller, D., & Watson, E. (1937). *Mechanics, molecular physics, heat and sound.* Boston: Ginn & Co.

Olenick, R. P., Apostol, T. M., & Goodstein, D. L. (1985). *The mechanical universe: Introduction to mechanics and heat.* New York: Cambridge University Press.

Olenick, R. P., Apostol, T. M., & Goodstein, D. L. (1986). *Beyond the mechanical universe: From electricity to modern physics.* New York: Cambridge University Press.

Riccobono, J. A. (1986). *Instructional technology in higher education.* Washington, D.C.: Corporation for Public Broadcasting.

Staff. (1987; Fall). Reports from the field, Meisner, G., Mogge, M. E., Rotter, C. A., *The Mechanical Universe and Beyond,* p. 3.

Yager, R. E. (1980). Tech. Rep. No. 21): University of Iowa, Science Education Center. Armes.

15

GRAPHER: A Case Study in Educational Technology, Research, and Development

Alan H. Schoenfeld
University of California, Berkeley

INTRODUCTION AND OVERVIEW

For the past few years, the Functions Group at Berkeley[1] has been developing computer-based microworlds, collectively called GRAPHER, for exploring aspects of mathematical functions and graphs. We began by asking how we could harness computational technology to help students understand some complex mathematical ideas, to help them "see" some things that are difficult to grasp conceptually. The result was a pilot version of a computer-based environment focusing on the multiple connections among and multiple representations of graphs and the algebraic formulas that generate them.

As always, the research took on a life of its own once we ran pilot studies of the system. Typically, the pilot sessions included a student, one of our staff who served either as tutor or experimenter, and the system itself. The nature of the student-tutor interactions in the presence of the system differed in interesting ways from those interactions in its absence. First, with the machine presenting some of the mathematics (and accepted as an unquestioned authority), the tutor took on the role of a colleague working with the student, to determine why the machine did what it did. This changed the dynamics of the tutorial interaction, suggesting new roles for teachers alongside the technology and a reconceptualization of "expert" teaching, both for humans and machines. Second, the nature of the dialogue between student and teacher changed. The mathematical

---

[1]Members of the Functions group whose work contributed to this paper are Abraham Arcavi, Kate Bielaczyc, Laurie Edwards, Marty Kent, Bill Marsh, Luciano Meira, Judit Moschkovich, Mark Nakamura, J. J. Price, and Jack Smith.

objects on the screen served as conversation pieces whose properties student and tutor explored in much more detail than is typical in most teaching interactions. This kind of interaction allows a microscopic examination into students' cognitive structures and the ways they come to understand a domain, and to ways they misunderstand it and what kinds of interactions might be necessary to remediate those misunderstandings. Our group is now working on a representational scheme to capture the richness of students' conceptual structures and the growth and change of those cognitive structures as students come to grips with a subject matter domain.

UNDERSTANDING FUNCTIONS AND GRAPHS

We start with the assertion that coming to grips with mathematical ideas—even in ostensibly straightforward domains—is an extraordinarily complex task. An appreciation for this complexity is essential for following the argument in this paper; we shall go into minute detail about a student's difficulties with one small corner of the world of functions and graphs. This "micro" approach and its ramifications only make sense if one is prepared to believe that learning in a domain consists of much more than the simple and straightforward accumulation of facts and procedures. In this section, as an example, we illustrate what it means to understand graphs of straight lines. Aspects of that understanding include the following.

1. General geometric knowledge about lines, whether or not a student thinks of those lines as being represented formally in the Cartesian plane. This knowledge includes a grasp of concepts such as point, angle, line, and parallel. It also includes knowing that two points determine a line, and that a point on a line and the direction of the line also determine it. It includes knowing that parallel lines do not intersect but that nonparallel lines intersect in precisely one point, and so on.
2. Knowledge of algebra and equation solving and knowledge of functions and the ways in which they are expressed algebraically. (We could expand this paragraph, but for the purpose of brevity, we'll stop here.)
3. Knowledge about the Cartesian plane: how to plot points, what a graph is, etc.
4. Knowledge of the relationships between functions and their graphs; for example, an understanding of the mystical phrase "a point (x_0, y_0) lies on the graph of $y = f(x)$ if and only if the pair (x_0, y_0) satisfies the equation $y = f(x)$.
5. Knowledge of algebraic conventions regarding lines and their entailments in the Cartesian plane. For example: (a) the slope/intercept form for writ-

ing the equation of a line is $y = mx + b$; (b) m is the slope of the line and b is its y-intercept; (c) b is the y-intercept which means that the graph of $y = mx + b$ crosses the y-axis at the point $(0,b)$; (d) the graph must cross at that point because when $x = 0$, the corresponding y-value is b in the equation $y = mx + b$; and (e) knowing what slope is and how different values of m result in graphs with different orientations.

6. Thinking simultaneously of the graph $y = mx + b$ as a collection of points (each of which corresponds to a value satisfying the equation) and as a single entity (as a function, which is itself a member of the class of linear functions, parametrized by m and b in the slope/intercept form). Knowing the graphical entailments of changes in m and b: How does a line change as the value of b increases, for example, or as m becomes a large negative number?

7. Knowing different algebraic representations for straight lines (e.g., the two-intercept form, the two-point form, the point-slope form) and how they are all related. Knowing that they are all alternate characterizations of the same line, which may be differentially useful to employ in different circumstances.

8. Knowing all of the above in such a way that knowledge of one aspect of straight lines (e.g., general knowledge, algebraic knowledge, graphical knowledge) constrains knowledge in any of the others. This redundancy serves as a powerful positive force and error detector in problem solving.

As one illustration of the interconnectedness of this knowledge, consider the statement made above that "two points determine a line." Although this is a real world statement, it has a Cartesian and algebraic interpretation as well: Given any two points $P_1(x_1,y_1)$ and $P_2(x_2,y_2)$, the student should be able to determine the equation of the line L and its graphical properties. The reader might wish to do so and reflect on the nature of the connections required to perform the derivation. We give one such derivation below.

If $x_2 = x_1$, then the line is vertical and its equation is $x = x_1$. If $x_2 \neq x_1$, the slope of the line segment from P_1 to P_2 is $m = [(y_2 - y_1)/(x_2 - x_1)]$. The point/slope form of the equation of L is thus $y = [(y_2 - y_1)/(x_2 - x_1)]x + b$, where b is the y-intercept of the line. Because (x_1,y_1) satisfies the equation $y_1 = [(y_2 - y_1)/(x_2 - x_1)]x_1 + b$, we can solve for b in this equation. Hence, the equation of L is $y = [(y_2 - y_1)/(x_2 - x_1)]x + [(y_1x_2 - y_2x_1)/x_2 - x_1)]$. Or if you prefer a neater algebraic form with different graphical entailments, this equation can be rewritten as $y = [(y_2 - y_1)/(x_2 - x_1)](x - x_1) + y_1$. (We can derive this latter form more directly if we think in terms of the graphical properties of the equation in the Cartesian plane.)

In the preceding discussion, I offered the barest outline of understanding the graphing of straight lines. I hope, however, that the main thrust of the discussion is clear. Understanding even this simple domain means (a) having a great deal of prerequisite knowledge at the student's fingertips; (b) having multiple perspectives on the objects involved (e.g., function as procedure and function as object); and (c) having multiple representations for them and a coordinated means of moving among the perspectives and representations. Finally, it means having all of this knowledge organized in ways that derive power from the redundancy (and mutual constraints) from all these links. Such connectedness does not come easily, and charting its development is the major thrust of our theoretical enterprise.

The example worked out above, graphing straight lines, is merely the tip of the functions-and-graphing iceberg. Consider parabolas for example. The coefficients in the standard algebraic form,

$$y = \mathbf{a}x^2 + \mathbf{b}x + \mathbf{c},$$

all determine the shape and location of the parabola, but not in clearly transparent ways; variations in the coefficients result in somewhat mysterious transformations of the parabola. Changes in **a** change the width and possibly the direction of the parabola and move the vertex; changes in **b** leave the shape of the parabola identical but move the vertex; changes in **c** translate the parabola vertically. Given the standard form, you have to do some algebra to determine either the vertex or the roots of the parabola. The coefficients in the vertex form,

$$y = \mathbf{a}(x - \mathbf{b})^2 + \mathbf{c},$$

provide different information. The width and direction of the parabola are determined by **a,** and the vertex is (**b,c**); changes in **a** affect only the width and direction of the parabola, while changes in **b** and **c** result in horizontal and vertical translations of the parabola, respectively. The third form, the root form,

$$y = \mathbf{a}(x - \mathbf{b})(x - \mathbf{c})$$

is less frequently used, but has its advantages and drawbacks. One can read the values where the parabola crosses the x-axis ($x = \mathbf{b}$ and $x = \mathbf{c}$) directly from the equation, but reading other graphical properties off the form is difficult, as are algebraic manipulations of this form.

One can think of these three forms, and the graph of the parabola, as different manifestations (or representations) of the same thing—as the same function seen from different perspectives. As one manifestation changes, all change in different ways. More importantly, given the student's goals at any particular time, it is frequently more advantageous to think of the function in one or another of these representations. Here, as in the case with straight lines, the student's knowledge of the domain is highly interconnected, with understandings about any aspects of these functions (in the general geometric world, the Cartesian plane, the algebraic realm) interacting with understandings of the others. Learning about functions and graphs means learning about all of these entities (linear, quadratic, etc.), their different manifestations and representations, and the connections among them.

THE SYSTEM: PRINCIPLES AND REALIZATION

Our goal is to provide the cognitive support structure to help students learn about functions and graphs from their introductory experiences with the Cartesian plane to the point where the students develop the kind of fluency in the domain— regarding the behavior of polynomials, rational functions, and circular functions—that was illustrated in the previous section. If this sounds ludicrously ambitious, I hasten to add that our intention is not to put all the intelligence in the machine. We do not intend to build a self-contained tutor for functions and graphs, a task which, given our conception of the desired goal state, would be extremely difficult and perhaps not possible on theoretical grounds. The idea is for the technology to serve as a catalyst facilitating conceptual growth and change. It does so in accord with the principles elaborated below.

The path we wish to make smoother with the help of the technology is a "learning trajectory" that results in the kinds of competence held by people who are fluent in the domain. This learning trajectory via the technology need not recapitulate the learning trajectory taken by those without it. With the help of the technology, students might (a) have different experiences with the mathematics and (b) be able to deal with it in a different order or in different ways. For example, students need not start by graphing very simple functions if the computer will display complex functions that the students can analyze. Different sequencing made possible by the technology may allow for different kinds, and orders, of "scaffolding."

A look at one such learning trajectory (in this case my own experience, which is typical) is a useful starting place. This trajectory indicates the kinds of conceptual development and change we wish to facilitate. In school, I learned to draw graphs such as $y = x^2 + x - 3$ by calculating the y-values for different x-values (usually $x = 0, \pm 1, \pm 2, \pm 3$, etc.; with more points if more detail was necessary), making a table of values, plotting the points from the table, and joining the plotted points with somewhat curvilinear segments. In the beginning the "overhead" for all the subsidiary operations was tremendous; after the trouble of calculating, plotting, and drawing, there was hardly the focus to reflect on the curves or their properties. When I was done plotting I didn't have the "real" curve; I had my clumsy approximation to it. That clumsy approximation wasn't an entity in and of itself; it was, rather, the complicated set of points and segments that resulted from the process of calculating, plotting, and drawing. Over time and with extensive experience, I came to abstract the idealized mathematical curve from the empirical procedure: After graphing many such curves, I came to see the real curve behind my actions as the background operations became less important (more routinized) and I could focus on the curve itself. Even so, this was only the first step in a long progression.

Consider what I saw when I compared the graphs of $y = x^2 + x - 3$ and $y = x^2 + x - 7$. The two looked similar. Moreover, having composed the tables for both, I know that the latter had the property that each of its y-values was precisely four units below the corresponding y-value for the former. Nonetheless, this was a point-by-point comparison of the following four-step type: (a) Take a representative sample of points on the first curve, (b) for each such point mark a point in the plane four units below it, (c) join the points, and (d) compare the composite sketches that result. This is part of the story, but not all of it. The mathematician also thinks of the two curves as representing the graphs of entities, where $y = x^2 + x - 7$ is the function (Note the singular; it is one object!) obtained by subtracting four from the function $y = x^2 + x - 3$, and the graph of $y = x^2 + x - 7$ (Once again, note the singular; it is perceived of as a whole single entity) is obtained by shifting the graph of $y = x^2 + x - 3$ four units downward in what is technically called vertical translation. At some point, I developed this understanding based on my ability to think of the functions and graphs themselves as conceptual entities—objects that, despite their complexity, I could think of as single, concrete manipulable objects. Later, I could begin to imagine the transformations as dynamic manipulations. Having graphed $y = ax^2$ for many values of a (often on the same sheet), I could eventually see the idealized family of parabolas $y = ax^2$ vary continuously as a varied. These understandings developed about imagined entities and transformations, because what I had in front of me were "snapshots" of members of the family of curves from which I imagined an animation consisting of a sequence of the snapshots moving in time. With more experience and through similar mental processes, I came to grips with the families of parabolas described in the previous section, their different forms, and the relationships among them. The same happened with other mathematical curves.

Our goal is for GRAPHER to make such development easier. As noted previously, GRAPHER is intended to be used in a social context—as a conversation piece facilitating discussion among students and between teacher and students. In the light of the preceding considerations, GRAPHER was designed with these six principles in mind.

1. GRAPHER should facilitate *conceptual concretizing,* the process by which complex concepts and relationships come to be conceived of as single, effectively concrete entities. Ideally, the microworld should help students move from the basics—learning about mappings in the Cartesian plane point-by-point—to the advanced understandings described above by helping them to negotiate every level transition along the way.

2. The dynamic and interactive nature of the computational medium should be exploited wherever possible to make tangible and explicit those mathe-

matical ideas that must otherwise by perceived or imagined indirectly. Even better, conceptual entities should be made meaningfully manipulable, so that students can learn about their properties by operating on them directly.

3. Wherever possible, the links the instructor wants the students to see should be made overtly. For example, multiple representations of the same concept should be simultaneously visible and, in light of the preceding principle, manipulations on one representation of a mathematical entity should result in instantaneous changes in the other representations as well.

4. The computational power of the medium should be exploited to perform drudge work for the student so that the student can focus on the conceptual issues being explored.

5. The medium should support meaning exploration rather than knowledge telling. That is, it should give students the opportunity to try ideas, see how they work, and develop their own understandings—rather than being told about something must be done and given drill and practice exercises.

6. The medium should encourage and facilitate students' reflection on their understandings and communication of those understandings.

In its first version, GRAPHER was implemented on Xerox 1109 LISP machines, which have large screens that allow for the simultaneous display of large multiple windows. It is now implemented on a Macintosh II. GRAPHER consists of three linked microworlds, which are called Point GRAPHER, Black Blobs, and Dynamic GRAPHER. At any time, students can move from one microworld to another and then return to the place they left in the previous world. Point GRAPHER is the entry-level microworld.

Point GRAPHER

In Point GRAPHER, students define (that is, specify the equations of) functions and explore their properties and graphs on a point-by-point basis. In its current state, Point GRAPHER requires students to have some familiarity with the Cartesian plane and knowledge of how to plot points in it. However, our next level of development includes a series of explorations (a curriculum) that lower the entry level to GRAPHER and will help students learn to plot points as well.

Point GRAPHER's screen (see Fig. 15.1) contains four major windows depicting function definition, graphing, point-by-point calculation of function values, and history. Additional interactions with the system are obtained through *dialogue boxes*. We explain each of these entities below.

The Function Definition Window

The function definition window occupies the top center portion of the screen. There the students can type in the equation of any function they wish to explore

288

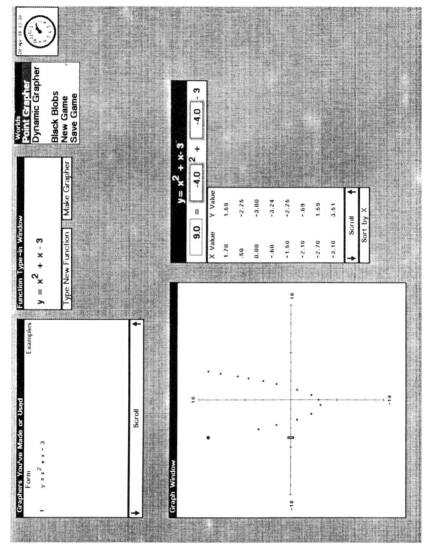

FIG. 15.1 Point GRAPHER's Screen.

(or recover any function that has been previously explored from the history window and either pick up from that point or modify the function). When the students click on the tab marked "make grapher," the system enters their function in the history window and starts the graphing process for them.

At this point, the function box and a table appear on the screen. The function box has two parts: a top white-on-black part giving the equation defining the function that the student has just produced and a box below it in which the slots occupied by variables are indicated by boxes.

Suppose the student decides to graph $y = x^2 + x - 3$. There are various ways to input values into the function. The simplest is the type-in mode. The student can, for example, type 2 in one of the x-boxes for the function. When the student hits a carriage return, Point GRAPHER swings into action. All the slots in the function box get filled in, showing that the value 2 has been substituted for x.

A dialogue box appears off the side of the graphing window with the following message.

$$
\begin{array}{l}
\text{When } x = 2.0 \\
\qquad y = (2.0)^2 + (2.0) - 3 = 3.0, \\
\text{so the point} \quad (2.0,\ 3.0) \\
\text{is on the graph.}
\end{array}
$$

The point (2,3) is then plotted on the graph, with arrows from the coordinate axes (in this case, at $x = 2$ and $y = 3$) indicating the point in question. Simultaneously, the values $x = 2$, $y = 3$ are entered in the function table. The process can be repeated as long as the student wants to continue in this fashion.

At any time, the student can switch to dynamic x-entry in which the values of x in the x-variable box change dynamically as the student moves a mouse. X-values are entered into the function box (an action equivalent to typing a carriage return in type-in mode) by clicking on a mouse button. There is a difference, however. As the student changes the x-values dynamically, the "moving point" (x,y) appears, dynamically, on the graph. It leaves no trace—until it is entered for a particular value of x—but its appearance foreshadows the appearance of the function.

The Graphing Window

Aspects of the graphing window were described in the preceding paragraphs. When it appears at the beginning of a session it offers the standard representation of the Cartesian plane, with x- and y-values ranging from -10 to $+10$. As (x,y) pairs of values for the function are computed, those values are plotted on the graph. When x-input is dynamic, a slider on the x-axis shows the x-value currently in the x-box, and a moving point on the graph indicates the corresponding (x,y) point of the function.

Students have various options in the graphing window. They can zoom in or out, changing the range of x- and y- values shown on the graph by a factor of 10. They can also connect the dots in their graph, getting a polygonal approximation to their function. By comparing that piecewise linear graph with the ghost image generated by the moving point, the students can see how close an approximation they can create to the real graph when they connect the dots. Finally, when they begin to understand more about functions, they can give the grapher hints. For example, students can tell the grapher that $y = x^2 + x - 1$ is symmetric about the line $x = -\frac{1}{2}$ and instruct it to plot "two points for the price of 1"—one they compute and its mirror image. (Note that if they make a mistake in instructing the Point GRAPHER to do so, the ghost image will provide immediate feedback indicating that they've goofed.)

The History Window

The history window provides a record of all the equations the student has explored and a means of access for regenerating any or all of the work that any student has done. It gives the teacher and students the opportunity to review the students' work. When someone clicks on any equation in the history window, that equation is reentered into the function definition box and the work on that equation to date (table, instructions to grapher regarding symmetry, etc.) are regenerated. Students and teachers can then go back over work done by the students, or students can continue work they stopped earlier. In addition, students can modify the function they've just entered in the function box and then explore the new function (e.g., they can change $y = x^2 + x - 1$ to $y = x^2 + x - 5$ and explore the new function).

Indeed, students can explore as many functions as they like at the same time. How do $y = x^2 + x - 1$ and $y = x^2 + x - 5$ compare? Have the machine build a table, plot the points, and draw the graphs simultaneously.

Options and Dialogue Boxes

Most of the options in the system (e.g., customizing the grapher, zooming, etc.) become available when students click at the top border of a window. A dialogue box opens up telling the students what the options are and, once they are selected, how to use them. For example, consider the student who has just examined the two functions in the previous paragraph and who wants to explore the behavior of functions of the general form $y = x^2 + x + A$. The student clicks on the top border of the function box and selects the option generalize. At that point, the system responds with a prompt of the following type. "Your function is of the form $[1]x^{[2]} + [1]x + [3]$. Click on the numbers you wish to generalize, and hit the carriage return." The student could click on any of the numbers, but suppose that she just clicks on the [3]. The machine then creates a new form $y = x^2 + x + A$, which becomes part of the student-defined list in the Dynamic GRAPHER (see next section). In this way, the system helps students to generalize the classes of functions they have explored and examine the properties of those general classes.

Black Blobs

The Black Blobs game was motivated by Sharon Dugdale's work with the Green Globs game (Dugdale, 1984) and is similar in spirit. With more powerful computers at our disposal, we have a much broader graphing capacity (i.e., we can graph larger classes of functions) and a nicer user interface. In addition, the history window in this microworld provides a means for students and teacher to explore prior work and reflect on the student's work to that point; the opportunity for students to call their shots in the game provides a nice diagnostic tool for students, teacher, and possibly the intelligent tutoring system which may ultimately be embedded in GRAPHER.

When students start a game, a coordinate grid appears on the screen with 13 almost randomly generated targets scattered over the grid. (Our screen is black and white while Dugdale's is monochrome green; hence the different but related names of the microworlds.) The idea is for students to hit the targets with graphs of functions. One enters the equation defining a function into the system, and the graph corresponding to that equation is drawn. The more targets a graph passes through, the higher the score one gets. Clearly, as students get better at seeing how a function will behave from its algebraic definition, they get more skilled at the game. Hence, playing the game should help students to develop the kinds of understandings about the domain (i.e., connections between the graphical and algebraic worlds) previously discussed.

Black Blobs also provides a simple utility. You can call up a clear screen (no blobs) or ignore the blobs that are on the screen and graph any function you care to define. Using this facility, you can see what graphs look like one at a time (as opposed to seeing the graph emerge point-by-point in the Point GRAPHER or as a member of parametrized classes of functions as in the Dynamic GRAPHER).

Dynamic GRAPHER

The structure of the screen for the Dynamic GRAPHER is similar to the structure for the Point GRAPHER in that it has a history window, graphing window (with zooming options, etc.), and function definitions. The difference here is that the objects that are defined are *classes* of functions. For example, one of the (predefined) classes of functions is linear functions in the form $y = mx + b$. If a student calls for this class (by clicking on it with the mouse), then a form comes up with boxes for the parameters that define the class. In this case,

$$y = [\ ]x + [\ ].$$

The student can then put values in the boxes, for example, 1 in the first box and 3 in the second, to obtain

$$y = [1]x + [3].$$

This identifies a specific member of the family, $y = x + 3$, which is graphed. But now the student may change either parameter dynamically. The system then graphs the family of functions in real-time as the parameter is changed. Hence, if the constant term is varied, the line moves up and down through the y-intercept; if the coefficient of x, which represents the slope of the line, is varied, the line rotates around the y-intercept accordingly. In short, the system displays the dynamic change—the animation—that I had to imagine for myself as a student when I struggled to learn the subject matter.

Here is a more complex example. The system can simultaneously display the three algebraic forms for the parabola described in the preceding paragraph, $y = ax^2 + bx + c$, $y = a(x - b)^2 + c$, and $y = a(x - b)(x - c)$, along with the graph; it will update all of the four forms if one of them is changed. For example, if you change the coefficients in the vertex form, the other algebraic forms change as well, and the graph moves accordingly. (If the graph moves so that it no longer crosses the x-axis, the root form changes to $y = a\ (x - ***)(x - ***)$ indicating the quadratic does not have real roots.) Similarly, you can change either of the other two algebraic forms or pick up the parabola and move it with corresponding changes in the remaining forms. See Fig. 15.2 which, alas, is a snapshot and cannot demonstrate the dynamic aspects of the Dynamic GRAPHER.

Finally, the various linear and quadratic forms discussed in this section are "canned" examples built into the system by the designers. The system is not limited to our examples, however. Instead, it is intended to support the students' explorations. Recall that the last option in Point GRAPHER is for students to generalize their own functions—to take any single functions in which they are interested and convert any of the constants in their functions into parameters. This leads to the general quadratic forms if one starts with a particular one. Of course that is only the beginning. A student who discovers the discontinuity in $y = 1/(x - 1)$, for example, can go on to explore functions of the type $a + b/(x - c)$ and then functions with more than one vertical asymptote, etc. (Of course, the hard part of curricular design is selecting tasks and creating the social contexts in which such discoveries are likely to take place.)

Brief Discussion

As previously noted, we tried to design GRAPHER to reflect a set of underlying principles. Given the space constraints under which I write, I can only hope that

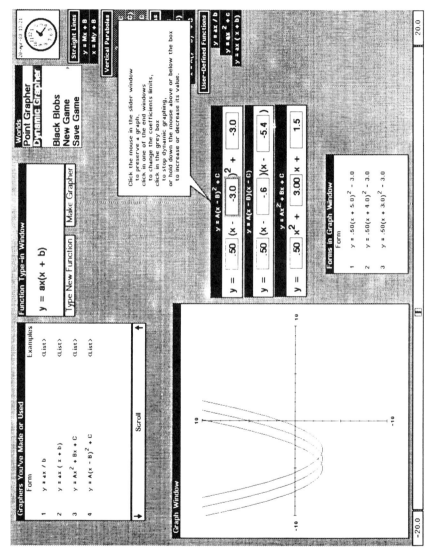

FIG. 15.2   Dynamic GRAPHER.

293

it is clear how the screen design reflects some of those principles. Of particular importance are: (a) making conceptual entities (graphs as rigid objects) computationally manipulable in the Dynamic GRAPHER; (b) making simultaneous multiple representations; (c) providing tables and filling them in in Point GRAPHER; and (d) using history windows. Another important principle is the main focus of the following section.

A SESSION WITH GRAPHER:
DETAILS AND IMPLICATIONS

In this section, we describe a student's interaction with GRAPHER as an illustration of the way the system works and the kinds of interactions it supports.

RB was the first person to pilot our system. She had recently earned a Bachelor's degree in psychology and claimed that she had once been familiar with all of the functions in GRAPHER; however, it had been 5 years since she had worked with functions, and she wanted to review them. We worked together with the system. Since my main purpose was to find out how she would interact with GRAPHER, I did not have a direct tutorial goal. Most of the time I served as a sounding board as RB explored the domain by herself. RB set her own agenda and got little direction or help from me unless she asked for it directly. In such cases I usually referred her back to the system with the suggestion that she explore something specific where GRAPHER would provide the relevant feedback. She had been introduced to Dynamic GRAPHER once before for about 10 minutes. This was our first extended session, described here in very telegraphic form. The part of the session discussed lasted about a half hour.

RB picked the $[y = mx + b]$ form of the straight line to explore. That form appears on the screen with default values $y = [1]x + [0]$. She began to change **m** dynamically by using the form $y = mx$. Her first goal was to obtain the value **m** $= 0$ in order to produce a horizontal line. When she achieved this she announced her intention to use the same form, $y = mx$, to obtain a vertical line. Of course this is impossible, and I could have told her so. However, my intention was to see what would come of her explorations; hence, I did not intervene.

RB increased the value of **m** dynamically until she got to $y = 10x$, the pre-set but changeable limit in Dynamic GRAPHER. She looked at the graph and said that the value of **m** would have to be larger in order to get a vertical line. How much larger? "Well, I guess maybe 12 or even as much as 13 . . . [and] maybe even as much as 15. I'm just talking about eyesight-wise [that is, one would be unable to distinguish the line from the y-axis] I think as much as 15 would get it." To check on her background knowledge, I asked about slope. She said correctly that the **m** in the equation gives the slope of the line, and that "slope is "rise over run," so slope is how far up the y-axis you are divided by how far across the x-axis you are." To check on her understanding, I asked her what a line with a slope of 1 would look like. In the context of our screen [in which the default configuration goes from -10 to 10 along both axes] she said "It's 10 up . . . divided by 10 over. So you have 10 this way divided by 10 that way. So that means you have a diagonal." So far, so good. To keep track of her guesses and allow for higher coefficients (options that are now in Dynamic GRAPHER), RB and I moved to Black Blobs.

RB typed in $y = 15x$ and was surprised that its graph did not appear vertical. When I asked her what integer-valued point on the coordinate grid it might go through, she (again, correctly) said it would be "15 up this way and 1 across on the x-axis right about here." Yet the contradiction between that statement and her desire that the line be vertical was not apparent to her. She suggested trying 14 and was surprised the line went down instead of up; we examined why. Then she suggested that $y = 20x$ might do it. I pointed out that the segment of the graph we were looking at only went as high as the value $y = 10$ and asked at what x-value she thought her equation $y = 20x$ might cross at the top of the screen.

RB: Well, half a unit I guess. [That is, the line would pass through (½, 10).]
AS: Why is that?
RB: I don't know. I see that there's a relationship there but as to why it's happening, I just don't know. . . . But it seems to me that if you double the coefficient of the x then it [the line] gets half-way closer to the y-axis.

When $y = 20x$ didn't work, she suggested $y = 80x$ in order to make the two half-way jumps at once. When that didn't work, she suggested that perhaps $y = 100x$ would be vertical. Supposing it was, I asked, what $y = 200x$ would look like.

RB: Well, I would think it would go over here [pointing] and that it would go diagonally into the negative quadrants . . . although that's not right either.

Acknowledging her confusion, we decided to explore one particular line, $y = 50x$, and see if we could figure out why the line appeared the way it did.

RB: All right. Let's try a straight 50. Let me see if I can predict where it's gonna be. . . . Well, [a slope of] 50 is 50 up and 1 over so it's gonna be way up there . . . so [pointing to the point (1,10) on the screen] it would get a little towards the 1 here . . .
AS: Is there any way of telling how close . . . towards the 1 it might be?
RB: Well, is there any way of telling? Well it should be 50, one 50th, right? 50 over 1? I mean . . .
AS: What does the 50 over 1 mean?
RB: Well, 50 over 1 means that it goes up 50 for every 1 that it goes over.
AS: So what does that tell you about what happens on the part of the screen you can see?

> *RB:* Well, we have 5 [pointing 5 units up on the y − axis], oh yeah, this is a tenth then, a tenth of the way where it needs to go.
>
> *AS:* Why is it a tenth of the way where it needs to go?
>
> *RB:* Because it would have to go up 50 to get over to the 1, and it's only gone up 5, which is $\frac{1}{10}$ of 50, so that means it's only gone $\frac{1}{10}$ of the way over towards it. Hey! Ok, so that means . . .
>
> Well, I knew this though, . . . that that's what the slope is—for each one up it goes that many over—but I'm still confused as to why I can't get it to be [vertical].
>
> Well, all right, if that's $\frac{1}{10}$ of the way, then I need to get it, if I want to get it parallel [to the y-axis], then . . . I don't know why I'm getting confused. It's $\frac{1}{10}$ of the way to the 1, but I don't want it to be on the 1, I want it to be on the zero. So it's infinity up for, Oh! *that's what it is,* it's any number up for zero over. So it's divided by zero so it is infinity.

I ask her to recapitulate.

> *RB:* I can see that the idea is that I can go up as much as I want but I don't want it to go over any, and you can't divide by zero . . . so it's undefined. So the slope of a vertical line is undefined. Hey! . . . Ok! I won't forget that anytime soon! Ok. So all that [i.e., a half hour with GRAPHER] and now I have a feeling for straight lines.

Discussion

The session with RB leads to the consideration of two major issues. First, it illustrates the point made in the opening section of this paper, that mapping out what it means for someone to understand a particular body of subject matter is an extremely delicate and subtle issue and that delineating the nature of a particular person's understandings (and how they change) is far more subtle and difficult. Superficially, it appears that at the outset RB had reasonable mastery of aspects of the domain, for example, the definition of slope (as rise over run) and its interpretation in graphical forms (a slope of 50 means the line goes up 50 units for each unit it goes to the right). A closer look reveals that her knowledge was quite fragile, localized, and often disconnected. Some of her knowledge was purely nominal, in that she could name or define terms but not trace through the consequences of those definitions. (For example, if line L_1 has larger slope than line L_2, then L_1 should rise faster than L_2. RB was surprised by this.) Some of it was incorrect or contradictory, some of it was stable, and some of it was transitory (evoked in our conversations, but context-dependent and not necessarily likely to emerge in other contexts). Some of her formal mathematics knowledge was tied to real-world meanings, some independent of it. To get an idea of what RB (or anyone else) understands of the mathematics, all of this—the knowledge, its connectedness, and its robustness—has to be laid out in very fine detail. To understand what it means for RB to come to grips with this (or any other) mathematics, one has to be able to explain how her knowledge structures grow and change, for example, how pieces of knowledge get represented, how they become stable, how connections get made, how actions in the problem space are constrained by those new connections, and how contradictions get resolved. A schematic outline of the cognitive territory is given in Fig. 15.3. For an extensive discussion of these issues and a sample analysis, see Schoenfeld, Smith, and Arcavi (1989).

The session with RB also serves to illustrate the point that GRAPHER should support meaning negotiation rather than knowledge telling. Note that RB was hardly a newcomer to this domain; she had years of experience with functions and graphs and some fairly well-engrained conceptions and misconceptions about them. In terms of the perspective outlined in the previous paragraph, she had a very stable (although sparse, often nominal, and largely disconnected) knowledge structure with regard to the domain. We view RB's learning in this domain as her engagement in a set of activities that results in reasonably stable changes in that knowledge structure.

Such changes do not come easily, although superficial appearances of subject matter mastery often give the impression that they do. Recall that my intention

Aspects of Domain-Specific
Knowledge Relevant to
Understanding Functions and Graphs

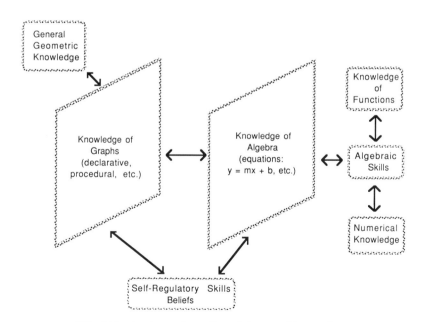

FIG. 15.3   A schematic outline of the cognitive territory

was not tutorial; I did not set out to teach RB about specific properties of straight lines. But suppose that I had. I might have laid out a nice curriculum, a set of examples that showed how the inclinations of various lines changed as their slopes changed. I could have led—naturally, clearly, and in much less time—RB to arrive at the idea that one cannot obtain a vertical line using the form $y = mx + b$. RB would have appeared to understand all this, but a week or two later nearly everything she learned would have vanished. The reason is that her prior set of conceptions was so strong and so well embedded that, unless the new information "took root" in it—that is, the new information became connected to her existing knowledge structures in a fundamental way—the superficial veneer of learning would fade, and the old structures would reemerge essentially unchanged. (There is nothing radically new in this assertion; consider the literature on "naive physics," for example.)

Our interaction over the subject matter, slow and uneven as it was, was rooted in RB's conception of the domain. GRAPHER made the mathematical objects tangible for her; she could create them and then reflect on their properties. This had two consequences. The first is that GRAPHER provided a different kind of feedback than the kind a student normally gets in mathematics. Normally, the student works a problem and arrives at an answer, after which one (by plugging numbers back into equations, essentially re-doing one's work, looking it up, or being told) figures out whether it is right or wrong. Here, the graphs spoke for themselves providing instantaneous and more meaningful feedback. If they didn't behave as expected, RB found out fast and could explore why. The second consequence is that GRAPHER's presence changed my relation (as tutor-of-sorts) with RB. In a standard tutorial role, I would provide feedback to RB on her conjectures. I'd have to produce the graphs myself, and the way I chose to graph different lines would be based on what I thought was important; hence the feedback necessarily contained my own biases. The mathematical authority would be mine in that the explanations would be perceived as coming from me rather than springing from the mathematics itself. In contrast, GRAPHER simply (almost magically and with authority that was completely accepted) produced the lines corresponding to RB's equations; it was then our task to make sense of them when the graphs didn't correspond to her expectations. The mathematics was "out there," rather than being tied up with me and my explanations. I was a much more neutral presence and could interact differently.

By way of analogy, consider what the discussion of friction in a physics lab would be like with and without equipment available for the discussion. The dialogue might start this way.

Tutor: Suppose you have a block that slides down an inclined plane. Now, suppose I increase the weight on the block. Will the block slide more slowly or more rapidly?

Student: There's a larger frictional force, so it should slide more slowly.

If there were no equipment available, the dialogue might have to proceed as follows.

> *Tutor:* Well, as a matter of fact, it doesn't. Why do you think that might be the case?

With the equipment available, it would go this way.

> *Tutor:* Try it.
> *Student:* (after the block doesn't do what was expected) Why did it do that?

There is clearly a big difference between the two. The apparatus is more than a source of feedback. It is a medium for exploration and a stimulus for dialogue—what we call a "conversation piece," allowing for a richer interaction over the substance of the subject matter. I contend that GRAPHER plays a similar role. The classic role proposed for microworlds by Papert (1980) among others. The kind of interaction that took place between RB and me is the kind of interaction likely to result in relatively stable changes in cognitive structures related to the subject matter (i.e., learning). (I am not proposing that exchanges be as unstructured as ours, merely as deep.)

CONCLUSION

This paper offers a characterization of subject matter learning as a highly complex enterprise, a characterization predicated on some underlying epistemological notions about the very localized nature of knowledge structures and the ways they grow and change. It illustrates that complexity in a particular domain—graphing elementary functions. We have described the computer-based system GRAPHER intended to help students come to grips with that domain. GRAPHER was designed as an instantiation of six cognitive design principles: (a) supporting conceptual concretizing, (b) making abstract entities concrete and manipulable, (c) linking multiple representations dynamically, (d) focusing conceptual attention by doing drudge work for students, (e) supporting meaning negotiation, and (f) facilitating reflection and communication. Preliminary explorations suggest that there are some significant benefits to the approach both for students (in terms of the kinds of interactions and understandings fostered by the environment) and researchers (who get to see some aspects of cognition at a very fine-grain size, a fact which we believe will lead to significantly increased understanding of the learning process). GRAPHER is one instance of educational technology that illustrates what Roy Pea (1987) has called the "transcendent functions for cognitive technologies in mathematics education," (p. 97)—the use of the computational medium to help people extend their own cognitive capacities. To the degree that we can come to grips with the issues sketched in this paper, we will be in a better position to design effective learning environments and to support the kinds of learning that take place within them.

ACKNOWLEDGMENT

The work described in this paper was supported by National Science Foundation grants MDR 8550332 and BNS 8711342. NSF support does not necessarily imply NSF endorsement of the ideas expressed in this paper.

REFERENCES

Dugdale, S. (1984). Computers: Applications unlimited. In V. P. Hansen (Ed.), *Computers in mathematics education (1984 yearbook of the National Council of Teachers of Mathematics)* (pp. 82–88). Reston, VA: National Council of Teachers of Mathematics.

Papert, S. (1980). *Mindstorms*. New York: Basic Books.

Pea, R. D. (1987). Cognitive technologies for mathematics education. In A. Schoenfeld (Ed.), *Cognitive science and mathematics education* (pp. 89–122). Hillsdale, NJ: Lawrence Erlbaum Associates.

Schoenfeld, A. H., Smith, J. P., & Arcavi, A. A. (1989). Learning. In R. Glaser (Ed.), *Advances in instructional psychology, Vol. 4*. Hillsdale, NJ: Lawrence Erlbaum Associates.

16

Social Niches for Future Software

Andrea A. diSessa
University of California, Berkeley

ACTIVITY STRUCTURES AND SOCIAL NICHES

Considering that computers have been with us for some time, I am surprised at the primitive state of our critical capability to judge their usefulness in educational settings. What I see often are arguments at very general levels: "Computers are dynamic and interactive and can respond patiently in an individualized manner." To be sure, one can rely to some extent on such generalizations, but most confidently in retrospect rather than in anticipation of good educational outcomes. Engineering and science are driven by expectations and promises like these, but anyone who has engaged in design or science knows how these can fail in the details. For some reason the promises just don't "get down to brass tacks;" general goals and possibilities don't really tell us what to do. Critical judgments of particular software are also too often driven by face validity. It is as if one can see quickly and easily the features that lead to in-depth learning. One hears that the software contains a lot of information; it offers the user immense freedom; the simulation is dynamic and interactive; ideas are rendered concrete and manipulable; the program satisfies a checklist of "good educational software" rules, and has a clever user interface.

I do not wish to dismiss any of these as unimportant. Yet at the same time, I wish to highlight a class of critical criteria that appears to be difficult or impossible to see in the software. I can introduce the topic with a set of questions that I compulsively ask students when they present to me a piece of educational software, say, a microworld, from which they are sure children will learn: What exactly is it that children will be doing with this software? What will they think they are doing with it? Why should they do what you think they should do? How is it exactly that they should learn from what they do? What happens if they have different purposes and if they should see the program in a different light than you intend? Why shouldn't the child get the wrong answers just to see what happens or turn your science microworld to artistic ends?

More compactly, I believe the single most important heuristic for evaluating software is to try to simulate the child's *activity structures* involving the full spread of his motivations, his capability to describe what he should be doing, what he can make of the things that do and don't happen in response to his initiatives, and so on. In fact, I claim that these activity structures are the things to be evaluated, not the software per se, and that it is in these structures that we can see not only how well children will learn, but also what they will learn.

This is evidently a knowledge-intensive evaluation technique. It must bring to bear everything we know about the consumers of our products, not only first impressions. This evaluation entails long-term trajectories and possibly subtle turns of meaning for people who may be very different from us. If you cannot answer the above questions about how students will perceive and interact with software, you have not even begun to evaluate it. Quick arguments about face validity and first impressions are intriguing, but it is proper to remain skeptical until activity structures can be analyzed.

The title of this paper emphasizes that student/machine couples are not the only level of analysis important to this kind of assessment. Any personal activity interfaces with social levels of activity. Group commitments and patterns of cooperative learning always contribute another level of dynamic to activity structures. At the very least, classroom social structure both among children and between the child and teacher must allow what we judge to be productive individual activities. More profoundly, activities are deeply affected by social concerns even if they are engaged in individually. This becomes more evident if we lift our gaze from the sanctioned school and classroom culture to the family and peer group (Eckert, this volume). Motivations of individuals reflect communal judgments of what is appropriate in various circumstances, including whether it is appropriate to think or try. Overall, we want to think about activity structures as fitting into niches defined by interest, need, capability, and judged appropriateness, both at individual and social levels.

It is not my aim to contribute theoretical or empirical results to this important area. For that I must defer to experts such as Saxe (this volume) and Lave (this volume). Instead, I take a designer's point of view and through these concerns introduce the general frame that we (the Boxer Project Group at Berkeley) have been developing to anticipate and, in part, to evaluate the effects of the work we are doing in developing a general purpose computational environment. What follows are some of our particular images, our simulations of future activity structures involving the system we are designing. They may be less certain than post hoc studies. But as designers, we cannot escape the fact that the likelihood of our success depends, in large part, on our ability to imagine what has never been.

THE CONCEPT OF A COMPUTATIONAL MEDIUM

The Boxer Project is aiming to test the feasibility of filling a rather grand and still hypothetical social niche, that of a computational medium. We are trying to produce a prototype of a system that extends—with computational capabilities— the role now played in our culture by written text. It should be a system that is used by a lot of people in all sorts of different ways, from the equivalent of notes in the margin, doodles, and grocery lists to novels and productions that show the special genius of the author or the concerted effort of a large and well-endowed group. In a nutshell, we wish to change the common infrastructure of knowledge presentation, manipulation, and development. More modestly, we want a general-purpose system to serve the needs of students, teachers, and curriculum developers, something that is so useful for such a broad range of activities that the community as a whole will judge it valuable enough to warrant the effort of learning a new and extended literacy.

Already the choice of target niche defines many of the properties of the system. Notably it will be complex and nontrivial to learn. Like written language, it will gain its foothold by being learnable in small chunks that are by themselves useful and by paying back the substantial investment made to learn it over a lifelong usefulness. My most compact description of the learning process is learnability through continuous incremental advantage; the system at each stage gives the learner an additional desired capability by learning a little bit more about the system throughout the whole course of intellectual development. The child gains by learning a word, say, cookie, and the adult gets a job by filling out the application competently or gains in professional prestige by eloquently expressing ideas.

What does a computational medium look like? Begin with written text, for it is certainly true that a computational medium would be forsaking a fundamental and established niche if it did not build on this already well-established medium. With only modestly enhanced capabilities, including automated ease in editing and formatting, computers in the form of text processors are already taking over in the workplace and in schools. This is, however, only the beginning of what we visualize as a computational medium.

The next jump comes when we free text from its purely linear form. Hypertext has captured the imaginations of media enthusiasts, and similarly we see a move in this direction as providing important new possibilities (as long as the extension is not so complex and specialized as to preclude use in the simple manner that a basic text processor represents). What hypertext supplies is the ability to chunk information into packages and to link, that is, provide nonlinear connectivity from package to package so that users can peruse a hypertext in a broadly flexible way and so that writers can collect and structure their ideas gradually, capturing in the physical layout an appropriate slice of the organization of those ideas, not just a linear slice.

For Boxer, our new system, we have added two simple hypertext capabilities to text. Text can be chunked into boxes, which are always part of text in some larger box. Thus, the basic structure of Boxer is a hierarchy of text and boxes (that contain more text and boxes) in arbitrary combination. Any box can be named if that is deemed appropriate.

Figure 16.1 shows some text and boxes that were merely typed directly into the system. When the user wishes to have a box-chunk of text, a key that makes a box is pressed. The box expands as text is typed or moved into it to accommodate whatever the writer wishes. From one point of view, boxes are just large characters that can be selected, deleted, cut, or pasted the way other text is in a word processor. The small grey boxes in the figure have been shrunk, hiding the detail contained inside. Shrinking and expanding are keystroke (or mouse-click) opera-

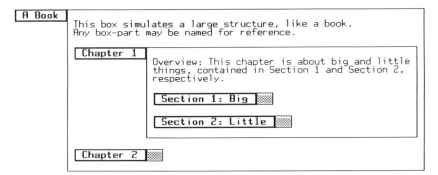

FIG. 16.1. Boxer documents are a simple form of hypertext chunked and hierarchically organized with boxes and cross-referenced with ports.

tions. The user can effectively zoom in on any visible box, temporarily forgetting its context, by expanding the box from its normal size; the frame of the box then occupies the full screen. We believe Boxer's capability to suppress or show detail with nothing more than a keystroke will be an important but easily understood expressive extension to ordinary written text. It can carry many important ideas—notably, representing levels of understanding and the process of zooming in on details, yet being able to shrink back to get an overview when necessary.

Linking, that is, direct access to remote corners of a Boxer document is provided by a kind of box we call a port. Ports are simply views of some box, the "target," which may exist anywhere else in the user's Boxer universe. An editorial change to either the port or its target results in changes to both. Ports make good cross-referencing structures, and one may trivially transport oneself to the place where the target of the port exists if one wants to see more about that context. The box at the bottom of Fig. 16.1 is a port whose target is the top-most box in that figure.

Of course, graphical images are important, and Boxer has graphics boxes that contain pictures. Graphics boxes also contain mobile and interactive entities we call "sprites," which can draw and simply be taught to respond to clicks from the mouse. These are the basis for Logo turtle activities, but also they add the capability to define modes of interaction with Boxer that were not built in. Thus we have extended text now to include images that can be both dynamic and interactive.

Finally, we add the capability to program. This already separates Boxer from essentially all hypertext systems. However, Boxer is unique in that not only does it contain a programming language, it *is* a programming language. Incremental learnability is provided in that anyone who can create and inspect a simple document in Boxer knows almost all the mechanics of creating and inspecting programs. Further, joining programming to hypertext processing adds utility and flexibility. For example, anything that is typed into Boxer is accessible data and can be manipulated at will. So, if we keep a personal journal as a box containing entries (more boxes), we can write a simple program to rearrange all the entries in the box into some other ordering, say, based on topic headings rather than chronological order. Programming is the glue that permits arbitrary recombination of the capabilities of the system or any documents written in it, giving as full control to the user of a document as to its author.

We caution the reader not to think of programming necessarily as an esoteric thing as computers were once considered. When programming becomes much more useful and incrementally learnable, it should, we hope, take its place beside written language as a complex skill, yet one which every literate person is expected to master to some appropriate level. Figure 16.2 shows some familiar Logo-like programming activities that are carried out by typing commands into a kind of box we call a "do-it" box.

The social niche for a computational medium is much too large and complex

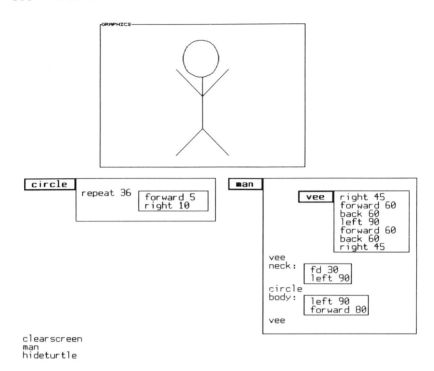

FIG. 16.2. Program definitions appear precisely where they are useful. A general utility such as CIRCLE appears in its environment of use. VEE is only useful in the MAN procedure, so it is located there. Units such as those identified as NECK: and BODY: can also be chunked with boxes.

to be described in any simple terms. In fact, it is misleading to think of it as a single niche at all, but rather we should think of it as occupying a large collection of niches. Indeed, a medium in our sense must justify its existence by being able to play chameleon for multiple roles.[1] We can usefully describe more specific social niches to first approximation as "genres," in analogy to various forms of written language. Accordingly, we are about to enumerate some new and (hopefully) important software genres that future learners will learn and recognize, taking care that we do not mistake superficial features of a genre for the network of social and personal activity structures in which the form works. Books may be

_____

[1]Here, common terminology fails us. One generally does not think of written language as a medium. Instead, its many physical manifestations are individually media, such as books, magazines, newspapers, etc. We should probably be describing written language and our proposed computational medium as abstract media, which have many physical manifestations.

pulp novels, which fit into a particular network of activities and interests or they may be technical monographs, which fit into quite another. It will be our obligation to make such distinctions where appropriate.

TOOLS

Some scientists have tried to define human beings as tool-using animals. While contemporary comparative psychology does not allow such simplifications, there is no doubt that important and special characteristics of humans and their civilizations can be seen in the tools they use. Especially relevant to us are technical and scientific cultures, which are rich not only in physical tools from pliers to computers to particle accelerators, but they are also rich in intellectual tools. Scientists or mathematicians may use little more than a pencil, yet their lives are full of intellectual instruments—well-worn patterns of thinking that accomplish particular frequently needed goals—handed down from previous generations of their cultures. In contrast, classrooms are substantially impoverished with respect to tools. Not only are they poor in physical resources, but children rarely come to the point that what they learn serves important functions for them in accomplishing what they wish to accomplish. Teachers are not much better off by these standards, though perhaps test booklets, mimeo sheets, blackboards and even "manipulatives" can count.

We imagine, then, that a computational medium can provide the substrate to support a new tool-rich culture in schools. Children will learn by using, inspecting, and even crafting tools to support activities they understand, and teachers will be much better supported for their own activities than at present. The transition to a culture that values its computational tools enough to spend substantial effort to support them has already happened in some corners of society— for secretaries and "just plain folk" as well as for programmers and scientists. The same can happen in schools.

With our emphasis on activity structures, we must say more about learning with tools. How might people learn, or fail to learn, with tools?

We can pass quickly through initial promises and expectations about tool use, which are widely extolled. Tools may give children and teachers capabilities in simple and concrete form that they need to engage in activities they want to engage in. In science, especially, tools can allow students to approach a different class of problems, say, more open-ended and exploratory ones (cf. Tinker, 1987). It is said that tools can take over the menial tasks, say arithmetical or graphical ones (such as point plotting), to allow students to work at the more important higher levels of deciding on approach and interpreting results. Now let us enrich these near-platitudes by taking a first pass at how using tools may fail to engender learning because of inadequate attention to the activity structures in which tools are embedded.

It is actually quite easy to generate examples of how learning with tools may fail. Elementary school children's use of arithmetic is a good example. First, arithmetic can, and perhaps should, be viewed and used as a simple but powerful intellectual tool. Yet it is almost never learned in contexts in which it can be seen as powerful. Instead, arithmetic is learned as an abstract practice, with the sole feedback to students on their capabilities at the *tool* level, rather than at the *use* level. Teachers say this sum is right, that multiplication is wrong. Not surprisingly, children do not build a real sense of number, or approximations, or other parts of understanding arithmetic that connect to effective use in context. We discovered in early use of Logo that children's use of number (in directives to the Logo turtle) was too fragile and failed them for the practical tasks they faced (Papert, Watt, diSessa, & Weir, 1979). They had no sense of a "little" or "a lot more" of scale or measurement capabilities of numbers. For example, students often considered 101 to be much bigger than 99; they chose numbers according to patterns—10, 11, 12—rather than according to meaning or need; and often they were unable to distinguish between distances and angles. At least some of this was quickly remediated in the turtle context where numbers did things for the children.

There are other ways to see the problems of learning arithmetic out of use-context. Children may learn to operate the tool, but when arithmetic turns to word problems, the tool is frustratingly difficult for students to use. It was learned in a way too disconnected from contexts of use.

Finally, even if children do learn to operate the tool in context, their understanding is likely to be fragile. The principles by which the tool works are never examined, alternatives are not understood, so children invent and perpetuate "bugs" in their procedures that they have no way of correcting on their own. Interesting corroboration of this mode of failure comes from research on how people understand calculators. Without inducing or being taught the underlying principles of operation of a calculator, but only teaching how to use it on some class of problems, users flounder in a sea of arbitrary button-pressing when faced with even modestly different problems. In contrast, if users are taught or induce a model of the internal workings of the calculator, they can much more easily adapt the tool to new uses (Halasz & Moran, 1983).

At higher levels of instruction, no one questions that algebra is an important tool in learning, say, physics. Yet we see the tool mistaken for the deeper principles of the domain. One of the most prominent and well-documented types of failures in physics instruction is precisely of this sort. Students think of physics as learning equations, hence they ignore the need for semantics beneath the terms in the equation (diSessa, 1985). In problem solving, not surprisingly students are stuck at the tool level and solve by exploring the combinatorics of terms in equations, rather than exploring the more fundamental space of physics concepts as they apply to the problem (Larkin, McDermott, Simon, & Simon, 1980).

Lillian McDermott (this volume) provides a final example of a similar failure.

In geometric optics, students are taught to use ray-tracing and a few standard methods that work for a relatively broad class of problems. But again, the tool seems to become the domain for students. The characteristics of the particular methods become the characteristics of optics, and one sees bizarre conclusions drawn about the world that are, in reality, reflections only of the tool provided to look onto the domain.

We can crystallize these reflections into an ideal scenario for the activity structure involved in deep and synergistic learning of tools and subject matter. In brief, we propose:

1. Tools in context
2. Understanding underlying principles
   - Bottom up: Tools through design
   - Top down: Inspectable tools

In more detail, tools should be introduced in a problem context that is, in some sense, thoroughly understood. In part this is to insure that the tools are clearly distinguished from the domain to be learned; this learning in context also insures that selection and application principles of tool use are clear. To the extent possible, the features of the situation that motivate the tool selection and the approach to solution that the tool represents (and alternate approaches to the problem) should also be understood. In simplified cases, students should be able to carry out manually the tasks for which the tool is designed. This is aimed, in part, at making clear the principles by which the tools work and to provide for alternative approaches when the tools themselves might fail. In the best case, students should build at least a simplified approximation to the tool or specify the basic properties of the tool.

When the bottom up approach—tools through tool design—is impossible, we would like to have tools built from pieces the student understands and inspectable tools that allow students multiple partial views of their construction. Such views illuminate operating principles and use or show (at an appropriate level of detail) the processes through which the tool accomplishes its work.

In a nutshell, students should be able to explore and hand-simulate tools; they should understand the usefulness and limits of tools in context; and they should be able to find alternatives when a tool fails. On a grander scale, we would like to encourage a community of teachers and students whose instincts are to build and share appropriate tools not only in the clean and widely applicable style that we are used to seeing in scientific tools, but also for more everyday and particular tasks.

Returning to arithmetic, the literature shows that arithmetic can be taught according to these principles. Gaea Leinhardt (1987) documents a wonderful case of teaching a subtraction algorithm by thoroughly exercising the problem context, making sure students understand the resources they have (principles), and, only after setting the scene, allowing students to invent the algorithm-solution. The inventive capability of students can be channeled through adequate preparation into deeper understanding and away from the prolific but unproductive invention of bugs.

Tools in Boxer

Figure 16.3 shows a very simple tool built in Boxer, a "turtle odometer," which keeps track of how far a turtle walks in its meanderings. Such a tool can be built from scratch in about 10 minutes by a competent Boxer programmer. The point-and-execute menu on the left is simply typed in place and allows the student to reset the odometer or dump its value (as DISTANCE: . . .) into a HISTORY box so as to keep track of data in experiments with the turtle. Here the student is investigating the length of a binary tree as the number of levels of branching is changed. The tree has been programmed and is being executed from the box in the upper right of the screen. Of course the student may type directly into the history as well. The menu item "clear-history" puts the current history into the box called "ancient history" providing a long-term log of activities. Capturing process in this way is extraordinarily easy in Boxer, and we expect it not only to ease and enhance use of tools generally, but as well to provide a better basis for reflection on the processes of investigation in addition to focusing on results. Note that the student has uncovered a very simple pattern in the length of the tree—100, 200, 300, 400—increasing linearly with number of branchings. Figure 16.3b shows the quite different pattern when one changes the FACTOR (ratio of size of branch to stem) from 2 to 3. Now the total length of the tree will converge with increasing level, never going higher than 300. Another interesting investigation would be the lengths of spirals of various kinds (some diverge in length when they wind inward, some do not). Or, a surprise is in store for a student who checks the length of the path of one of four turtles situated at the four corners of a square and directed to walk constantly toward their clockwise neighbors.[2]

The turtle odometer illustrates some of the reasons a computational medium can contribute toward establishing real examples of our ideal scenario of learning with tools. A collection of tools in computational media can be easier to learn and more flexible at the same time since they are all similar, that is, they use the basic capabilities of the system to define how one interacts with and modifies them. A menu is just a list of commands waiting to be executed. Similarly, basic capabilities of the system such as text editing (e.g., typing directly into the

---

[2]The curved path of each turtle until it meets its neighbor is the same length as the sides of the original square, illustrating that the motion of the chased turtles, orthogonal as it is to the motion of the chasing turtle, does not affect the chase at all.

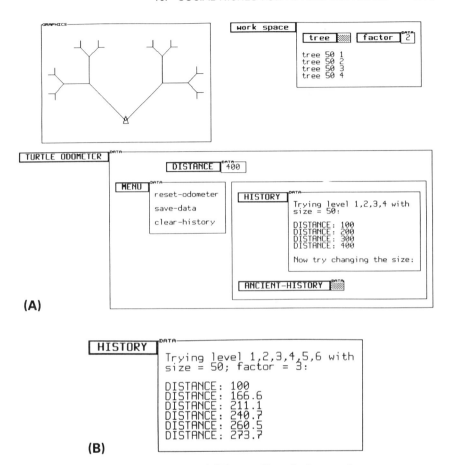

FIG. 16.3.   A turtle odometer. (a) The total length of a tree diverges as number of branchings increases; (b) The length of the tree converges in a different experiment.

history box and/or editing the menu) replace idiosyncratic special-purpose facilities.

Note as well that this tool is so simple as to be entirely under the control of the student. One is in very little danger of mistaking the tool for the task itself. Indeed, if the student does not build the tool, it can certainly be inspected and modified if necessary (say, to adapt to the needs of the four turtles problem). Tools do not need to be complex to be useful and productive in a learning context, and we hope many small tools may contribute as much as a few complex ones to the tool-rich culture we envision.

More centrally, a computational medium allows us incredible flexibility in crafting the context for learning tools and for learning with them. It is easy to generate and modify microworlds, like the turtle in this case. And these contexts can all be interfaced trivially to tools because everything lives in the same environment. Indeed, a tool built for use with one microworld can easily be moved to another in contrast to the extremely problematic interfacing of different applications (editor, database, spreadsheet, statistical package, programming language, etc.) in the present state of computer use. Tools can even extend the microworld itself by being programmed to act back on the microworld (e.g., some features of turtle control can be added to the turtle odometer menu).

Figure 16.4 shows a screen calculator built in Boxer. This slightly more complex example emphasizes how Boxer's hierarchical inspectability can provide students with models of how the tool works by simply allowing its inspection through expanding and shrinking. Figure 16.4b shows the insides of the calculator. We see boxes that define the individual keys of the calculator, their graphical appearance (SHAPE), and their responses to mouse clicks (L-CLICK). We also see the central processing unit (CPU). Inside the CPU are the various commands that manipulate the registers on the surface of the calculator, ANSWER, ENTRY, OPERATION and HISTORY. Any piece of the calculator can be executed with a point and click to see how it works in isolation.

Like the calculator, any Boxer tool provides a model of itself and, if we take some care, these models can be nearly as informative as we wish. Again, modifications and extensions can be added by students and teachers to extend the usefulness of the tool beyond the range for which it was originally intended.

In our examples, we have slighted the importance of tool-use for teachers. However, at this stage it should be easy to imagine how Boxer can be used for teachers. In the first instance, learning tools are quite relevant for teacher training prior to and during service. But I also think more mundane tools will be perceived by teachers as being important and will provide important learnability through incremental usefulness. Simply to keep track of and organize notes, plans, and grades in a convenient hierarchically organized book is substantial help. One can add simple tools to, say, warn the teacher of students doing poorly (at least those getting poor grades), or automatically to grade and store results from quizzes. Indeed, it is relatively easy to make programs to generate simple quizzes, perhaps tuned to students' levels of achievement. Making or modifying premade interactive worksheets for students would also be, we project, common activities for teachers. We have produced examples of all of these genres, but classroom use by the teacher is yet to come.

The Boxer group has some preliminary empirical results with students learning in this tool-oriented manner. Ploger and diSessa (1987) show a 12-year-old student, in his first hour and a half contact with Boxer, learning some central ideas about probability and statistics by using and building simple Boxer tools. We were extraordinarily pleased to see how well this scenario matched our ideal scenario of learning tools in context, from the bottom up (through design) and

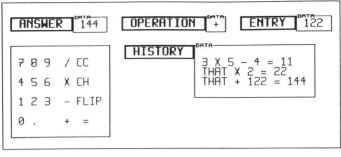

(A)

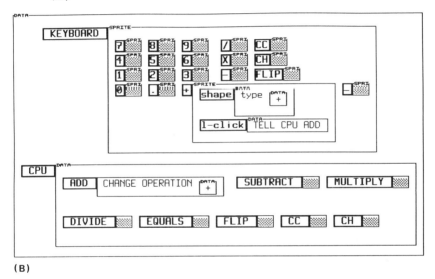

(B)

FIG. 16.4. **(a)** A Boxer screen calculator; **(b)** The calculator mechanism has many pieces, but each is very simple.

from the top down (through inspecting and modifying supplied "transparent
tools.")

KNOWLEDGE SPACES

Boxer's hypertext capabilities augmented by programming provide an opportunity to represent knowledge externally in ways that until now have been inaccessible to most humans. The point is simply that with a much more expressive
medium, extended beyond the linear and static representational capabilities of

text, we may be able to present, so as to be effectively and reflectively accessible, much more of the complex weavings of relations and possibilities inherent in the knowledge of some domain. I call such presentations *knowledge spaces*. So, for example, biology might to some seem to be simply a swarm of facts and definitions. Yet with an appropriate presentation, we could highlight relations more than isolated facts, or highlight large-scale organizational principles. In another model, we might aim to take the burden of fact-recall at least partially away from students, and thereby support focus on the use of that knowledge in problem solving and on problem solving strategies themselves. Again, it is facile to pretend one can simply lay out the knowledge of some domain for inspection and appropriation, but if we pay attention to the structure of activities built onto such displays, new learning patterns may be created.

An early model in my own thinking about knowledge spaces occurred several years ago when my then 9 year old son demanded that I teach him about algebra. Without having time to think out a curriculum, the two of us sat on the couch with a pad and pencil, and I started asking him about how much he already knew. It turned out that the idea of attaching a variable number to a symbol was in his repertoire. So I began trying to see if the axioms of algebra made any sense to him. Did he know, for example, that $A \times B = B \times A$? Yes, indeed, that seemed right to him. When I asked how he knew, he announced that $A \times B$ was the area of a rectangle, and if you turned it on end, that would not change the area of the rectangle, but you would write it $B \times A$.

The distributive law, $A \times (B + C) = A \times B + A \times C$, was more problematic. At first he had no idea if it was right or not. But we tried a few examples, and it seemed to work. Although he considered this clear evidence, he did not consider it definitive proof. After a bit more work, he announced spontaneously that if B and C were the same, then he could see how it worked. $A \times (B + B) = A \times 2B = 2 \times AB = A \times B + A \times B$. Apparently, he could see reason behind the law by collapsing $B + C$ into one term $2B$ and thus eliminating the "distribution."

I won't continue the recounting, but would like to build this small episode into a plausibility case for the following activity. Imagine a group of children at a relatively early stage in trying to understand some domain like algebra. Let us put to them the task of organizing into a knowledge space all they know and their reasons for believing those things. We should, of course, provide some examples and some schematic structure on which to hang their developing ideas.

Figure 16.5 incorporates slightly idealized bits from the above scenario into a partially built knowledge space. Note that we have made room for developing knowledge and even the history of idea-development in categories like TRUE, ???, and INTERESTINGLY FALSE! (say, $1/(a + b) = 1/a + 1/b$). Each true item needs evidence or reasons, possibly from examples or from categories of reliable knowledge, like geometry facts such as $A \times B$ is the area of a rectangle. In the figure, we have connected pools of reliable knowledge to the items they support using ports. The boxes containing those pools reify important epistemological

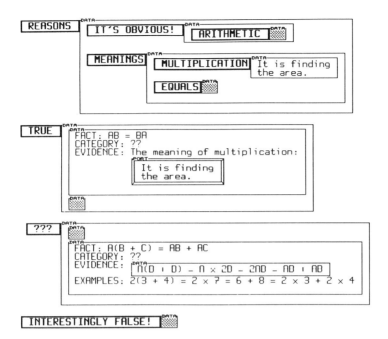

FIG. 16.5.   A hypothetical knowledge space on algebra constructed (mostly) by children trying to understand the domain.

categories, while ports permit reasons also to be distributed where they are needed in the logic of the subject. The evolution of reliable pools, possibly seeded concretely but subtly by teacher insertions in the knowledge space, could play an important role in the evolution of the presentation. Could some problematic "fact" be understood in the same way some previous fact was understood? For example, is there a geometric presentation of the distributive law? Try it!

Perhaps it is entirely fanciful to imagine that students would build something like an axiomatic deployment of some domain in this way. But I don't think it is fanciful that they would learn a lot about the domain, what is true and false, why and how one can organize and think about the whole system. I have no doubt such work could at least contribute to the appreciation of the difficulty of achieving such a compact representation of a field as an axiomatic deployment. Similarly, it could highlight the possibility of multiple organizations as opposed to the presumption built into most mathematical presentations that there is only one way of thinking of a domain (a presumption, incidentally, that students seem to learn all too well).

There are several reasons why I imagine this as an activity of a group of students over an extended period of time. First, it is a difficult task and can well absorb the insights of a group of children. I am quite certain an individual child would soon stall for lack of ideas. In addition, I imagine that the community effort involved in arguing about and negotiating organization for the community knowledge space would prompt a lot of reflection and learning that an individual child would not meet even if he were capable of completing the task.

There is much to learn before implementing a knowledge space curriculum (sequence of activities). How much teacher intervention and seeding of the knowledge space should be done? Which seeds in various domains will actually flower in children's hands? What is the appropriate relative emphasis on constructing, reorganizing, and reflecting? We intend to explore these issues as part of a project to teach elementary school children about the topic of motion.

Another model for knowledge spaces is the work of a graduate student, Steve Adams. He is studying the knowledge and capability of child dinosaur experts (between ages 9 and 12) for the purpose of enhancing the learning of even these children and figuring out how to engage children who would not ordinarily come to such expertise. It turns out such child experts have a surprisingly large range to their expertise and interest. To be sure, this includes an extraordinary range of facts. But these are driven by interest and competence at multiple levels of knowledge, from empirical generalizations ("It seems to me most two-legged dinosaurs are meat eaters.") to sophistication and interest in the progress of science ("I wouldn't buy a book with a copyright before 1986; things are changing too fast." "What's really interesting is when scientists get things wrong, like the apatosaurus . . .").

At the core of the microworld Adams is building in Boxer is a more or less conventional database, (shown in Fig. 16.6). This, of course, allows some usual database activities like querying to find or verify correlations such as the two-legs/meat-eating one mentioned in the preceding paragraph. But since it is in the midst of a computational medium, all sorts of other activities can be envisioned. Indeed it is the design and support of these activities that makes the database really interesting. *The transition from database to knowledge space rests on the medium providing support for specific activities in which children will learn.*

Perhaps the simplest activity is to peruse and add to the database. This may be done, of course, using only basic Boxer editing capabilities. Provision has been made to encourage introducing items of special interest, for example, in a STO-RIES field in the entry template. Another activity that was used as a knowledge assessment instrument proved surprisingly engaging and productive. That is to take the entries of the database and reorganize them according to any principle the child wishes. One could sort according to appearance, geological period, or how fast they could run (a suggestion of one of the child experts who then proceeded to invent a theory of how to tell how fast a dinosaur could run). Adams has seeded other activities with simple programs suggesting ways to go. One is a quiz program that is easily extended to different classes of questions. Students can invent a class of questions and then build a question and answer key. The key itself may be literally the result of a simple query to the database. (The query

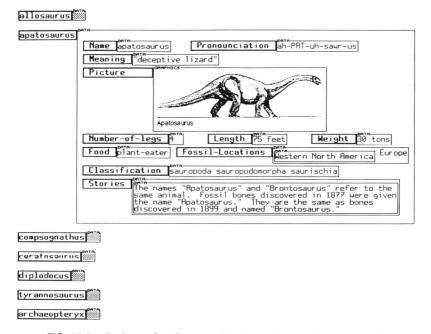

FIG. 16.6.   A piece of a dinosaur database—the activity center of a dinosaur microworld.

provides the answer to the question for each dinosaur.) Students should learn by inventing their own classes of questions and by taking their own and other students' quizzes. We know that this is highly motivating for students from previous work with Logo. Boxer's inherent database capability makes such activities easier to begin (e.g., perusing and directly modifying), more broadly engaging (e.g., including pictures), and more extensible (through better inspectability of existing program and simpler programming more generally).

I will close this section with an example of children's spontaneous activity that highlights one of the special features of Adams' designs, a feature that challenges most views of the instructional designer's task. It also challenges our scientific competence to understand activity structures and even our common instincts for the direction of scientific work in instruction and learning. Two of Adams' dinosaur experts took to drawing cartoons for each other based on corruptions of real dinosaur names. So, for example, a dimetrodon became a dime-meter-odon (Fig. 16.7), with a dime slot for a mouth and a meter needle on its fin.

A frenzy of activity of this sort over a several hour period resulted in dozens of such cartoons. It is difficult to say exactly what is learned in such an activity, though some after-the-fact pronouncements might be made. Certainly this would be among the better ways to generate mnemonics for dinosaur names, and to

straightasaurus

Dimemeterodon

Apaddedsaurus

Try to bite

Proceratops

Broccoliasaurus

FIG. 16.7. A spontaneous social activity—trading dinosaur cartoons: straight-osaurus (stegosaurus); dime-meter-odon (dimetrodon); a padded-saurus (apatosaurus); try-to-bite (trilobite); pro-ceratops (pro-toceratops); broccoli-saurus (brachiosaurus).

become acquainted with some species partners know, but you don't. But the point is precisely that such narrow cognitive views of judging the value of activities may be beside the point. Instead, what seems more important is that the activity provides an opportunity for enjoyment based directly on their expertise and an opportunity to share ideas, factual or whimsical. I believe such activities played a central role in what made these children dinosaur experts in the first place almost independent of particular knowledge outcome.

I suggest our emphasis on activity structures should add a new direction to a constructivist view of learning: Not only must we design for continuity of knowledge, but we must design for *continuity of goals* as well. What we expect as personal goal-directed engagement should lead only gradually from what learners initially consider interesting and meaningful enterprises toward what we set as educational targets, say, acting in the world as little scientists and mathematicians. Designers should be chastened that without due consideration to activity structures and continuity of goals, they might never consider dinosaur cartoons a legitimate activity or that they might design such activity out of instruction by default.

MIND TOYS

Children and adults engage in word and language play from babbling to cross-word puzzles, and they are almost always better off for having done so. Any respectable medium should have such possibilities. In the case of computational media, such play can serve double duty in gaining familiarity with and power over the media, but because the medium itself has mathematical-structural properties, it can be play with mathematical and scientific ideas at the same time.

The Logo community has witnessed an interesting social phenomenon. If we look through books and newsletters on Logo and listen at conferences we find a tremendous fascination with the topic of recursion (a program that uses itself as a subprocedure). Recursive designs, in fractal, space-filling, and other forms, occupy an impressive amount of attention, as does the topic of understanding the recursive programs that generate them. In the abstract, it is hard to imagine that such an exotic and specialized mathematical topic should become a central concern in any nonmathematically-sophisticated population, let alone among school teachers and children. Yet this is the case. The intrinsic interest in this structure, which creates elaborate patterns seemingly out of thin air, coupled with the fact that simple programming puts one in the position of experimenting with it, has nearly made a popular craze out of it.

There are even simpler structures in Boxer that may play the role of mind toys for future generations of computationally literate people. As mentioned earlier, ports make excellent cross-references to related material in a Boxer world. But what happens in the case of self-reference? What happens when a reference in a chunk of Boxer references the chunk itself? Figure 16.8 shows a box that contains two references to further material. The first is an ordinary reference, but the second is a reference to the object that is referring. That is, the second port is a port to the box containing the port. Before looking at Fig. 16.9, try to imagine what will happen if you merely shrink the first port and expand the second.

Behind this almost familiar effect (which one sees on the evening news when the monitor behind the anchor contains the feed from the camera, hence an image of the monitor, which contains . . .) is a deep mathematical idea: Any structure

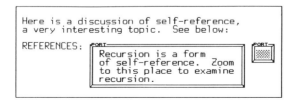

FIG. 16.8. An item on self-reference with port cross-reference to recursion.

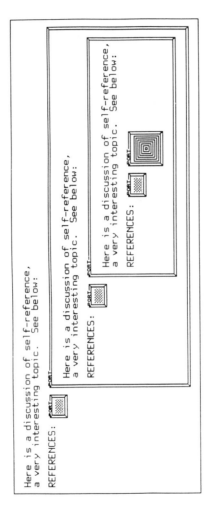

FIG. 16.9.   The item references itself leading to an infinite structure.

that contains itself (say, in a port), plus some other stuff, must be infinite.[3] Infinity is an always-intriguing phenomenon, especially when it seems so near to us, having been created simply with everyday objects and operations—a bit of mystery and simplicity in an elusive but nearly-graspable package. Part of the root attraction of this phenomenon is that our common sense says it cannot happen. It is the same common sense that led Kant to claim that the whole must be bigger than the part. Yet here it is, a part that is the whole and vice versa—all made with a few keystrokes.

Designing for Rare Events

I do not claim that toying with such a phenomenon, by itself, leads to deep mathematical understanding. Though I believe there are many such experiences (perhaps not so spectacular as this, but which may be just as personally meaningful) lying in wait in a computational medium, I think the importance of such experiences goes beyond their cumulative direct effect in developing mathematical sophistication. I am looking to rare but profound experiences that children may have with something that they can see as mathematical (or merely intellectual) that can change their minds about what is interesting and exciting to do. I recall a handful of such experiences in my childhood that, in retrospect, I believe were important to my entering a scientific career. I would conjecture that 7 plus or minus 2 profound experiences of this sort might entirely turn around a student's perceptions of what to spend time on. Of course such a conjecture is far beyond our capability to prove or even genuinely to understand at this stage of the scientific understanding of learning. But again I wish to point to patterns of activity, this time the relation between possibly rare but important events and life-long learning patterns, as deserving more attention than we give them. Designing learning environments for profound but rare events may be more than we can imagine carrying out competently at this stage. But designing so as to exclude them by default may be more dangerous than we should bear.

SUMMARY

The central objects of our concern in the design of educational artifacts ought not to be the technology itself but, instead, the specific activities of students and teachers using the technology and the social niches into which these activities fall. If we do not realize this, we are in danger of perpetuating old and ineffective patterns (or worse) at the same time that we introduce dazzling new machines. In

---

[3]This corresponds to the standard Dedekind definition for infinity, a set that contains itself as a proper subset.

this context, I have examined ways we can most profitably use coming powerful extensions of written text as computational media.

I have argued that tools can be sterile objects unless we provide for appropriate activities with them including designing them and using them in contexts that exercise and measure the tool's power. On the other hand, new computational media can support such activities in ways that until now have been inaccessible; they can provide better learnability, flexibility, and extendibility. There is hope that schools may be substantially changed into tool-rich cultures that simultaneously empower and enlighten. I have argued that the extended representational capabilities of computational media provide important new opportunities, such as knowledge spaces, in which to present and reflect on the nature of a field and on one's own development of knowledge. Again, this can only happen if we investigate and design appropriate activity structures to properly connect the student's mental life to what appears on the computer screen. Finally, I have argued for both the importance of rare events that may play a substantial role in intellectual development, but also for the importance of concern for the continuity of goals and activity structures in their own terms.

ACKNOWLEDGMENTS

I wish to thank the following for helpful comments on drafts of this paper: Steve Adams, Melinda diSessa, Don Ploger, and Alan Schoenfeld.

This work was supported, in part, by the National Science Foundation. Opinions expressed are those of the authors and not necessarily those of the foundation.

REFERENCES

diSessa, A. (1985). Learning about knowing. In E. Klein (Ed.), *Children and computers*. San Francisco: Jossey-Bass Inc.

Halasz, F. G., & Moran, T. P. (1983). Mental models and problem solving in using a calculator. *Proceedings of CHI '83*, 212–216.

Larkin, J., McDermott, J., Simon, P., & Simon, H. (1980). Expert and novice performance in solving physics problems. *Science, 208*(20), 1335–1342.

Leinhardt, G. (1987). The development of an expert explanation: An analysis of a sequence of subtraction lessons. *Cognition and Instruction, 4*(4), 225–282.

Papert, S., Watt, D., diSessa, A., & Weir, S. (1979). *Final report of the Brookline Logo Project* (Memo 545). Cambridge, MA: M.I.T. A.I. Laboratory.

Ploger, D., & diSessa, A. A. (1987). *Rolling dice: Exploring probability in the Boxer computer environment* (Boxer Technical Report E1). Berkeley, CA: University of California, School of Education.

Tinker, R. F. (1987). Educational technology and the future of science education. *School Science and Mathematics, 87*(6), 466–476.

Summary: Establishing a Science and Engineering of Science Education

Marcia C. Linn
University of California, Berkeley

ESTABLISHING A SCIENCE AND ENGINEERING OF SCIENCE EDUCATION

Those concerned about science education are now, more than ever before, building a science of science education (complete with an array of methodologies, paradigms, and underlying purposes), and an engineering of science education (complete with technological tools, models, and a commitment to trial and refinement). Participants representing the broad range of disciplines concerned with science education have come together at two conferences supported by the National Science Foundation to foster the science and engineering of science education. This chapter identifies key issues raised at the most recent conference, communicates a sense of the discussion that ensued, and summarizes recommendations that will make such discussions more integral to the field and will encourage investigations likely to lead to major improvement in the practice of science education.

At the first conference, held in 1986, 45 leaders in science education assembled to identify common concerns, synthesize diverse efforts, and recommend future actions. Attendees represented those concerned with (a) science curriculum design and evaluation, (b) teacher preparation, and (c) research on learning and instruction. Some described themselves as science educators, others described themselves as cognitive scientists, and still others described themselves as curriculum designers. Many participants met for the first time at this conference. These individuals reached consensus on some issues, as summarized in the conference report (Linn, 1987) and recommended that future meetings be held to analyze research and development related to these issues, to involve additional leaders in science education, and to work towards implementing the recommendations.

The second conference for approximately 200 participants was held in January, 1988 and was jointly sponsored by the Lawrence Hall of Science, the Xerox Palo Alto Research Center, and the School of Education. It brought together many of the same individuals as well as other leaders in science education to discuss research studies and technological developments in science education, to refine the recommendations from the first conference, and to identify avenues for further progress.

ELABORATING THE RESEARCH BASE
FOR SCIENCE EDUCATION

The presentations and discussions at the second conference elaborated, refined, and extended the developing science of science education articulated at the first conference. The report from the first conference described "a new thrust in science education research" that included a growing consensus about the nature of the learner, increased understanding of the process of teaching, a new view of the curriculum, and creative efforts to exploit the new technologies. Presentations at the second conference addressed each of these areas reinforcing some issues and introducing some new themes.

The Learner

Discussion of the nature of the learner elaborated on (a) the active nature of the learner, (b) the origin and purpose of the learner's conceptions of scientific phenomena, and (c) techniques for meeting the needs of all learners. Emerging themes included the effect of social context of the learner and the developmental influences on the learner.

The Active Nature of the Learner

Learners are empowered by their active inquiring approach to learning. Echoing discussions at the first conference and building on the views of Dewey (1902), Piaget (1972), and Schwab (1973), participants discussed how the learner's propensity to construct understanding from experience could be channeled to increase scientific proficiency. It was widely agreed that much instruction fails because it does not tap into the active nature of the learner. Thus, J. L. Schwartz (conference communication, January 17, 1988) argued that mathematics and science are not spectator sports but involve the active participation of the learner. However, participants cautioned that hands-on activity is not enough. "Minds-on" learning, in addition to hands-on learning, is needed. Schwartz warned that unless we think beyond hands-on learning to the cognitive activities of students when they manipulate materials, we lose sight of the objective of instruction and keep students busy but unchallenged.

Schoenfeld (this volume) echoed the sentiment and indicated that students need to become involved in the questions of science in order to persevere in problem solving. This general principle was extended broadly. Lunetta (this volume) cautioned that much effort to encourage cooperative problem solving has failed precisely because students are active but do not jointly engage in the learning process. di Sessa (this volume) argued that technology becomes more effective for learning if the actions that students perform when they interact with technology are carefully designed. Many indicated that technological tools need redefinition to ensure that the actions of students lead to integrated understanding.

Directing the active nature of the learner to appropriate cognitive goals involves capitalizing on the learner's desire to construct a view of the world. Newman (this volume) suggested that cooperative group learning experiences might actively involved students in jointly solving problems and thereby distribute the effort required for complex problems. By carefully channeling the activities of students, cooperative experiences could involve students in more complex problems than they could solve on their own. Brophy (1986) argued that students need to be aware of the educational objectives of classroom activities in order to benefit from instruction. Linn and Songer (1988) report that student groups learn far more science and understand it better if they are required to predict the outcomes of experiments prior to conducting them, to reconcile their predictions with the experimental outcomes, and to use this reconciled understanding to predict the results of subsequent experiments. In this example, students' actions are directed to cohesive understanding of science.

In summary, the active nature of the learner requires direction. Hands-on activity may engage students but not promote understanding of science. Computer-based materials offer many opportunities for interaction, but, as discussed further in the section on technology, may provide hands-on experiences that do not teach science. The recommended remedy is careful analysis of the cognitive demands of activity-based learning.

Student Conceptions of Scientific Phenomena

The nature of students' ideas of scientific phenomena continues as a central area of study. Early studies by Piaget identified characteristics of learners that generalized across problems. This approach has given way to the realization that students' ideas are strongly tied to specific subject matter (e.g., Larkin, McDermott, Simon, & Simon, 1980). As a result, presenters summarized students' conceptions in a variety of scientific disciplines and suggested how these views of science might arise.

Students' ideas about scientific phenomena differ from those of experts in content, coherence, and stability. McDermott (this volume) summarized conceptions of a broad range of scientific phenomena that teachers might encounter in science classes and argued that more detail is needed before the origins of these ideas can be understood. Herron (this volume) described some of the conceptions students hold about chemistry and discussed how these influence problem solving. Lunetta (this volume) contrasts ideas developed from naturally-occurring experiences to those presented in schools.

The coherence of students' ideas about science has attracted considerable recent interest. Reif (this volume) showed how students' unintegrated ideas prevented them from using information they have learned to solve new problems. McDermott (this volume) pointed out that students often lack understanding of how scientific ideas are related and therefore cannot apply them to naturally-occurring phenomena. In her research, McDermott (1984) has shown how students' numerous conflicting ideas about mechanics cannot be accurately applied to realistic problems. The emerging questions are: (a) What are reasonable expectations about the coherence of students' scientific ideas, and (b) what might be techniques for encouraging students to continuously seek greater coherence for their understanding?

These questions lead to discussion of the degree and nature of understanding appropriate for science students. It is possible to consider a continuum starting with isolated ideas that arise from limited experiences, but which are adequate for explaining those experiences, to integrated ideas that have coherence with other beliefs and may be adequate for explaining unfamiliar phenomena. An empirical study reported at the conference illustrates points on this continuum. Saxe (this volume) studied Brazilian adolescent candy sellers. He contrasted those with and without formal schooling. He found that both groups were successful in selling candy, but that those who received schooling developed more powerful and generalizable strategies than those without schooling. Schooled candy sellers had a coherent corpus of understanding that applied to school and candy problems. Unschooled candy sellers also had a coherent corpus, but it was limited to candy selling.

Researchers are starting to understand how coherent scientific ideas develop throughout the lifespan (e.g., di Sessa, 1988). Even expert scientists often discover that ideas they previously thought were unrelated require reconciliation. Many well-educated participants at this conference remarked that they had incoherent ideas about certain scientific and mathematical phenomena. Thus, it is desirable to identify appropriate intermediate goals along the continuum towards integrated understanding.

Techniques for encouraging coherence grow out of the general view that the learner is actively constructing a view of the world. Participants suggested techniques for channeling the constructing nature of the learner to achieve integrated understanding. Reif (this volume) argued that instruction often fails to emphasize the coherent nature of scientific information and called for programs that help students organize information systematically, so its structure and coherence become apparent. He showed how instruction featuring hierarchical organization of information can actively involve students in building coherent understanding. Linn (1989) pointed out that instruction using complex technological tools like programming often fails to emphasize the coherence of a discipline like computer science. She argued for instruction focused on the thinking skills used to build integrated understanding. Schoenfeld (in press) illustrated why mathematics instruction often ignores common sense and suggested techniques for helping students make sense of mathematics. Many participants echoed a conclusion from the first conference—arguing that cohesive understanding can best be imparted through systematic coverage of a few scientific topics rather than superficial coverage of many topics.

Serving the Needs of All Learners

Participants at the first conference stressed that differential access to learning experiences and the expectations of parents, teachers, and society at large influence participation and performance in mathematics and science. At the second conference, participants noted that science instruction often appeals to only a small segment of the population. For example, science teaching that stresses the logical positivist perspective—that science consists of facts and that there is one correct method for scientific investigation—often discourages girls and also does not describe the work of practicing scientists. Furthermore, many students complain that science courses are irrelevant to their lives. Science and mathematics courses would interest many more students if they seriously addressed scientific problems relevant to those who are expected to enroll and if they stressed that scientific progress is not a smooth path, but a series of ups and downs. The lack of coherence and lack of relevance of scientific ideas presented in science classes discourages many students from enrolling.

The Effect of Social Context of the Learner

The social context of much science learning exacerbates the lack of coherence of ideas presented in class. The effect of social context on learning emerged as a theme of the second conference. Participants argued that the social environment influences which students participate in science classes, how students view scientific information, and how students approach cooperative learning opportunities. For example, an in-depth study of one high school social environment led Eckert (this volume) to identify two opposing social groups and show how these groups condoned different school activities. One group involved students who saw themselves in a hierarchical structure and engaged in activities and science classes in part to move up in the hierarchy. These students competed with each other to learn science and to gain social status. In contrast, another group involved students who supported each other, rejected all school-based rewards, and preferred cooperative learning environments more characteristic of art or vocational classes. Once the boundaries between these groups are drawn, according to Eckert, they are rarely crossed. As a result, some students avoid science because of their preferred approach to learning. Offering a wider range of science learning environments could increase participation.

Developmental Influences on the Learner

The question of what develops as students get older has several answers. Piagetian-inspired research has identified qualitative changes in scientific ideas while Vygotsky (1978, 1986) has focused on developmental constraints in learning.

Vygotsky (1978, 1986) coined the term *zone of proximal development* to describe the learning students achieve with appropriate instructional support. Vygotsky differentiated progress in learning new information from prior knowledge. He suggested that learners were constrained both by their initial state of understanding and by the conditions under which they were obliged to learn new information. The zone of proximal development referred to the potential information students could incorporate given their learning skills and the nature of instruction.

Finding ways to extend the zone of proximal development was a focus at the second conference. Newman (this volume) described how teachers could attend to this "construction zone" to enhance learning. Newman pointed out that students will appropriate new, more complex understandings if those understandings are within their zone of proximal development. Brown and Campione (this volume) sought techniques to fully extend the earners' zone of proximal development. They stressed that intermediate stages in learning must be understood to encourage maximal development. As a result, researchers wishing to improve instruction need to define the next level of cognitive skills and teach that, rather than defining only the final level of cognitive skill desired. This involves careful analysis of the cognitive task. For reading comprehension, Brown and Campione identified questioning, summarizing, clarifying, and predicting as activities within the zone of proximal development of middle school students who are having difficulty with reading comprehension. They showed how instruction attentive to these skills can enhance learning.

Efforts have also been made to distinguish prior knowledge from the amount of new information that students can acquire from appropriate instruction. The former is frequently called intelligence, while the latter is the zone of proximal development. Linn (1970) found that both the initial state of the learner and the learner's ability to use developmentally appropriate instruction contributed to proficiency.

Guided by Vygotsky's notion of the zone of proximal development, developers have sought a scope and grain size of instruction relevant to the initial state of the learner and to the skills available to the learner for processing new information. Researchers have focused specifically on individual learners and on the dynamics between the learner and the teacher that make this process effective in realistic situations (e.g., Brown & Campione, in press).

In summary, the second conference built upon the first to create a more integrated and cohesive view of the science learner. This process of jointly constructing a broader and more robust view of science education constitutes a start on the science of science education.

The Curriculum

Reconceptualization of the curriculum to reflect the information explosion, to incorporate technological advance, to motivate the full range of students, and to impart a robust view of science concerns science educators. Participants at the second conference elaborated this conceptualization, focusing on (a) imparting deep understanding of scientific concepts, (b) emphasizing the process rather than the product, and (c) balancing explicit instruction and discovery learning.

Deep Understanding

The information explosion has increased pressure to provide fleeting coverage of a broad range of scientific principles and concepts in each precollege science class. As a result, students memorize more and remember less. Furthermore, students do not develop the coherent, integrated ideas deemed essential to effective understanding of complex scientific phenomena. The result of this approach has been poor performance on numerous assessments (e.g., Dossey, Mullis, Lindquist, & Chambers, 1988; International Association for the Evaluation of Educational Achievement, 1988; National Commission on Excellence in Education, 1983; National Science Board, 1988). To improve scientific understanding, participants generally agreed that less material should be covered, but also argued that the nature of instruction must be changed.

One major problem in achieving deep understanding is that students do not realize what they do not know. Brown and Campione (this volume) discuss what has been called the illusion of comprehension that arises when students overestimate their level of understanding. They stress that curricula should provide insight into the learning processes so students recognize what they know and what they do not know. Schoenfeld (this volume) shows how technological tools can help students recognize loopholes in their knowledge and engage learners in resolving inconsistencies. Newman (this volume), Linn and Songer (1988), and Brown and Campione (this volume), show how students gain understanding by making predictions or hypotheses and then reconciling the results of experiments with their anticipations.

Reif (this volume) argues that current knowledge of the nature of the learner, the teacher, and the subject matter makes it possible to teach students the skills that they need in order to deal with new scientific information. Reif claims that knowledge organization is crucial to understanding. He recommends instruction that provides an effective organization and representation for complex information, thus ensuring that students will be able to retrieve it in the future.

Given the call for depth of coverage of scientific concepts, participants noted that some choices must be made concerning what information should be covered. Rutherford (1988) has been seeking a systematic understanding of appropriate topics for science and mathematics curricula. Other efforts are needed to specify the degree of understanding appropriate for the topics selected.

In selecting topics for science instruction, many noted that students often lose interest in science because curricula do not seem relevant to their lives. Reif (this volume) called for instruction that prepares students to function effectively in our society. Saxe (this volume) showed that schooling could influence performance in informal settings and Eckert (this volume) demonstrated that some students come to reject school and avoid it because they do not see any relevance of the material or the instructional environment to their intended life activities. Participants pointed out that some curricula have emphasized the relevance of concepts, but that so far such curricula have succeeded more with students already interested in science than with those who need to be motivated to pursue science (e.g., Rutherford, Holton, & Watson, 1970). Relevant, socially responsible curricula would help attract and retain students in science.

Thus, the current science curriculum fails to impart the understanding students need. Both the delivery of instruction and the topics taught require change before we can expect effective learning.

Process versus Product

Much of science and mathematics instruction places emphasis on the product of learning rather than the processes required and, thus, may lead students to believe that the processes are unimportant and that getting the right answer constitutes scientific understanding. Curricula often reinforce this view by teaching students to decode words before they are taught to comprehend text, to learn computational algorithms before learning to solve real mathematical problems, and to master the mechanics of writing before learning to communicate. Stewart (this volume) illustrates this argument for genetics instruction, pointing out that students often learn mindless algorithms for solving genetics problems, rather than understanding the process of genetic transmission. Goodstein (this volume) demonstrates how extensive trial and refinement of the Mechanical Universe television series resulted in programs that communicated the processes involved in learning scientific concepts. He noted that creative use of dynamic graphic representations of scientific phenomena was effective in illustrating these processes. Others pointed out that in some cases, even when the process is emphasized, students fail to learn when to apply the process or how to select processes for particular problems. Furthermore, some courses discourage students from developing their own techniques for solving problems by teaching the rote application of a given process and by penalizing students who invent their own short-cuts or use more advanced techniques (e.g., Schoenfeld, in press). Curricula that focus on the processes students need provide students with lifelong learning tools.

Balancing Explicit Instruction and Discovery Learning

Designing curricula to help students construct a powerful view of science and of themselves as learners was more a focus of the second conference than the first. Participants discussed the balance between discovery learning and explicit instruction in helping students understand the nature of scientific learning. Explicit instruction identifies the reasoning processes students use, breaks them down into manageable parts, and provides support while these processes are practiced, refined, and applied to more and more complex problems, but must be directed to appropriate goals.

Researchers differed in their views of explicit instruction. Reif (this volume) argues that identifying a reasoning process, finding a clever way to teach it so the student can use it, and providing feedback so that the student masters the subparts of that process, is a powerful way to teach scientific problem solving. Brown and Campione (this volume) argue that the curriculum must make the reasoning process visible and can channel the active nature of the learner. Goodstein (this volume) argues that technological tools can be used to make the steps of a reasoning process accessible to learners. Others argue that explicit instruction often conflicts with what students already know. Schoenfeld (this volume), di Sessa (this volume) and Lave (this volume) argue that students need to be encouraged to develop their own ideas and that the explicit approach will stifle students and interfere with the development of coherent understanding.

These discussions centered on helping students become responsible for learning new concepts and principles versus helping students understand complex phenomena. Most agreed that both goals are appropriate. Some believed both goals could be achieved with explicit instruction or discovery learning, while others felt different methods were needed.

Summary.   Recent insights into the nature of the learner combined with the proliferating scientific information has led to a demand that the science curriculum be restructured. Most agree that fleeting coverage of the ever-increasing amount of scientific information cannot be accomplished and that, instead, instruction should focus on deep understanding of a few scientific topics, along with developing the strategies and procedures that students need to analyze new information. As a science of science education develops, these themes will remain central.

The Process of Teaching

Discussions of the process of teaching built on the concern with teacher workload and preparation and also addressed apprenticeship as a model for teaching.

Teacher Preparation

Preparing teachers to implement the curricula envisioned by participants raised many questions and yielded few answers. Most agreed that teachers need understanding of the subject matter as well as skill in using a wide range of instructional strategies.

Few teachers are prepared for this responsibility. Goodstein (this volume) contrasted the number of precollege physics teachers (25,000) with a number of such teachers who have degrees in physics (2,000). Larkin (this volume) and Goodstein both called for techniques to amplify the role of the few qualified teachers as well as technological tools to improve instruction for the many students who wish to enroll in precollege classes.

In addition, suggestions for improving teacher preparation arose. di Sessa (this volume) suggested that powerful technological tools would allow teachers to devise their own instructional materials. Larkin (this volume) argued that teachers should work collaboratively with curriculum developers and cognitive researchers to come up with curriculum materials that incorporate effective teaching strategies.

Apprenticeship

Participants examined the apprenticeship model for teaching, contrasting several views. Apprenticeship typically involves a cooperative learning environment where apprentices and masters work side by side; masters can be observed and questioned; and apprentices gradually acquire the skills of masters through trial of appropriate activities, discussion of problems, and a cooperative approach to task completion. Lave (this volume) describes apprenticeship for Liberian tailors. Collins, Brown, and Newman (in press) have argued that a variety of effective interventions that involve some form of cooperative learning can be thought of from the perspective of apprenticeship.

Brown and Campione (this volume) point out that the similarities between reciprocal teaching (Brown & Palinscar, 1987) and the other approaches labeled "apprenticeship" by Collins, et al. (in press) are compelling, but also note differences between the various programs. First, Brown and Campione argue that their approach involves a systematic analysis of the domain not typical of apprenticeship. Second, they criticize those who have interpreted their reciprocal teaching in terms of the Skinnerian notions of modeling behavior and then fading or withdrawing support as students become more proficient in implementing the model. Brown and Campione argue that a main component of their approach is to identify the strategies students need to learn and to devise instructional techniques that teach these strategies. Extensive trial and refinement is required to identify these strategies. Furthermore, they noted that the strategies they identified for helping students with reading comprehension did not generalize to the domain of mathematics or science. Domain-specific analyses of learning must be undertaken to promote the discussion, argumentation, explanation, and reflection that fosters student understanding. Brown and Campione contrast the readily externalized processes typical in learning tailoring or candle-making to the quite internal processes in mathematics and scientific problem solving. They suggest that the apprenticeship approach does not directly apply to abstract domains.

Summary.   Discussion of the process of teaching raised many questions and suggested the need for more work. Effective models of instruction are difficult to generalize to new domains, suggesting that our knowledge in this area remains fragmented.

The New Technologies

Participants at the first conference noted that the new technologies had permeated scientific advances and offered promise for science education. The second conference focused on using technology to support student learning and to implement explicit versus discovery-oriented learning strategies.

Supporting the Learner

Technological tools can help learners achieve cognitive goals but, as participants pointed out, such tools can also engage learners in irrelevant or unproductive activities. di Sessa (this volume) argued that developers are sometimes satisfied when students interact with computers and fail to evaluate the benefit of the interaction. Schoenfeld (this volume) described technological environments for mathematics instruction that encourage students to discover important mathematical principles, but pointed out that similar tools used incorrectly could reinforce incoherent views of mathematical phenomena. Similarly, Lepper (1985) has shown that computer tools generally motivate students, but often reinforce the wrong goals. As argued in the discussion of the active nature of the learner, the computer-learner interaction engages the student, but not necessarily in learning science.

The cognitive benefits of technological tools must be carefully evaluated. For example, technological environments can easily implement graphic, dynamic, and verbal representations of the same concept, and many argue that multiple

representations of scientific concepts may encourage learners to construct cohesive ideas. J. L. Schwartz (personal conference communication, January 17,
1988) encouraged systematic analysis of these environments, cautioning that
students may perceive each representation as a new phenomenon and end up
completely confused. Lunetta pointed out that similar concerns surrounded use
of laboratories in science instruction (e.g., Hofstein & Lunetta, 1982). On balance, students may integrate information that is unrelated and reach unintended
conclusions.

Discovery versus Explicit Instruction

Technology has been used to implement the full gamut of instructional provi-
sions from reconstructible environments to rigid tutoring systems. Participants
viewed several of these at the conference. Technological environments support
both explicit- and discovery-oriented instruction.

di Sessa (this volume) demonstrated an environment that allows relative
novices to harness technological power to solve their own problems in discovery
mode. He focused on the active nature of the learner, arguing that this environment could engage the learner. The teacher and the curriculum are responsible
for channeling energy to relevant activities.

Larkin (this volume) described effective uses of computer-assisted instruction,
intelligent computer-assisted instruction, and intelligent tutoring systems. She
argued that these form a continuum and have no clear boundaries. In fact, due to
the primitive nature of text analysis tools, she argued that most such tutors are
quite explicit in their instructional strategies.

At present, the nature of the technology places constraints on both discovery-
oriented and explicit instruction. Explicit instruction tends to be quite brittle
because of difficulties in analyzing open-ended responses. Stewart (in press)
argued that technology-based instruction can achieve more general goals if used
appropriately. Thus, discovery-oriented tools require an effective curriculum to
harness their power to desired educational goals.

All argued that extensive trial and refinement is necessary to make any technological environment effective. J. Minstrell (personal communication, January
17, 1988) described how a physics teacher refined the curricula to incorporate
technology. Many goals were revised or dropped as the result of adding technological environments. J. Kilpatrick (Personal communication, January 17,
1988) pointed out that since technology changes the goals of instruction, educators need to experiment with alternative goals and uses of technology to
achieve those goals.

Given the reliance on trial and refinement for effective use of technology in
instruction, technological environments that facilitate such efforts seem highly
desirable. Two were described. Larkin (this volume) described cT, an authoring
system running on the Macintosh Plus, which allows individuals to design their

own instruction after a fairly short period spent learning the programming language. cT speeds up development and increases the scope of trial and refinement because refinements are easy to implement. di Sessa (this volume) described Boxer, running on a Sun workstation, which provides a flexible and readily reconstructed environment.

The benefits of trial and refinement are also apparent in other curriculum projects. Goodstein (this volume) described the extensive trial and refinement required to make effective use of television. Linn and Songer (1988) described revisions to a curriculum using real-time data collection. White (in press) described improvements in the ThinkerTools environment.

Summary.   Recent efforts to use technology for science education suggest that the process is more of an art than a science. Trial and refinement is the key to incorporating technology into instruction. The same principles that have guided curriculum design without technology apply when technology is introduced. It is important to pay attention to the nature of the learner and to ensure that the activities students engage in contribute to their understanding. At present, the integration of technology with effective teaching is frequently overlooked. The goals of instruction require refinement in light of technological advance, since technological tools supplant some educational goals and make others more crucial. Overall, technology does offer considerable promise to science education, but that promise will only be realized if an integrated and coherent perspective on technology, the science curriculum, the science learner, and the science teacher is achieved—essentially an engineering of science education.

Conclusions

This conference chronicled progress in developing a shared perspective on science education as the illustrative discussions reported in the previous paragraphs suggest. This process bears similarity to the process of integrating ideas described for individual learners. Researchers in the field have varied perspectives on science education, each warranted by some reasonable evidence. Integration of these diverse perspectives is complicated because the methods researchers use to gather evidence yield information that is not always readily reconciled and because many questions remain unanswered. Fortunately, there is strength in diversity of methods.

Researchers in science education employ a broad diversity of methodologies. Recent work in the field of cognitive science has relied heavily on analyses of the thought processes of individual students, often studied by examining videotapes of individuals solving problems while talking out loud. In addition, ethnographic approaches have been used to study the context of classroom learning. While some researchers rely on analysis of individual learners, others only feel confident about conclusions when they have data on larger groups of individuals. This diversity led Ann Brown to remark somewhat ironically, we are not cognitive scientists, so we gather *d*ata (with a hard d) on groups of students (Brown, this volume).

Are there methodological perspectives that yield both *d*ata and details? Progress is being made toward combining methods involving protocol analyses and clinical interviews with those involving collection of data on groups of students. Several investigations of student learning in realistic settings provide examples. Brown and Campione (this volume), Pea and Kurland (1987), Linn and Songer (1988), and Leinhardt and Greeno (1986) have all attempted to integrate studies of individual student reasoning with systematic collection of data on groups of students.

Both because of differences in methods and because of differences in perspective, researchers in science education hold diverse views. Some believe that curriculum design should flow from instructional theory, while others would base innovation on the insights of creative teachers. Some focus on class and societal context influences on learning, while others focus primarily on how individuals construct meaning. Some note that cultural and gender influences are potent factors in performance, while others believe that socialization and educational influences override these factors. Some participants in the field place their greatest confidence in the skills of the teacher, while others feel efforts should be made to improve the curriculum, and still others focus primarily on improved uses of technology.

Clearly, opportunities for larger more integrated investigations would greatly contribute to coherence in this field and to the emerging science of science education. Because principles are just emerging, and their generalization is still an issue, studies involving trail of curriculum innovations followed by refinement in subsequent trials, have great promise. Such integrated investigations, however, require sustained work of multidisciplinary teams. A mechanism is needed to encourage more efforts of this sort. One approach is found in the recommendation for centers for collaboration in science education in the next section.

Recommendations

The four recommendations put forth at the first conference were elaborated at the second conference. The first recommendation, that Centers for Collaboration in Science Education be established, remains paramount. Procedures for implementing this recommendation were a major focus of the conference. Progress and priorities for each of the four recommendations are discussed in this section.

Establish Centers for Collaboration in Science Education

Subsequent to the first conference, other groups have also concluded that centers are needed to advance the field (e.g., Pea & Soloway, 1988). Centers for

Collaboration in Science Education respond to the critical need for (a) greater interaction among professionals involved in science education, (b) the large scale projects such as exemplary schools and alternative curricula now necessary to advance the field, and (c) the sustained intellectual leadership likely to foster the science and engineering of science education. Current isolated projects rarely achieve multidisciplinary perspectives on educational problems, because they cannot support sustained, thoughtful interchange of ideas. Although electronic communication has greatly expanded the informal networks in the field, a mechanism for long-term formal collaboration is needed.

Centers for collaboration in science education also have the potential of attracting talented individuals to address the complicated issues discussed previously. Stable centers for collaboration will attract and sustain interest from a broad range of disciplines including science education, psychology, anthropology, physics, chemistry, mathematics, biology, teaching, cognitive science, philosophy, and administration. It will not be necessary to raid universities to establish centers. It will be essential, rather, to broaden the base from which staff for such entities could be recruited.

Centers for collaboration have the potential to implement cooperative approaches to solving the problems of science education. They can support graduate students embarking on a career in the field as well as professionals who wish to retrain to participate in the emerging science of science education. In addition, centers provide opportunities for long and short term visits from those interested in collaborating with center members. These might include individuals who have conducted investigations relevant to those conducted at the center and who wish help in interpreting their findings, as well as those who are interested in data collected at the center but want to examine it from a different perspective.

Thus, centers could provide opportunities to carry out large-scale collaborative projects that would greatly advance the field. They could contribute to the intellectual power and stability of the field by attracting and sustaining talent, and they could train students and others who wish to collaborate with science educators. A question is how to ensure that centers achieve these objectives.

Establishing a center that becomes more than a loose consortium of separate projects requires working relationships that lead to collaboration across disciplinary boundaries. Effective models for such collaboration exist, but are difficult to implement in environments that generally reward competition and discourage cooperation among equals. Some suggest considering models found in research on the family. Cooperative families deal with conflict, acknowledge alternative views, and build on each other's perspectives. Collaboration means that participants in centers (a) consider perspectives that could be lost because their originators come from the wrong background, (b) listen to the views of individuals who speak with a different accent, and (c) pay attention to ideas that are presented with uncertainty. The trust characterizing effective cooperative families can help to encourage those from disciplines relevant to science education to communicate. Cooperation is likely to arise when all the participants address a common goal such as establishing an exemplary school, or designing an innovative learning environment, or generalizing an approach to instruction to a new domain. A center needs a few people who can see things from various perspectives and can provide leadership in constructing a new understanding.

These models can only succeed, however, if individuals who contribute to them are properly rewarded. Universities often discourage multidisciplinary cooperation because faculty are situated in departments. Thus, in order to establish effective centers, the participation of those establishing the reward structure is required.

To secure funding, centers must be accountable to the organization providing support. Nevertheless, it is likely that centers will provide a better balance between accountability and creativity than individual research projects because centers involve a longer commitment and potentially greater project diversity than is likely when individual projects are evaluated separately. Still, structures are needed to guard against the possibility that the long-term commitment strays in a direction not anticipated or desired by the funding agency, and to ensure that allocating funds to centers does not disenfranchise individuals with promising views.

In summary, the potential of centers for collaboration in science education can be realized when effective collaborative structures are established, when substantial funding is available, and when integrated research programs are realized.

Expand Opportunities for Sharing Information

This conference was a first step in implementing the recommendation that opportunities for sharing information concerned with science education be expanded. Additional conferences, especially others that encourage collaboration across discipline boundaries, offer great promise. For example, many science education researchers were exposed to the Mechanical Universe physics curriculum for the first time at this conference. Collaboration between those developing innovative approaches to instruction such as this one and those conducting research on the learner and on instruction offer promise. Efforts are needed to increase communication between those who design curriculum, those who train teachers to use it, and those who study learners and instruction. Projects that integrate advances in understanding learning with advances in design of instruction, such as that found at the Center for Design and Educational Computing reported by Larkin (this volume), seem promising. In addition, alliances between school systems, the National Education Association, and those involved in research and development in science education are needed.

In addition to conferences, there are other opportunities for communication. Several suggested increasing the number of publications by cognitive-oriented researchers in journals such as *Physics Today* or *The Biology Teacher*. Others suggested that discipline-oriented investigators publish articles in cognitively oriented journals, and still others called for multidisciplinary articles in multidisciplinary journals.

Increase Fundamental Understanding of Science Learning

This conference served to communicate increases in understanding of science learning and provide opportunities for informal discussion of their implications. Such increases form an important component of the proposed Centers for Collaboration in Science Education and a rich on-going activity in the field.

Strengthen Response to Instructional Needs

Research in science education should reflect and respond to real instructional needs. The emphasis at this conference on the context of science learning, including reports on formal and informal learning and on peer influences on performance, provide an important step towards understanding and responding to instructional needs. Rich ethnographic studies of teaching add information at a comparable grain of analysis to the protocol analyses of problem solving. It is equally important to conduct interventions in realistic settings and analyze their impact on learning.

Efforts to respond to instructional needs are most effective when they involve teachers as collaborators rather than when they set out to "fix" the problems of schools. Yet, the immediate concerns of those involved in schools often prevent them from considering the sweeping changes that those developing new curricula or investigating technological innovation feel are essential. Efforts to integrate these different cultural perspectives on education are needed in order to ensure that the enterprise of science education meets instructional needs.

To develop and sustain the new thrust in science education research, we must avoid the chronic amnesia that often characterizes research in education. New efforts are most effective when their findings are compared and contrasted to efforts initiated in the past. Thus, concerns with the zone of proximal development are better understood when they are analyzed in conjunction with other developmental perspectives. Efforts to encourage activity-oriented hands-on or minds-on learning are best understood when analyzed in conjunction with the large projects of the 1950s and the 1960s which attempted to accomplish many of the same objectives. Efforts to incorporate the new technologies into education are best understood when analyzed in contrast to other technological innovations that have contributed to science education. A rich science and engineering of science education will benefit from a coherent perspective incorporating both past efforts and current concerns.

ACKNOWLEDGMENTS

This material is based upon research supported by the National Science Foundation under grant No. MDR-8470514. Any opinions, findings, and conclusions or recommendations expressed in this publication are those of the authors and do not necessarily reflect the view of the National Science Foundation.

The author appreciates comments on an earlier draft from Andy di Sessa, Marjorie Gardner, Jim Greeno, Julia Hough, Fred Reif, and Alan Schoenfeld.

REFERENCES

Brophy, J. (1986). *On motivating students.* Unpublished manuscript, Michigan State University, The Institute for Research on Teaching, East Lansing.

Brown, A. L., & Palincsar, A. S. (1987). Reciprocal teaching of comprehension strategies: A natural history of one program for enhancing learning. In J. G. Borkowski & J. D. Day (Eds.), *Cognition in special children: Comparative approaches to retardation, learning disabilities, and giftedness.* Norwood, NJ: Ablex Publishing.

Brown, A. L., & Campione, J. C. (in press). Academic intelligence and learning potential. In R. J. Sternberg & D. K. Detterman (Eds.), *What is intelligence? Contemporary viewpoints on its nature and definition.* New York: Ablex Publishing.

Collins, A., Brown, J. S., & Newman, S. E. (in press). Cognitive apprenticeship: Teaching the craft of reading, writing, and mathematics. In L. B. Resnick (Ed.), *Cognition and instruction: Issues and agendas.* Hillsdale, NJ: Lawrence Erlbaum Associates.

Dewey, J. (1902). *The child and the curriculum.* Chicago: The University of Chicago Press.

di Sessa, A. (1988). Knowledge in pieces. In G. Forman & P. Pufall (Eds.), *Constructivism in the computer age.* Hillsdale, NJ: Lawrence Erlbaum Associates.

Dossey, J. A., Mullis, I. V. S., Lindquist, M. M., & Chambers, D. L. (1988). *Mathematics: Are we measuring up?* Princeton, NJ: National Assessment of Educational Progress.

Hoftstein, A., & Lunetta, V. N. (1982). The role of the laboratory in science teaching: Neglected aspects of research. *Review of Educational Research, 52*(2), 201–217.

International Association for the Evaluation of Educational Achievement (IEA). (1988). *Science achievement in seventeen countries: A preliminary report.* New York: Pergamon Press.

Larkin, J., McDermott, J., Simon, D. P., & Simon, H. A. (1980). Expert and novice performance in solving physics problems. *Science, 208,* 1335–1342.

Leinhardt, G., & Greeno, J. G. (1986). The cognitive skill of teaching. *Journal of Educational Psychology, 78,* 75–95.

Lepper, M. R. (1985). Microcomputers in education: Motivational and social issues. *American Psychologist, 40,* 1–18.

Linn, M. C. (1970). *Effects of a training procedure on matrix performance and on transfer tasks.* Unpublished doctoral dissertation, Stanford University, Stanford, California.

Linn, M. C. (1987). Establishing a research base for science education: Challenges, trends, and recommendations. *Journal of Research in Science Teaching, 24*(3), 191–216.

Linn, M. C. (1989). Science education and the challenge of technology. In J. Ellis (Ed.), *Informal technologies and science education* (The Association for the Education of Teachers in Science [AETS] yearbook). Washington, DC: ERIC Clearinghouse for Science, Math, and Environmental Education.

Linn, M. C., & Songer, N. B. (1988, April). *Cognitive research and instruction: Incorporating technology into the science curriculum.* Paper presented at the annual meeting of the American Educational Research Association, New Orleans, LA.

McDermott, L. C. (1984). Research on conceptual understanding in mechanics. *Physics Today, 37,* 24–32.

National Commission on Excellence in Education. (1983). *A nation at risk: The imperative for reform.* Washington, DC: Government Printing Office.

National Science Board. (1988). *Science and engineering indicators—1987.* Washington, DC: National Science Foundation.

Pea, R. D., & Kurland, M. D. (1987). On the cognitive effects of learning computer programming. In R. D. Pea & K. Sheingold (Eds.), *Mirrors of minds: Patterns of experience in educational computing* (pp. 147–177). Norwood, NJ: Ablex Publishing.

Pea, R. D., & Soloway, E. (1988). *The state of the art in educational technology R & D: Issues and opportunities.* Report to the Office of Technology Assessment, Washington, DC.

Piaget, J. (1972). *Science of education and the psychology of the child.* New York: The Viking Press.

Rutherford, J. A. (1988, April). *Science education in the United States: A report from the Longitudinal Study of American Youth.* Symposium presentation at the annual meeting of the American Educational Research Association, New Orleans, LA.

Rutherford, F. J., Holton, G., & Watson, F. G. (1970). *The Project Physics course handbook.* New York: Holt, Rinehart & Winston.

Schoenfeld, A. H. (in press). On mathematics as sense-making: An informal attack on the unfortunate divorce of formal and informal mathematics. In D. N. Perkins, J. Segal, & J. Voss (Eds.), *Informal reasoning and education.* Hillsdale, NJ: Lawrence Erlbaum Associates.

Schwab, J. (1973). The practical 3: Translation into curriculum. *School Review, 81*(4), 501–22.

Stewart, J. (in press). Potential learning outcomes from solving genetics problems: A topology of problems. *Science Education.*

Vygotsky, L. S. (1978). *Mind in society: The development of higher psychological processes* (M. Cole, et al., Eds.). Cambridge, MA: Harvard University Press.

Vygotsky, L. S. (1986). *Thought and language* (A. Kozulin, Ed. and Trans.). Cambridge, MA: MIT Press.

White, B. Y. (in press). ThinkerTools: Causal models, conceptual change, and science education. *Cognition and instruction.* Hillsdale, NJ: Lawrence Erlbaum Associates.

Author Index

International Association for the
 Evaluation of Educational
 Achievement, 329, *340*
Isom, F., *51*
Itakura, K., *138*

J

Jackson, I., 198, *201*
Jansen, W., *54*
Joerger, K., 141
Johnson, D. W., 197, *201*, 238, 239,
 240, 241, *247, 248*
Johnson, R. T., 197, *201*, 238, 239,
 240, 241, 242, *247, 248*
Johnson, W. L., 172, *179*
Johnstone, A. H., 34, 41, 42, *49, 51*
Joshua, S., 15, 16, *28*
Jung, W., 10, *28*
Jungck, J. R., 57, 58, 61, 65, *67, 68*

K

Kahnemann, D., 95, *108*
Kane, M. J., 118, 122, *137*
Kant, E., *179*
Kaplan, B., 232, *234*
Karmiloff-Smith, A., 43, *51*
Kärrqvist, C., 10, *27*
Kass, H., 41, *54*
Kellington, S., 41, *52*
Kempa, R., 35, *51*
Kenealy, P., 8, *28*
Kerner, N. K., 39, *51*
Klainin, S., *51*
Kleinman, R., 39, *51*
Klopfer, L. E., 5, *27*
Koedinger, K., 57, *68*
Koran, J. J.,*51*
Koran, M. L., *51*
Krajcik, J. S., 243, *248*

Kramers-Pals, H., 37, *51, 52*
Kran-Gandapus, T., *51*
Kulik, J. A., 78, *84*
Kulik, S., 74, *84*
Kunkle, D., 79, *84*
Kurland, M. D., 336, *341*

L

Laboratory of Comparative Human
 Cognition, *234*
Labudde, P., 103, *108*
Laird, J. E., 172, *179*
Lambrecht, J., 37, *51*
Lampert, M., 122, 131, *138*
Landers, R., 122, *138*
Lansdown, B., 196, *201*
Larkin, J. H., 5, 24, *28*, 163, 166, 171,
 172, *179*, 308, *322*, 325, 332, 334,
 338, *340*
Laudan, L., 66, *67*
Lave, J., 76, *84*, 219, 232, *234*, 251,
 302, 331, 332
Lawson, A., 34, *51, 52*
Lawson, R. A., 3, *28*
Lazonby, J., 33, *51*
Lehman, J. R., *51*
Leighton, R. B., 269, *279*
Leinhardt, G., 309, *322*, 336, *340*
Lepper, M. R., 173, *179*, 333, *340*
Lesh, R., 79, *84*, 122, *139*
Lewis, M. W., 118, *138*
Lindquist, M. M., *83*, 329, *340*
Linn, M. C., 73, *84*, 106, *108*, 235,
 248, 323, 325, 327, 328, 329, 335,
 336, *340*
Lipson, J., *67*
Lochhead, J., 69, 72, 73, *83, 84*
Lunetta, V. N., 235, 236, 237, 243,
 247, 248, 252, 257, 259, *263*, 325,
 326, 334, *340*
Lynch, P., 42, *51*

Subject Index